SUSTAINABLE MATERIALS FOR TRANSITIONAL AND ALTERNATIVE ENERGY

Modern Materials and Sensors for the Oil and Gas Industry Series

SUSTAINABLE MATERIALS FOR TRANSITIONAL AND ALTERNATIVE ENERGY

Volume 2

Edited by

MUFRETTIN MURAT SARI
Texas A&M University, Commerce, TX, United States

CENK TEMIZEL
Saudi Aramco, Dhahran, Kingdom of Saudi Arabia

CELAL HAKAN CANBAZ
Ege University, Izmir, Turkey

LUIGI A. SAPUTELLI
ADNOC
Frontender Corp, United States

OLE TORSÆTER
Department of Geoscience and Petroleum,
the Norwegian University of Science and Technology (NTNU),
Trondheim, Norway

ELSEVIER

Gulf Professional Publishing
An imprint of Elsevier

Gulf Professional Publishing is an imprint of Elsevier
50 Hampshire Street, 5th Floor, Cambridge, MA 02139, United States
The Boulevard, Langford Lane, Kidlington, Oxford, OX5 1GB, United Kingdom

Notices

Knowledge and best practice in this field are constantly changing. As new research and experience broaden our understanding, changes in research methods, professional practices, or medical treatment may become necessary.

Practitioners and researchers must always rely on their own experience and knowledge in evaluating and using any information, methods, compounds, or experiments described herein. In using such information or methods they should be mindful of their own safety and the safety of others, including parties for whom they have a professional responsibility.

To the fullest extent of the law, neither the Publisher nor the authors, contributors, or editors, assume any liability for any injury and/or damage to persons or property as a matter of products liability, negligence or otherwise, or from any use or operation of any methods, products, instructions, or ideas contained in the material herein.

British Library Cataloguing-in-Publication Data
A catalogue record for this book is available from the British Library

Library of Congress Cataloging-in-Publication Data
A catalog record for this book is available from the Library of Congress

ISBN: 978-0-12-824379-4

For Information on all Gulf Professional Publishing publications
visit our website at https://www.elsevier.com/books-and-journals

Publisher: Joe Hayton
Acquisitions Editor: Katie Hammon
Editorial Project Manager: Alice Grant
Production Project Manager: Sojan P. Pazhayattil
Cover Designer: Miles Hitchen

Typeset by MPS Limited, Chennai, India

Dedications

To my lovely wife, Gulhan, for her valuable support and encouragement along with my academic career... And, to my kids, Ece, Emre, and Kaan for their patience and understanding even when I was very busy. I am truly blessed to have each of them in my life and their love and devotion made everything else possible.

Mufrettin Murat Sari

To my daughters, Ada and Ela who brought joy and happiness to my life.

Cenk Temizel

My effort that went into the creation of this book is dedicated to my wife, Ezgi Canbaz who assisted me and laid the way open for maximizing my concentration, to my parents Fusun and Kaya Canbaz who gave their true love without any expectations and supported me with patience in any circumstances, and to my brother Serkan who made me feel lucky to have a honest brother like him. Additionally, special thanks to my coauthors, Yildiray Palabiyik, Hakan Ozyurtkan, and Fatma Bahar Hosgor who trusted me and take sides with me to prepare one of the fb-non-chapters of this book.

Celal Hakan Canbaz

For my family who provides me the energy and inspiration, all the time.

Luigi A. Saputelli

Thanks for the support from PoreLab Center of Excellence and Department of Geoscience and Petroleum, Norwegian University of Science and Technology.

Ole Torsæter

The Editors also would like to thank the following lead and coauthors for their valuable assistance: Cengiz Yegin; Fatma Bahar Hosgor; Fatma M. Yurtsever; Hayriye Merve Yurdacan; Irfan Celal Engin; Jorge Saldana; Mesut Yurukcu; Mustafa Hakan Ozyurtkan; Mehmet Ozdogan; Mustafa Salman; Nirup K. Nagabandi; Omer Karakoc; Serkan Demirel; Yagmur Yegin; Yildiray Palabiyik.

Contents

List of contributors

Celal Hakan Canbaz
Ege University, Izmir, Turkey

Serkan Demirel
College of Computer and Information Sciences, Regis University, Denver, CO, United States

Irfan Celal Engin
Mining Engineering Department, Faculty of Engineering, Afyon Kocatepe University, Afyonkarahisar, Turkey

Fatma Bahar Hosgor
Petroleum Software Ltd., London, United Kingdom

O. Karakoc
Department of Materials Science and Engineering, Texas A&M University, College Station, TX, United States

Mufrettin Murat Sari
Texas A&M University, Commerce, TX, United States

N. Nagabandi
Essentium Inc, Pflugerville, TX, United States; Incendium Technologies LLC, Round Rock, TX, United States

M. Ozdogan
Department of Mechanical Engineering, Binghamton University, Binghamton, NY, United States

Mustafa Hakan Ozyurtkan
Department of Petroleum and Natural Gas Engineering, Istanbul Technical University, Istanbul, Turkey

Yildiray Palabiyik
Department of Petroleum and Natural Gas Engineering, Istanbul Technical University, Istanbul, Turkey

Jorge Saldaña
Department of Political Science, University of Houston, Houston, TX, United States

M. Salman
Department of Electrical and Computer Engineering, Binghamton University, Binghamton, NY, United States

C. Yegin
Incendium Technologies LLC, Round Rock, TX, United States

Y. Yegin
Department of Nutrition and Food Science, Texas A&M University, College Station, TX, United States

H.M. Yurdacan
University of Southern California, Los Angeles, CA, United States

Fatma M. Yurtsever
Department of Chemistry, University of Arkansas at Little Rock, Little Rock, AR, United States

Mesut Yurukcu
Department of Physics and Astronomy, University of Arkansas at Little Rock, Little Rock, AR, United States; Incendium Technologies LLC, Round Rock, TX, United States

About the authors

Mufrettin Murat Sari is a Chemistry Professor at the Department of Chemistry, Texas A&M University, Commerce, TX, United States, and Life and Health Science Department, University of North Texas at Dallas. He has 20 years of experience in the fields of Materials Chemistry, Applied Biochemistry, and Nanotechnology. His research interests are micro/nanomaterials, nanofluids, nanopharmaceuticals, modified surfaces, and their biotechnological, engineering, and environmental applications. Previously, as main academic appointments, he was a teaching assistant at Hacettepe University, researcher at Criminal Department Research Center, Ankara, Turkey, professor at Military Academy, and visiting professor at Texas A&M University, College Station. He received a PhD degree from Hacettepe University in 2005. He has around 40 scientific articles/proceedings and other publications with hundreds of citations. Also, he joined Interfacial Phenomena in Nanotechnology and Biotechnology, IPNB, Research Group at Texas A&M University for postdoctoral study in 2015, and studied Design of 2-Dimensional Next Generation Thermal Interface Nanomaterials.

Affiliation: Texas A&M University, Commerce, TX, United States
Contact Details:
Mufrettin Murat Sari
Department of Chemistry, Texas A&M University-Commerce, Commerce, TX, 75428, United States
e-mail: Mufrettin.Sari@tamuc.edu, mufrettin@gmail.com

Cenk Temizel is a Sr. Reservoir Engineer with Saudi Aramco. He has around 15 years of experience in the industry working on reservoir simulation, smart fields, heavy oil, optimization, geomechanics, integrated asset modeling, unconventionals, EOR in the Middle East, the United States, and the United Kingdom. He was a teaching/research assistant at the University of Southern California and Stanford University before joining the industry. He serves as a technical reviewer for petroleum engineering journals. His interests include reaction kinetics/dynamics of fluid flow in porous media and enhanced oil recovery processes. He served as a session chair and member of organizing committees for several SPE conferences. He has around 100 publications and patents. He holds a BS degree (Honors) from Middle East Technical University–Ankara (2003) and an MS degree (2005) from University of Southern California (USC), Los Angeles, CA, United States, both in petroleum engineering.

Affiliation: Saudi Aramco, Dhahran, Kingdom of Saudi Arabia
Contact Details:
Cenk Temizel
2855 Pinecreek Dr. A328, Costa Mesa, CA, 92626, United States
Phone: +1-650-3195742
e-mail: cenktemizel@gmail.com

Celal Hakan Canbaz is a Senior Reservoir Engineer with 16 years of experience. Previously, he had Reservoir Domain Champion position with Schlumberger in Middle East, Researcher position with The Petroleum Institute (former ADNOC Institution), and Research/Teaching Assistant position with Istanbul Technical University, Turkey. He is an expert in SCAL analysis, reservoir wettability characterization, well testing analysis, perforation and testing design, multiphase flow meters, CO_2/oil/water interactions, wellbore flow dynamics, and PVT data interpretation. He involved more than 30 projects with Shell, BP, Chevron, ExxonMobil, and Petrochina in Middle East. He is the first place winner of several paper contests in different SPE organizations (Turkey, UAE, and Oman), author/co-author of more than 30 conference/journal papers, two books, a US Patent, and technical reviewer of several energy related journals. He holds a BSc (2005) and MSc (2008) degrees in Petroleum and Natural Gas Engineering from ITU, and PhD in Energy from Ege University, Turkey.

Affiliation: Ege University, Izmir, Turkey
Contact Details:
Celal Hakan Canbaz
Geoscience and Reservoir Engineering Consultant
Phone: +90 5425906970
e-mail: canbazhakan@hotmail.com

Luigi A. Saputelli is a Reservoir Engineering Expert Advisor with 30 years of experience as reservoir, drilling and production engineer in PDVSA, ADNOC, Hess and Halliburton.

He is a researcher, lecturer, and active volunteer in the Society of Petroleum Engineers (SPE) where he serves as JPT editor since 2012, Production and Operations Advisory Board since 2010, founding member of Real-time Optimization Interest Group and Petroleum Data-driven Analytics technical sections. He is recipient of the 2015 SPE International Production and Operations Award.

He has published more than 100 industry papers on digital oilfield, reservoir management, and real-time production optimization. He holds a BSc in Electronic

Engineer from Universidad Simon Bolivar (1990), with a MSc in Petroleum Engineering from Imperial College (1996), and a PhD in chemical engineering from University of Houston (2003). He is also serving as managing partner in Frontender, a services firm in Houston.

Affiliation: ADNOC
Frontender Corp, United States
Contact Details:
Luigi A. Saputelli
Frontender Corporation, 8558 Katy Freeway Suite 103, Houston, TX 77024, United States
Phone: +1 281 217 2783
Phone: +971 55 930 4236
e-mail: lsaputelli@frontender.com

Ole Torsæter is professor in reservoir engineering at the Norwegian University of Science and Technology (NTNU), research associate at PoreLab a Norwegian Centre of Excellence and Adjunct Professor at the University of Oslo. He has been researcher or visiting professor with SINTEF, Phillips Petroleum, ResLab, New Mexico Tech, Texas A&M, University of Bordeaux and A*STAR, Singapore. Torsæter has supervised 220 Master- and 25 PhD-candidates, and he has published 200 research papers and the most recent are on nanofluids for EOR. Ole Torsæter has a Dr. degree from NTNU with a thesis on water imbibition in chalk where he showed that the Ekofisk Field was a good candidate for water flooding. Torsæter received the Darcy Technical Achievement Award from the Society of Core Analysts (2014), Distinguished Achievement Award (2016), and Management and Information Award (2018), both from SPE. Torsæter is member of the Norwegian Academy of Science and Technology.

Affiliation: Department of Geoscience and Petroleum, the Norwegian University of Science and Technology (NTNU), Trondheim, Norway
Contact Details:
Ole Torsæter
Department of Geoscience and Petroleum, Norwegian University of Science and Technology (NTNU),
S.P. Andersens v. 15 A, 7031 Trondheim, Norway
Phone: +47 91897302
e-mail: ole.torsater@ntnu.no

Preface for volume 2

In the context of emerging and converging technologies, formulated as Nano-Bio-Info-Cogno, nanotechnology provides us encouraging alternatives and permanent solutions for better sustainability of human health and our ecosystem. Maintaining environmental sustainability and protecting human health are vital concerns over the last decades, and therefore, nanotechnology and particularly nanomaterials have attracted great attention along with a rapid increase in research efforts on them. One of the biggest streamlines in nanotechnology is synthesis, characterization, fabrication of nanomaterials, and material-based sensors and their use in state-of-the-art, environmental, and engineering applications. Besides their valuable contribution to the other areas of science and technology such as electronics, aerospace, automotive, industrial chemicals, biomedicals, pharmaceuticals, manufacturing, textile and dye industry, next generation nanomaterials and sensors have been increasingly used to solve growing problems and meet expanding necessities in the energy area.

Energy needs in today's fast moving world with many advanced technologies have become much greater than ever seen before. However, current capacities of conventional energy production technologies and limited energy sources are far beyond the rising expectation in parallel with the increasing world population and energy consumption. Although there is a growing demand for energy resources, discoveries of new energy fields and reservoirs are declining, and the energy industry has significant challenges to increase energy sources for sustainability. Hence, cleaner, cheaper, and more reliable energy and energy sources have become a crucial issue. At this point, the energy industry is one of the powerful candidates for extensive use of materials and sensors but a late adapter of new generation materials-based emerging technologies throughout the years. For instance, the petroleum industry has been capable of adapting and utilizing the latest technologies in a wide spectrum from exploration to production. Using advanced materials and sensors in the energy have achieved a breakthrough and opened up a new path in a wide variety of high-tech applications. Likewise, many revolutionary improvements and adaptation have been achieved by using both materials and sensors in different fields of the petroleum industry, such as enhanced oil recovery, exploration, drilling, production, refining, and distribution.

The number of researches that aim to improve eco-friendly and renewable energy solutions including next-generation materials and sensors increases significantly and they present promising alternatives. However, their current level of use in related technology is not satisfying and far behind than expected. Literature lacks a comprehensive reference where the advanced materials used in the energy areas are thoroughly covered and explained. Hence, the main aim of the book study is to close this gap providing a strong reference by the experts in respective subjects in different energy areas including relatively new, renewable, and greener technologies recently being adopted by the industry.

This volume in the Advanced Materials and Sensors for the Oil and Gas Industry Book Series includes five chapters.

Chapter 1, Smart and State-of-the-Art Materials in Oil and Gas Industry, provides a general introduction about the importance, emergence, and potential growth of the state-of-art materials such as nanoparticles, nanoadditives, supramolecular assemblies, piezoelectric and shape-memory materials, and then reviews their latest applications in the energy, oil, and gas industries including sonic logging-while-drilling, ultrasonic borehole imaging, precision pressure measurement, marine seismic survey, subsea pumps, reconfigurable seals, the underwater connectors, deep water valves, and other adaptive components.

Chapter 2, Advanced Materials for Geothermal Energy Applications, includes comprehensive information and update about latest advances in geothermal industry materials and utilization of new generation materials and tools in this field such as fiber coatings, dressing materials, and composites as well as advanced drilling fluids, advanced cement materials, improved reservoir monitoring in harsh environments, and thermal infrared remote sensing tools. Comparison of the latest technology with their conventional counterparts have also been presented in terms of their efficiency and cost-effectiveness.

Chapter 3, Functional Green-Based Nanomaterials Toward Sustainable Carbon Capture and Sequestration, focuses on the use of functional green-based nanomaterials and covers a wide range of studies including synthesis, functionalization, characterization, and CO_2 adsorption/desorption performance of CO_2 nanosorbents originated from green materials in the energy area. A comprehensive background for nanocatalysts and sensors in the coal gasification process has been discussed in Chapter 4, Nanocatalysts and Sensors in Coal Gasification Process, with the extensively used examples such as Cu-ZnO-Al_2O_3/HZSM-5, Ni-CeO_2, Fe-Co-Mn, Fe/ZSM-5, Ni/Al_2O_3-MgO nanocatalysts, and $LaFeO_3$ and LaTiO.4FeO.6O_3 perovskites nanosensors.

Finally, Chapter 5, Advanced Materials for Next-Generation Fuel Cells, reviews the methods of fabrication and next-generation fuel cells' applications with the current technology of the advanced inorganic and organic materials and summarizes the advantages of nanostructures such as core-shell, nanorods, nanoframes, and nanoparticles over conventional methods.

The second volume in the series, Sustainable Materials for Transitional and Alternative Energy, presents a list of processes across the energy industry coupled with the latest research involving advanced nanomaterials, helping engineers get up to speed on the field of nanoparticle applications beyond the petroleum industry. Topics include green-based nanomaterials toward carbon capture, the importance of coal gasification in terms of fossil fuels, and advanced materials utilized for fuel cells. This volume includes the chapters all showcasing the outstanding efforts of prominent researches from different countries having very diverse academic and industrial affiliations. It is intended to access a wide range of readers including academicians, researches, graduate, and undergraduate students from various backgrounds such as petroleum engineers, petroleum researchers, nanotech researchers in the oil and gas industry, chemical engineers, and material scientists. We hope that the chapters of this volume will provide readers with valuable insight into materials and sensors for state-of-the-art translational and alternative energy applications with respect to the production, design, fundamentals of architecture, and applications.

CHAPTER ONE

Smart and state-of-the-art materials in oil and gas industry

O. Karakoc[1], Y. Yegin[2], M. Ozdogan[3], M. Salman[4], N. Nagabandi[5,6], C. Yegin[6], Mesut Yurukcu[6,7] and Mufrettin Murat Sari[8]

[1]Department of Materials Science and Engineering, Texas A&M University, College Station, TX, United States
[2]Department of Nutrition and Food Science, Texas A&M University, College Station, TX, United States
[3]Department of Mechanical Engineering, Binghamton University, Binghamton, NY, United States
[4]Department of Electrical and Computer Engineering, Binghamton University, Binghamton, NY, United States
[5]Essentium Inc, Pflugerville, TX, United States
[6]Incendium Technologies LLC, Round Rock, TX, United States
[7]Department of Physics and Astronomy, University of Arkansas at Little Rock, Little Rock, AR, United States
[8]Texas A&M University, Commerce, TX, United States

1.1 Introduction

Conformance problem essentially is production of unwanted water of gas and can be divided into two broad categories; one, poor sweep efficiencies of oil driving fluid resulting in fingering or seeping through high permeable rock matrix and two, excessive water or gas production via coning, leaks and flow behind pipes. Both the above problems can have multiple reasons and nanoparticles (NPs) have high potential to provide solutions in each of the problem that could drive the next generation of technology. We will review current and emerging techniques in nanotechnology for conformance problems. Although shape memory alloys (SMAs) are widely used in a numerous commercial application in the aerospace, automotive, and medical industries, the focus of this chapter will be centered on gas and oil industries. The primary advantage of SMAs in energy exploration is the durability and simplicity of SMAs and their capability to operate in severe and harsh corrosive environments including high pressures and temperatures. Additionally, this chapter covers brief introduction of piezoelectric materials and their application in oilfield services.

The main focus of this chapter is on enabling a brief summary of additives, NPs in conformance problems, mechanism of shape memory effect (SME) and superelasticity, most widely used SMA alloys, supramolecular assembly solutions, and their application in energy, gas and oil industries are compiled and presented.

Sustainable Materials for Transitional and Alternative Energy.
DOI: https://doi.org/10.1016/B978-0-12-824379-4.00005-7

1.2 State-of-the-art materials

1.2.1 Additives

Chemical "additives" are also added to the drilling muds. The purpose of an additive may be to change the viscosity or other properties of the mud to enhance the drilling mud's performance. Chemicals are used in the drilling process for multiple purposes, including the protection of the expensive drill shaft and bits. This equipment would otherwise be destroyed if drilling fluids were not available for lubrication, corrosion control and cooling. Chemicals are also used to reduce damage to the exposed formation rock and maximize the rate of penetration. It is crucial that the chemicals used in the drilling process be compatible with the rock being drilled. The drilling process can encounter all types of rock, including limestone, sandstone, granite, dolomite or a composite. The chemical used must not, for instance, react with the limestone and damage it. In many cases, the precise composition of the drilling fluid is determined in a lab, where field conditions are reproduced, and the appropriate fluid is formulated.

These drilling fluids are referred to as *muds* and are in liquid form. Their purpose is to cool the drill bit and provide a medium for transport of the rock and other debris to be lifted to the top of the well. The drill muds used today are water-based, diesel oil-based and synthetic-based (made from mineral or oil extracts). Due to environmental issues, oil-based muds (including synthetics) are not allowed to be discharged which means additional clean-up charges for the operators of offshore rigs, but even with the extra cost, oil-based muds are still favored due to their superior performance compared to alternative water-based muds.

1.2.1.1 Bacterial control additives

Bacterial growth and biofilm formation may cause high costs because of hydrogen sulfide (H_2S) formation in reservoirs. It can also result health and environmental hazard issues. Acids produced by bacteria can cause corrosion and biofilms can also plug the oil-bearing formations. Sulfate-reducing bacteria can anaerobically grow in the oil and gas environment and produce biofilms. It is very difficult to remove biofilms after they are formed. They can cause corrosion on metallic surfaces and result failures at production facilities. Therefore, bacterial control and prevention of biofilm formation are necessary for oil industry.

It is an essential step to check the efficiency of biocide treatment procedures in oil industry [1]. If our bacterial evaluation methods are not effective very high or very low number of biocides might be introduced in oil field systems. Inefficient control of bacterial growth and sulfate-reducing bacterial activities may create big environmental, safety, and production issues. Therefore, efficient and fast biocide treatment is important to control rapid bacterial growth. Microorganisms can produce H_2S through their metabolic activities and result microbiologically influenced souring. It is much easier to control microbiologically influenced souring at the early stages by adding biocides [2].

There are many different bacterial detection methods such as analytical profile index (API) serial dilution method, enzymatic assay, colorimetry, most probable number technique, colorimetry, electrochemical determination, and DNA sequencing. API serial dilution method is one the most commonly used methods for bacterial detection. Enzymatic assay for adenosine triphosphate (ATP) is a useful method for biocidal control [3]. Bioluminescence measurement is a trustable method for ATP determination. Electrochemical determination method is an excellent method for monitoring of biofilm formation. Most probable number technique is a traditional technique for bacterial enumeration [4]. DNA sequencing method is used for identification and enumeration of bacteria. Depending on an investigation has been completed at diesel pipelines in India, 11 different bacteria were identified. Bacterial species identified in the pipelines are *Bacillus cereus* ACE4, *Serratia marcescens* ACE2, *Pseudomonas aeruginosa* AI1, *Pseudomonas stutzeri* AP2, *Bacillus subtilis* AR12, *Bacillus megaterium* AR4, *Klebsiella oxytoca* ACP, *Bacillus litoralis* AN1, *Bacillus pumilus* AR2, *Bacillus carboniphilus* AR3, and *Bacillus* sp. [5]. They were not identified sulfate-reducing bacteria in their samples. The main bacterial species in the samples were *B. cereus* and *S. marcescens*. Sulfate-reducing bacteria can be a big problem in the oil fields. These bacteria can easily contaminate the oil reservoirs via water injection. Sulfide production by these bacteria can cause quite expensive issues for the industry [6]. Biocides are used for controlling sulfide-reducing bacteria [4]. Cheung et al. [7] investigated the effect of pressure and temperature on growth rate of sulfate-reducing bacteria. Their study showed that temperature plays a bigger role on their growth than the pressure.

Bacterial metabolic products can result microbial corrosion and influence the corrosion of materials. Understanding and solving the issue related to microbial corrosion require a multidisciplinary approach including microbiology, chemistry, and metallurgy. Different types of biocides have commonly been used in water treatments. Properties of the biocides should be known before applying in the industry. Boivin [8] showed some guidelines for the proper selection of biocides. Microbiological problems should be taken under control at the early stages of microbial growth to prevent any costly outcomes.

Main requirements for bactericide selection for drilling fluid are summarized by Zhou [9]:

1. It should be able to kill wide range of bacteria;
2. It should be nontoxic and should not cause any danger for environment and humans;
3. It should not be corrosive;

4. It should not damage to the drilling fluid;
5. Its killing efficiency should not be decreased by bacterial environmental adaptation.

Some of the biocides proposed for bacteria control in petroleum industry: 2, 6-Dimethyl-m-dioxan-4-ol acetate [10], 2-Bromo-4-hydroxyacetophenone [11], formaldehyde [12], Nitrate [13], thiocyanomethylthio-benzothiazolea [14], dimethyl-tetrahydro-thiadiazine-thione [15], diammonium salts of tetrahydrophthalic acid [16], monochloroamine [17], and di-(tri-N-butyl)-(1,4-benzodioxan-6,7-dimethyl) diammonium dichloride [18].

1.2.1.2 Corrosion inhibitor additives

Corrosion is one of the costliest issues in petroleum industry. There are many corrosion surfaces in oil industry such as oil storage tanks, production wells, pipelines, etc. It is almost impossible to prevent the contact of air with fluids in oil industry. Even though the industry uses reducing agents to remove oxygen from the system, still some extra agents remain after oxygen removal process. Deposition conditions can also cause corrosion problem. Some solid particles and iron sulfide can be accumulated on the surfaces and decrease the efficiency of corrosion inhibitor additives. These accumulated solid particles can also serve as a harbor for some anaerobic bacteria. Bacteria can produce biofilms and these biofilms can protect bacteria and help them grow underneath of them [19]. Sulfate-reducing bacteria can contribute corrosion problem by producing H_2S which reacts with iron to generate iron sulfide.

Corrosion may occur in numerous places in the oil industry such as steam generators, cooling units, pipelines, drilling muds, oil production parts, refinery units, etc. Most of the corrosion inhibitors may cause environmental hazards such as imidazolines, fatty amines, chromates, etc. The use of some of the other alternative anticorrosion additives is limited because of their high costs. In general, corrosion inhibitory additives can be classified as inorganic and organic; anodic and cathodic; and nonfilming and filming [20].

Polypeptides can be considered as a good corrosion inhibitory additive due to their more environment friendly nature [21]. Polyaspartate is the strongest corrosion inhibitory additive as compared to other polypeptides [22]. Its efficacy increases at high pH and calcium ion conditions. Silverman et al. [23] searched the efficiency of polyaspartic acid as a corrosion inhibitory additive. Depending on their study, polyaspartic acid at $pH > 10$ prevents the corrosion on steels. Silicate based inhibitors can also be advantageous to use in oil industry because of their low toxic effect on environment and low cost [24].

Sulfate-reducing bacteria can easily grow in water systems. Chlorine dioxide (ClO_2) can be added to stop their growth and biofilm formation; however, it is very corrosive to the metal surfaces. Chromates can be a good alternative to ClO_2 corrosion inhibitory additives. Chromates are not preferred due to their high toxic effect to the environment. As an alternative to chromates, a mixture of water, fatty imidazoline, alcohol, ethoxylated fatty diamine, and acid can be preferred [25]. This type of mixture is both more effective than chromates and safer to the environment.

Here are the examples for some additional corrosion inhibitors: 3-phenyl-2-propyn-1-ol [26], Hydroxamic acid [27], Mercaptoalcohols [28], Acetylinic alcohol [29] 2,5-bis(N-pyridyl)-1,3,4-oxadiazoles [30], Tall oil fatty acid anhydrides [31], Quaternized fatty esters of alkoxylated alkyl-alkylene diamines [32], Sulfonated alkyl phenol [33], Thiazolidines [34], and 1-Hydroxyethylidene-1,1-diphosphonic acid [35].

Corrosion inhibitory additives are commonly added to the system either as a batch addition or continuous injection addition. Encapsulated types of corrosion inhibitors have the advantage both batch and continuous addition methods. They have showed a good effect on prevention of CO_2 corrosion in oil fields. Corrosion inhibitory additives can be released on a prolonged time manner. Encapsulated types of corrosion inhibitors significantly improve the efficiency of the corrosion treatment since they can be applied as a batch and delivered in a continuous manner [36]. For example: Amido amine can be used for corrosion prevention on metal surfaces in the well. It can be dispersed in a polymer matrix for slow releasing.

1.2.1.3 Fluid loss additives

Fluid losses in petroleum industry may occur when the fluid contacts with porous structures. Amount of fluid losses highly depend on porosity and permeability of the structures. Extensive fluid loss should be eliminated because of its cost and possible damage to the environment. Therefore, it is a very important issue to prevent fluid loss.

Bacterial cultures can also be used as an alternative to polymers. They can also form polymers and prevent fluid loss. Anderson et al. [37] selected a bacterial culture because of its exopolymer production ability. The researchers evaluated the addition of the bacterial culture can replace polyanionic cellulose (PAC) without affecting the fluid retention and viscosity when added in drilling mud. Addition of bacterial cultures was not as effective as PAC in terms of viscosity and fluid loss. However, the addition of these bacterial cultures can decrease the required amount of PAC.

Viscoelastic surfactant fluids can be made by combination of different surfactants. Surfactant concentration should not exceed certain level since they may accumulate and form micelles that may cause the network formations that show viscoelastic properties [38]. Some type of salts such as potassium chloride, sodium isocyanate, ammonium chloride can be added to create viscoelasticity in surfactant solutions. The electrolyte content of surfactant solutions plays an important role on their viscoelastic properties [38]. Fluids can be gelled by using viscoelastic surfactants and used in hydraulic fracturing. However, viscoelastic surfactant fluids may cause big amount of fluid loss into the reservoirs. Therefore, it is essential to add fluid loss additives into viscoelastic surfactant fracturing applications [39]. Fluid loss can be decreased by the addition of calcium hydroxide or magnesium oxide into a fluid which is gelled with viscoelastic surfactants [40]. As a result, the addition of these additives can significantly minimize the fluid losses into the reservoir pores.

1.2.1.4 Lubricants

The protection of the drill bit bearing wear is an important issue for petroleum industry. Therefore, it is a necessity to use lubricants for drilling applications. The usage of synthetic greases has some advantages compared to hydrocarbon-based greases [41] such as capable to function at lower temperatures, pumpability with high viscosity, and high oxidative stability. Synthetic greases show many advantages and many different usages because of their controlled synthesis. For example: greases are used for the lubrication of rock bit bearing. In general, greases are prepared by the addition of thickeners into lubrication oil. A typical grease contains many different additives [41]: corrosion, oxidation protection, antiseize protection, extreme pressure agents, solubility aids, and antiwear additives.

There are many different compositions used as lubricants such as molybdenum disulfide, ellipsoidal glass granules, phospholipids, alcohols, polarized graphite, olefins, paraffins, polymers, calcium-sulfonate-based greases, and esters etc. (Table 1.1) [42].

1.2.1.5 Fluid viscosifiers

Viscosifiers are one of the major additives of the drilling and fracturing fluids in petroleum industry. The main function of the viscosifiers to enhance the carrying capacity of the fluids by increasing the viscosity of the fluids. Commercially available viscosifiers are mainly polymer-based materials while newer technologies such as viscoelastic surfactants have been in development.

Table 1.1 Grease compositions [42].

Ingredient	Amount (%)	Examples
Thixotropic base material	40–90	Metal salt complex greases
Metal complex grease	50–90	Lithium complex grease, aluminum calcium complex grease
Antiseize agents	5–50	Metal fluorides
Boundary lubricant	5–50	Metal borates, molybdates, carbonates, acetates, stearates, etc.
Friction adjusters	0–12	Polytetrafluoroethylene (PTFE), graphitic materials, natural or powders and synthetic fibers, molybdenum disulfide, fibers etc.
Antiwear additives	0–5	Sulfurized isobutylene, phosphate esters, dithiocarbamates, dithiophosphates, naphthanates, or the like
Antidegradant additives	0–2	Antioxidants and antiozonants

1.2.1.6 Synthetic-based muds

Synthetic-based muds were used to minimize the negative environmental effects of oil-based muds. The main reason for their development is the legislation which prevents the discharges into marine environment of oil-based muds. They are made similar to oil-based muds. Synthetic-based muds have been made to keep the performance of these additives as like oil-based muds while decreasing their negative environmental effects. Instead of mineral oil or diesel oil, organic fluids have been used such as acetal, esters, ether, linear alkyl benzenes, and polyolefins [43]. Synthetic-based muds are costly. Therefore, companies are looking for balancing between costs and improve the rheology while still staying within the environmental limits.

1.2.1.7 Clay stabilizers

It is essential to take care of maintaining wellbore stability during drilling in petroleum industry. Swelling of the clays may cause a collapse in wellbores. Therefore, clay stabilizers need to be used as additives during drilling. Some of the examples used as clay stabilizers are polymer lattices [44], polyacrylamide [45], dimethyl diallyl ammonium chloride [46], polyols and alkaline salt [47], hydroxyaldehydes or hydroxyketones [48], quaternary ammonium carboxylates [49], polyvinyl alcohol, potassium silicate and potassium carbonate [50], and potassium salt of carboxymethylcellulose [51].

1.2.1.8 Antifreeze additives

Antifreeze chemical agents are used to reduce the freezing point of the fluids. Some of the antifreeze additives are calcium chloride, methanol, ethylene glycol, propylene glycol, and sodium chloride.

1.2.1.9 Additives for odorization

Odorization is important for gas transmission company in terms of safety. It is a type of warning in the detection of gas before it reaches dangerous levels. A successful odorization program involves accurately injection of odorant, monitoring, and record keeping. Some examples of the odorants used in the industry are tetrahydrothiophene, thiophene mercaptans with pyridine and picoline [52], and mixture of ethyl-, propyl-, butyl-, and amylmercaptans [53]. Companies should consider which pipelines may require odorization, selection of odorants, detectible levels of the gas, and their odorization can successfully meet the regulatory requirements [54].

1.2.1.10 Defoamers

Foam formation may occur in many parts of the process in oil industry. Use of defoamers or antifoaming additives might be necessary for efficient operation. Foam occurs because of the generation of bubbles when contaminant or any pressure drop in fluids. Defoamers can help the separation of gas from water and oil; therefore, they can prevent the problems may occur in expensive equipment. Hydrocarbons, silicones, fluorocarbons, and polyesters are some of the chemicals used as antifoaming agents in oil industry.

1.2.2 Nanoparticles in conformance problems

NPs are class of materials which have at-least one dimension in the range of 1–100 nm. These materials typically possess different chemical and physical properties compared to their bulk counterparts because of their high surface area to volume ratio leading to high surface energy which forces them to search for low energy states and thus impart the unique properties. Due to their unique properties, NPs are used in several fields including medicine, materials, catalysis, space and others for various applications that are not possible using bulk materials. Such benefits of nanotechnology can also improve the profitability and oil recovery if appropriately employed and in this respect, NPs have found a way in reservoir confirmation, enhanced oil recovery, and other oil related operations and applications in various ways. While the oil is classified in to primary, secondary and tertiary based on their recovery stage, most emphasis currently is on improved oil recovery (IOR) which includes secondary and tertiary recovery since just primary recovery is not profitable and moreover cannot meet the growing global energy demand. Multiple challenges are faced during such oil production processes and with enough resources, nanotechnology can provide breakthrough solutions in near future. A right step in this direction is the amount of research taking place in nanotechnology to solve different oil production related problems at the fundamental level and, conformation problems are most prevalent in oil wells across the globe.

1.2.2.1 Improving sweep efficiencies

Oil driving fluids used in IOR are carbon dioxide (CO_2), water, and polymer loaded water which can end up having low sweep efficiencies for various reasons. Improving the sweep efficiency in these systems could be as simple as improving mobility control of the oil driving fluid but can get complicated when other factors like displacement in porous media, mobility control, and particle stability. As of today, many several lab scale studies involving nanofluids (NPs in liquid) are on-going which are directed at improving the sweep efficiencies and soon can find themselves in the real world.

1.2.2.2 Nanoparticles stabilized foam

Foam is a suspension of gas in a liquid and can be generated by providing mechanical energy to a gas in water system. Natural gas, nitrogen, and most widely CO_2 are used in gas flooding operations during EOR. They typically work by reducing oil viscosity and oil-water IFT but end up having much lower viscosities than the gas and can lead to viscous fingering. This is solved by using/generating foam during these flooding operations which have high apparent viscosity of the gas [55,56]. Foam can be efficiently generated and stabilized by foaming agents like surfactants, but NPs are finding applications as foam stabilizers or even replace surfactants in certain cases which will prove beneficial at high salinity and high temperature well conditions.

Silica NPs are often explored for this purpose due to their vast availability, well understood behavior and low cost. For a water—CO_2 system stabilized by NPs, a study by Hamed et al. [57] found that a lower NP size is effective to stabilize foam in high NP concentration solutions while a high NP size is better for low NP concentration solutions. A study by Emrani et al. [58] found that silica NPs used with an alpha-olefin sulfonate surfactants generated stronger foams, improved high temperature stability, and had fine texture. In another study, Kim et al. [59] used smaller sized silica NPs to create stable CO_2-in-brine and decane-in-brine foams. Other studies by Mo et al. found that NP concentration as low as 100 ppm can be effective to change the foam properties for CO_2 in water and the foam got viscous with increase in NP concentration. In most cases, NPs adsorbed at the gas-water interface forming a barrier which hindered bubble coalescence. Also, it was found out that the dilutional viscoelasticity of the gas-water interface is improved by NPs in studies done by Lu et al. [60].

Studies conducted by Nguyen et al. [61] found that light, medium, and heavy oils showed considerable oil recovery improvement when NP-stabilized CO_2 foam was used. They also indicated that such results were possible due to foam leading to less oil-in-water emulsions. Amorphous and crystalline silica behaved similarly foam stability, mobility, morphology, and resistance factor as per studies conducted by Yu et al. [62]. More important consideration was hydrophobicity of the NPs, where the increase in hydrophobicity of the NP, significantly reduced the CO_2 bubble size. In this regard, several studies were conducted using polymer coated NP (PNP) which have different surface properties compared to bare silica NPs. PEG-coated NPs were used by Espinosa et al. [63] to stabilize CO_2-foam. Studies by Worthen et al. [64] present overall approach in designing NP for CO_2 EOR, which could be used to alter the NP surfaces by coatings. Polymer/surfactant-coated NPs can generate stable foams in certain cases where the precursors individually cannot [65—68]. More details about PNPs and the influence of coating, coating ratios, particle sizes, and surfactant type are nicely summarized in the work of Jazeyi et al. [69].

Fly ash NPs are another interesting class other than silica to use as foam stabilizers [70]. Fly ash is a byproduct of coal combustion and is typically large, which can be ground into NPs by several means like balling milling. Typical chemical composition of fly ash includes SiO_2, Al_2O_3, Fe_2O_3, CaO, MgO, Na_2O, K_2O, TiO_2, BaO, SO_3, P_2O_5 and few amounts of other metallic oxide [71]. Work done by Singh et al. [70] demonstrations that using fly ash NPs with a nonionic surfactant, strong CO_2 foam with fine texture can be produced. Eftekhari et al. [72] used 200 nm fly ash NPs obtained via ultrasonic grinding along with AOS to stabilize nitrogen, which proved to be stronger than using AOS alone.

In addition to the above NPs, aluminum oxide [73–75], iron oxide [76–79], nickel oxide [77,80,81], titanium oxide [73,82–85], zinc oxide [86–89], zirconium oxide [80,90–93], and carbon nanotubes (CNTs) [90,94–99] are also used for EOR methods but most studies with these particles was concentrated on influencing the wettability, relating to conformation problem only indirectly.

1.2.2.3 Nanoparticles polymer flooding

Mobility ratio (M) needs to be less than one for effective polymer flooding, since higher M leads to viscous fingering and thus create conformance problems. In polymer flooding, they are used to increase the viscosity of the displacing fluid (typically water) which reduce the mobility ratio. Polyacrylamide (PAM) and variants of it are widely used for such applications but this is limited when a high-salinity and high-temperature reservoirs are encountered since the polymer hydrolyzes in these conditions leading to lower viscosity of the polymer solution. NPs can be employed as additives in polymer flooding to improve performance under harsh conditions [100–102]. Silica NPs form chemical bonds between them and the polymer to improve rheological properties and thus achieve more stability in harsh conditions.

Increasing NP concentration increases the viscosity of the hybrid solutions of polymer and NPs and a rapid viscosity increase was observed beyond a certain critical NP concentration (CNC) [101]. In the studies performed by Zhu et al. [100], it was found that hybrid NP solutions show shear-thinning behavior and the viscosities of the solutions with high NP concentrations are relatively high. Within all these studies, it is understood that below CNC, most NPs are unavailable since they are hydrogen bonded on to polymers and beyond CNC, there are free NPs available for cross-linking different polymer chains leading to rapid viscosity spike up. Hu et al. [101] found that at high temperature conditions, aging of the hybrids led to improved polymer stability possibly due to better bridging of NPs with polymer chains. Overall, the polymer/NP hybrid can enhance polymer flooding efficiency by improving mobility ratio under reservoir conditions.

Other NPs like FeO, CuO, NiO, Al_2O_3, Ni_2O_3, MgO, Fe_2O_3, SnO, Nanoclay, and CNT [73,74,78,79,82,83,94,102–108] are reported to be used in EOR but not specifically only to increase viscosity of the nanofluid but to improve other EOR factors like change in IFT, wettability to increase oil production.

1.2.2.4 Nanoparticles for less water production

Several issues like fingering, coning, high permeable rocks, fissures, alternate paths, and more could lead to higher than required water production in IOR. One efficient way is to plug these areas cleverly to cut the amount of water that enter the production lines. The power of nanotechnology could be used here in targeted delivery, innovate methods to seal only needed parts of the reservoir, smart materials that behave differently under various conditions and thus blocking water efficiently, and in other ways. Traditionally the solution to this problem has been gelling agents placed as plugs deep in the well or at the well bore or in the zone where the problem surfaces. But, recently, studies are showing the environmental challenges involved in the chemical being used for gelling agents and NPs can present a bright solution here.

NPs for this purpose can be divided into three use cases: (1) injection-fluid-based, on-command blockage and opening of reservoir zones; (2) precision injection into reservoir zones; and (3) improved stability of conformance control fluids under harsh reservoir conditions. Recently developed NPs with active cross-linking functionality with HPAM under reservoir conditions will enable such placement of gel [109]. In another study [110], silica NPs were functionalized with organic/inorganic nanostructured hybrid polymers, which can be used as cross-linking agents under reservoir conditions instead of environmentally nonacceptable chemicals. It is also possible to engineer in a way that the gel is more hydrophobic and hence confine water even better.

In more studies, Huang et al. [111] described a colloidal silica gel system where the gelation is activated by salt. Suleimanov and Veliyev [112] showed that the polymer gel strength is improved by adding metal NPs. Rankin et al. [113] also showed use case for colloidal silica for gelling applications in reservoirs with fractures or high permeability layers. Lakatos et al. [114] describes a use case of silicate system for gelling purposes which was enhanced by addition of nano silica yielded a precise placement of the gel and stability up to 150°C, permeability dropped down by 4–5 order of magnitude and improved the oil production. All the above results were achieved while still using environmentally acceptable systems. Suleimanov et al. [115] also discuss about similar nanogel system which increases the gel strength and thus eventually leading to high oil production. Goel et al. [116] reported studies of using nonspherical silica NPs (rod-like and string-of-pearl like) for nanofluid systems has yielded superior gelation activity in saline conditions. In all the above systems, NPs are expected to form super structures by cross linking with polymer systems to create a superior gel. These cross-links can be activated with various stimuli like salinity, temperature, pH or others and are expected to last at high temperatures and for prolonged periods of time.

More studies conducted by Kalgaonkar et al. [117] have shown that positively charged NPs are not suitable for gelling purposes and activator concentrations are important in controlling the gelling time. It was also understood in these studies that, at higher temperatures, gelation time decreases due to the increase in particle collision. Lafitte et al. [118] reported a single particulate NP additive for gelling applications, where cross-linking is not involved and rather the plugging is achieved by physically plugging of the NPs at the site. Further, cellulose nanocrystals (CN) were used as NPs for improving emulsion stability by Pandey et al. [119] in which they found that, water required larger pressure gradient compared to oil to established flow and thus could be potentially applied for zone blocking and could be used as permeability modifiers and thus achieve plugging zone wise. This was also expected due to formed cross-linked network. The huge implication of this work is usability of naturally occurring bio-degradable materials for flooding purposes. In another study by Esmaeilnezhad et al. [93] ferro nano fluids are employed, in which the viscosity of the fluid changes based on the applied magnetic field. Magnetite NPs were used and by applying a magnetic field, drastic improvement in gel strength leading to solid-like plugs. The results of these studies could be used to apply in zonal controlling permeability by controlling local magnetic fields. Another interesting study is done by Panthi et al. [120] where they used magnetic hyperthermia to locally heat up thermo-sensitive polymer to form local soli-like gels for blocking [121].

Overall, use of NPs for gelling purposes is exciting and carryover of the lab scale techniques to field studies could drastically change the productivity and profitability in oil industry due to several innovative ideas for targeted zonal blocking and use of environmentally friendly system.

1.3 Smart materials

1.3.1 Shape memory materials and piezoelectric materials for oil and gas industry

The oil and gas industries need a full spectrum of services from exploration of hydrocarbons to production of hydrocarbons. Smart systems with intelligent and adaptive features and functions have become important part of technology, necessitating the use of microcontrollers, actuators and sensors. The development of smart and high functional density applications must overcome difficulties coming from commercial and technical limitations including operating environment, allowable cost, available space and response time [122–124]. Even though there are many types of active materials as shown in Fig. 1.1, shape memory materials and piezoelectric materials have been found many applications in the oil and gas industries.

Active materials mostly demonstrate shape changes under the external fields such electric, optical, thermal and magnetic. Two significant criteria actuation energy density and frequency of active materials determine their suitability in the actuation application. Actuation energy density refers to work-output per unit volume. Comparison of actuation energy density and frequency of popular active material are presented in Figs. 1.1 and 1.2. SMAs have largest actuation energy density and low actuation frequency. Magnetic SMAs have higher actuation frequency density as compared to conventional SMAs. Shape memory polymers (SMPs) have lower actuation energy

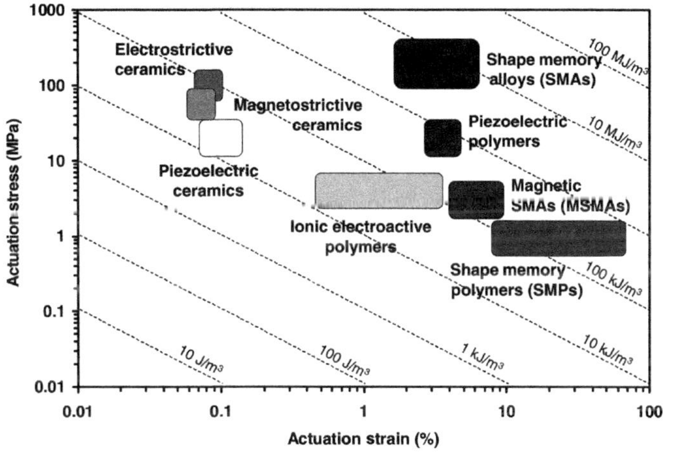

Figure 1.1 Actuation stress versus actuation strain exhibiting actuation energy density of various active materials [125].

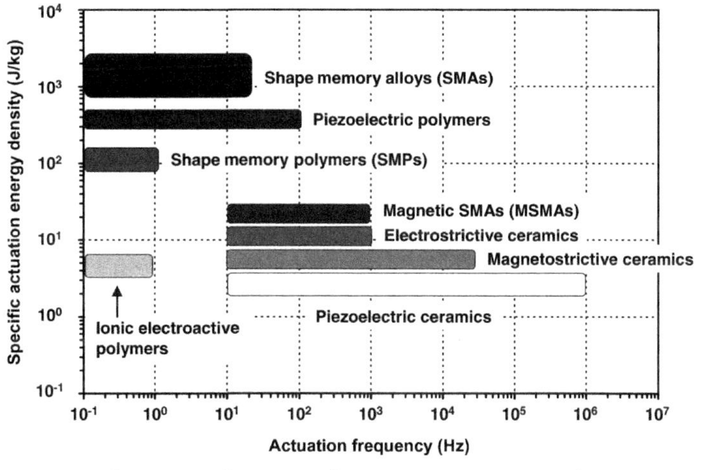

Figure 1.2 Comparison of actuation frequency for various active materials [125].

density and slower actuation frequency as compared to SMAs. Piezoelectric ceramics have intermediate actuation energy density and high actuation frequency.

SMA was firstly discovered in 1932 by researcher Ölander [126]. Smart alloy gained popularity after Buehler and his coworkers in 1962 exhibited SME in a titanium-nickel (TiNi) alloy [127]. After the discovery of SMAs, extensive progress has been performed in the applications of these multifunctional alloys and the fundamental understanding of the shape memory capabilities. SMAs have become promising candidate in many applications due to their unique SME and superelasticity. Other than transformation temperatures, suitability of SMAs for any application depends on many factors including magnitude of strain and their reversibility which are represented by recoverable strain (ε_{rec}) and irrecoverable strain (ε_{irr}) [128]. Recoverable strain is the magnitude of strain that material can recover through complete superelastic and shape memory cycles. Magnitude of recoverable strain depends on testing condition such as applied stress level and upper cycle temperature (UCT). ε_{rec} has shape strain and elastic strain components. Shape strain may stem from the martensitic transformation from austenite to martensite for superelasticity, or reorientation/detwinning of martensite for SME. Irrecoverable strain (ε_{irr}) may come from either formation or propagation of dislocations, or stabilized martensite which does not transform to austenite even after sample heated above austenite finish temperature (A_f). The former is irreversible process, whereas retained martensite can transform to austenite when heated to elevated temperature greater than A_f temperature.

1.3.1.1 Shape memory materials

1.3.1.1.1 Mechanism of shape memory effect and superelasticity

Martensitic transformation occurs in a military manner which atoms move in a coordinated manner over short distances (less than one atomic spacing) instead of random diffusion of atoms [128]. It is a diffusionless solid to solid phase transformation between high temperature austenite phase and low temperature martensite phase. It can be activated by stress, magnetic field and temperature. Composition does not experience any changes during transformation, while transformation is the change of crystal structure one form to another. Fig. 1.3 demonstrates mechanism of SME during martensitic

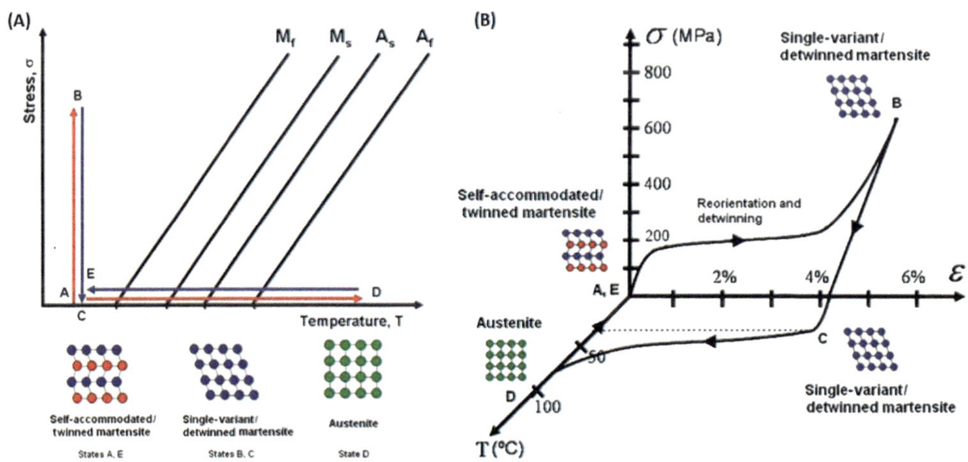

Figure 1.3 Schematically illustration of (A) stress – strain data and (B) stress – strain temperature graphs for SME. M_f: Martensite finish temperature, M_s: Martensite start temperature, A_s: Austenite start temperature and A_f: Austenite finish temperature [128].

transformation. Self-accommodating twins are observed in the microstructure at temperatures below M_f [128]. These self-accommodating twins can transform into single variant detwinned structure, resulting in shape changes.

In the presence of stress, when material is heated above A_f, materials undergo from martensite (low temperature phase) to austenite (high temperature phase) transformation which restores the original shape of a plastically deformed specimen. These phenomena are called as thermoelastic martensitic transformation (SME). Fig. 1.4 indicates mechanism of superelasticity where material deforms slightly above A_f temperature (fully austenitic state) [128]. Superelastic behavior is a result of stress-induced formation of some martensite above its normal temperature. Upon unloading, immediately martensite reverts to undeformed austenitic phase, providing "rubberlike" elasticity in this class of materials.

1.3.1.1.2 Influence of alloying on the shape memory properties of shape memory alloys

Initial studies have started with AgCd and AuCd in 1930s and followed by discovery of NiTi alloy in 1963, large variety of newest materials have been researched over the last eight decades. Various allaying elements are added to existing compositions to produce new compositions, giving rise to variety of properties for SMAs. Such a variety in the behavior of SMA provides great flexibility to meet requirements of a given engineering application. This section will cover most commonly known SMAs and their properties.

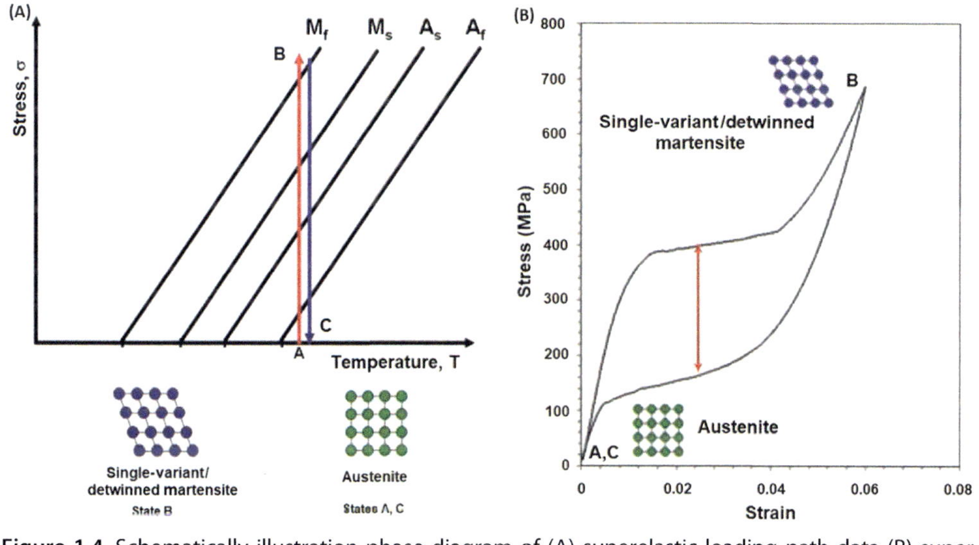

Figure 1.4 Schematically illustration phase diagram of (A) superelastic loading path data (B) superelastic loading cycle [128].

1.3.1.1.3 Ni-Ti-based Alloys

NiTi Alloys: Nitinol are most widely known SMA and find greatest number of commercial applications due to their good mechanical properties, resistance to corrosion and exhibiting characteristics of SME and superelasticity (pseudoelastic behavior) [125]. This pseudo-elastic alloy can display incredible elasticity that is almost 10–30 times greater than that of conventional metal [129]. NiTi alloys can also fully recover actuation strain of up to 8% [125]. Its operating temperature is below 100°C. It can be found in various forms of tubes, sheets, wires, and springs.

Ni–Ti–Cu: Copper addition at the expense of nickel element to Ni–Ti alloy reduces both thermal hysteresis and transformation strain of Ni–Ti alloys [130]. Even though any change in the composition of binary NiTi alloy greatly affects transformation temperatures, Cu addition to binary NiTi alloys significantly reduces the sensitivity of transformation temperatures [125]. Optimum composition range is found as 5 at-% ≤ Cu ≤ 10 at-% in the Ni–Ti–Cu alloys. Further increase in Cu content embrittles the alloy [125].

Ni–Ti–X (X: Pt, Pd, Au, Hf, or Zr): Ternary alloying elements (X: Pt, Pd, Au, Hf or Zr) into NiTi are adopted to increase transformation temperatures for high temperature applications [125,128,131–138]. Fig. 1.5 exhibits transformation temperatures and actuation strain of some high temperature shape memory alloys (HTSMAs). NiTiHf materials have advantages of lower cost and higher shape memory capabilities compared to Pt, Pd, Au and Zr alloy base NiTi alloys [128,134,139,140]. Focus of the

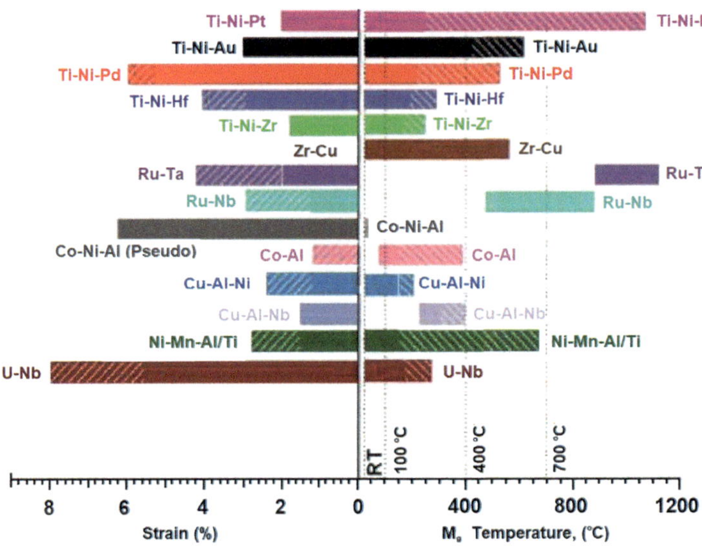

Figure 1.5 Transformation temperatures and corresponding transformation strains for different high temperature SMAs [128].

earliest studies was centered on Ni-lean NiTiHf alloys due to rapid decrease in transformation temperatures when Ni content is above 50 at-%. However, recent research has shifted towards Ni-rich NiTiHf alloy (Ni > 50 at-%) since precipitation hardening by aging treatment strengthens materials against plastic deformation and permits control of transformation temperatures [134,139–144]. Furthermore, investigations on the actuation fatigue properties of NiTiHf display them as viable candidate in the high temperature applications [145,146].

1.3.1.1.4 Copper-based shape memory alloys

Cu–Zn–Al and Cu–Al–Ni alloys are commercially available for shape memory applications [147]. Even though Ni–Ti materials have better shape memory response and corrosion resistance than Cu-based SMAs, finished form of Ni–Ti materials has higher cost than Cu-based alloys due to fabrication difficulties during forming and melting. Cu-based SMAs have far more complex structure of martensite than those of Ni–Ti alloys [147]. Copper-based alloys have good thermal and electrical conductivity with high formability, making them a perfect alternative to NiTi in many commercial applications [147].

Martensitic transformation temperature of Cu-based SMAs is greatly affected by even small variation in the alloy composition similar to those of NiTi-base SMAs. Thus, fabrication process requires precise control during melt process. Quenching rate plays an important role in obtaining SME in Cu-base alloys. Rapid cooling is necessary to retain metastable β phase which is responsible for SME. Rapid cooling prevents decomposition of β phase into the equilibrium phases.

Cu–Zn–Al: The upper limit for transformation temperatures of Cu–Zn–Al are recorded as 120°C [147,148]. Overheating or aging can cause decomposition of parent phase of Cu–Zn–Al into equilibrium phases, restricting their use at operating temperature of approximately 100°C. Low critical stress for formation of slip makes mechanical behavior of Cu–Zn–Al restricted to 200 MPa. The alloy has excellent SME and pseudoelasticity within the operational range of stress level, while maximum achievable transformation strain is 3%–4% [149]. Cu–Zn–Al is mostly used Cu-based alloys in the commercial applications due to their high ductility.

Cu–Al–Ni: The upper limit for transformation temperatures is 200°C [147,148]. Transformation temperature can be altered by changing content of nickel and aluminum. Thermal hysteresis is nearly constant and independent from content of elements. Cu–Al–Ni is relatively less sensitive to aging phenomena and stabilization [125]. Due to difficulty in production of this alloy, addition of manganese is employed to improve its ductility and titanium to refine its grains [125]. Main reason behind limited use of this alloy is low ductility caused by intergranular cracking [150]. Thus, this alloy fails under around 280 MPa and achievable transformation strain is around 3% [150]. Besides, Cu–Al–Ni is correlated with poorer cyclic stability [150].

1.3.1.1.5 Iron-based alloys

Fe—Mn—Si and Fe—Ni—Co—Ti are the most widely known ferrous SMAs. Specific thermo-mechanical treatment results in SME in $FeNi_{31}Co_{10}Ti_3$ which have a thermal hysteresis of around 150°C [125]. Fe—Mn—Si is another commercially available ferrous alloy. Addition of Si improves the SME and increases the critical stress level for slip in austenite. Fe—Mn—Si alloys are able to exhibit transformation strain of 2.5%–4.5% [149].

1.3.1.1.6 Additional shape memory alloys

Co—Ni—Al: Addition of Ni to CoAl or Co to NiAl causes the production of Co—Ni—Al systems. They demonstrate perfect oxidation and corrosion resistance at elevated temperatures. The material exhibits high temperature cubic structure- low temperature tetragonal structure transformation. Alloy was not gained popularity due to its brittle nature after the discovery of martensitic transformation in 1971 [125]. However, recent developments in the formation and control intermetallic phases through aging has improved the ductility of alloy [122,151]. $CoNi_{33}Al_{29}$ has A_f of −26°C and transformation strain of 4%. The material is mostly known for its magnetic properties along with pseudoelastic and SME behavior. Co—Ni—Al exhibits a magnetic field activated reorientation strain when sample is subject to a magnetic field. High actuation frequency (~1 kHz) is observed for Co—Ni—Al, while thermally induced SMAs are known with low frequency responses. However, transformation strain of $CoNi_{33}Al_{29}$ is limited to 0.06% when subject to magnetic field [152].

Ni—Mn—Ga: Ni—Mn—Ga systems are most largely studied magnetic SMA. Martensitic transformation was discovered in Ni_2—Mn—Ga in 1984 [153]. Magnetic field induced reorientation strain was recorded as 0.2% for Ni_2—Mn—Ga single crystal. Further investigation after development in the understanding of structure of this alloy pave a way to obtain larger magnetic field induced strains of 6% [154] and up to 10% [155,156] in the single crystals. Low blocking stress (6−10 MPa) is primary limitation of magnetic shape memory alloys (MSMAs) Complete suppression of magnetic reorientation strain takes place at blocking level. It is found that pseudoelastic transformation coupled with magnetic field results in transformation strain of 0.5% and raises blocking stress of the Ni_2—Mn—Ga to 20 MPa [157], Fe—Ni—Co—Ti, Ni—Mn—Al, Co—Ni—Ga, Co—Ni—Al [152,158−160], Fe—Pd [161−164], and Fe—Pt [162] are other SMAs studied. Their magnetic-field induced strains are lower than that of Ni—Mn—Ga. Maximum field-induced strain is 3.1% for Fe—Pd which is also more ductile than Ni—Mn—Ga systems [165].

1.3.1.1.7 Challenges in the development of shape memory alloys

Durability is a measure of the stability of actuators, and how the actuation capabilities of material degrade with cycling due to deformation or plasticity in the samples.

Actuation fatigue is a measure of actuators life cycle capability before failure of samples. During thermo-mechanical cycling in the fatigue/durability testing, there are a loss of actuation strain and shifts in the shape memory characteristics such as hysteresis and transformation temperatures. To make SMA alloys stable, thermomechanical training (e.g., reverse loading, cycling), alloying and microstructural control (e.g., control of Ni control, precipitation hardening) [166].

Work output is generally defined as applied stress level multiplied by average actuation strain and the how much work is performed per cycle. Due to operating temperature limitation (≤100°C) in commercial NiTi alloys, particular ternary alloying elements (X:Pd, Pt, Au, Zr, and Hf) are shown to increase transformation temperatures [128,131–137,145,167,168]. The Ni–Ti–Pd and Ni–Ti–Pt alloys with transformation temperatures of 100°C–300°C have work output of 11 and 8 J/cm^3, respectively (Fig. 1.6) [169].

However, work output of NiTiPd and NiTiPt alloy with transformation temperatures above 300°C drops off dramatically due dislocation slip at elevated temperatures. At 500°C, they show almost zero work output [169]. In designing of SMA materials for high temperature applications, some underlying characteristics must be determined to achieve optimum work output with reasonable dimensional stability at high temperature. It is imperative to have low stress level for deformation by detwinning, and strong resistance to dislocation slip at high temperatures [169].

1.3.1.1.8 Shape Memory Materials (SSM) applications in oil and gas industry

The SMAs can retain their permanent shape when its original shape is manipulated by heating/cooling cycles that enables the SME or superelasticity to be activated. This

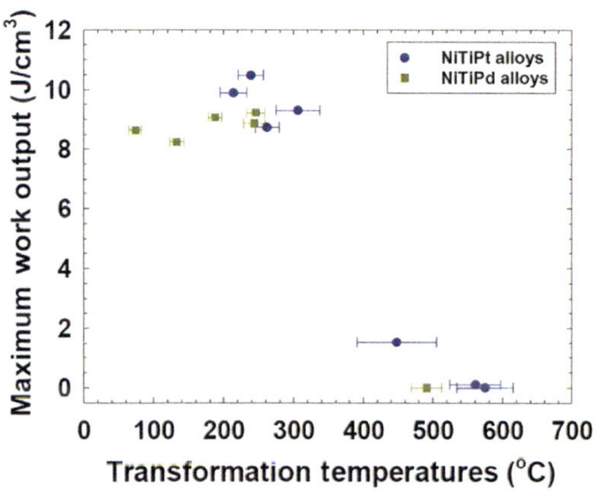

Figure 1.6 Work output of NiTiPt and NiTiPd alloys with respect to transformation temperatures [169].

effect has been utilized in wide range of applications such as aerospace, medical, consumer and oil industries [124].

In the last few decades, there has been growing industrial demand and academic exploitations for the employment of SMAs in offshore oil and gas exploration-production industry to improve the yield, durability and functionality of the oil extraction process. One of the main reasons of the increase in the demand is the continuously improving technology which pushes the current products to be smarter, smaller, cost effective and more powerful. And this demand has yielded numerous promising patents, journals and real-life applications of SMAs in oil and gas industry because they are very strong candidate for offshore oil and gas industry.

While designing a SMA based device there are couple design considerations that needs to be addressed carefully. One of the main considerations is material selection. Oil exploration mostly occurs at high temperatures which is sometimes up to 250°C. Thus, the selected SMA material should be able to operate well at this temperature ranges [170]. In addition, designed SMA devices should withstand harsh subsea environments that requires high corrosion resistance, light weight, easy maintenance and pseudoelastic behavior. Another important design consideration are cost, durability, size, form, etc. Some specific examples to SMA applications in oil and gas industry are actuators, sealers, connectors etc. [171,172].

While the SME occurs through the martensitic phase transition in metallic alloys, it is achieved in SMPs on the molecular level via cross-links and switching segments. Cross-links, which can be either covalent bonds or physical interactions, are used to determine the permanent shape of the polymer and switching segments are used to fix the temporary shape with T_{trans}.

The history of shape-memory polymers goes back to 1940s [173]. However, the importance of SMPs is recognized in the 1960s when self-shrinkable tubes and films are made using cross-linked polyethylene [174]. Remarkable attention is paid to SMPs in the 1980s and growing today with the number of publications in emerging applications.

SMP is a polymer that can memorize its permanent shape. Through an external stimulus such as temperature, SMPs can switch from temporary shape to permanent shape. Strain recovery rate and strain fixity rate are two important quantities describing shape-memory effects. The strain recovery rate is the ability of SMP to memorize its permanent shape and strain fixity is the ability of a material to fix mechanical deformation which can be realized through switching segments. SMPs have advantages of high elastic deformation compared to their counterparts SMAs. Due to their low cost, low density, biocompatibility and biodegradability with human body, they found great potential in many applications, for example, biomedical applications. SMPs can be used at different temperatures and stiffness. Temperature is one of the commonly used stimuli for SMPs. SMPs are multistimuli responsive materials [175], the stimuli could be in form of direct heating, radiation, and laser heating, microwaves [176,177], pressure [178], moisture or solvent vapors [178], and change in pH values [179].

SMPs have several advantages over their SMA peers [180]: (1) wide range of external stimuli e.g., light, magnetic field, electricity, solution, (2) flexible programming through single and multistep processes, (3) possess adaptable characteristics that can be mixed with composites, (4) biodegradability with human body and tissue, and (5) light and can occupy large volume which is useful for aerospace applications, for example, NASA and The Air Force Research Laboratory developed lightweight self-deployable SMPs for potential space applications [181].

Superior recovering capability of SMPs, up to 100%, enables them to be used in many engineering applications. Many application areas are evolved by utilizing distinct characteristic features of SMPs. Even though few products based on SMPs are currently available in the oil and gas industry, there is a growing trend towards SMM based publications and products in the offshore gas and oil industry [172]. For example, SMPs are ideal for sand management and development of packers in downhole due to large strain recovery during SME [172]. Furthermore, main reason behind the failure of metal part is harsh and corrosive environment of the sea. SMPs is ideal material to protect devices against corrosive and erosive environment due to its strong corrosive resistance compared to traditional metals in the sea [172].

Combinations of SMA and polymer are used to design a catheter distal tip, actuator [182]. A smart surgical suture is fabricated from a shape-memory thermoplastics monofilament fiber, and feasibility is tested by elongating fiber by 200% [183]. Furthermore, remote control can be achieved to switch between different phases of SMPs, which can be applied by selecting wavelength, polarization direction and intensity [184]. While most of the SMPs are programmed to be two way (one permanent and one temporary shape), recent developments show that triple SMPs have also been developed [185]. SMPs are still in the development process. Researchers are still looking for the improvement of SMPs using shape memory composites which can solve the problems that SMPs can not.

1.3.1.1.9 Shape memory alloy actuator applications

SMA-based actuators can provide desirable characteristics for oil and gas industries compared to the conventional hydraulic, pneumatic or motor-based actuator systems. Table 1.2 shows that the SMA materials are able to provide relatively more work output and strain compared to various active materials, however they exhibit low frequency response [187]. NiTi, NiTiHf, CuAlNi, and CuZnAl are some examples to SMA materials used to produce high force and strain which are important design considerations for an actuator [188]. In addition to exerting large forces during the phase transition, their stiffness and natural frequency could be tuned through heat treatment (changing the length and thickness of the wires) [189]. Another significant properties

Table 1.2 Property comparison for various materials [186].

	Stress (MPa)	Strain (%)	Efficiency (%)	Bandwidth (Hz)	Work (J/cm^3)	Power (W/cm^3)
Piezoceramic	35	0.2	50	5000	0.035	175
Single crystal piezoelectric	300	1.7	90	5800	2.55	15000
SMA	200	10	3	3	10	30
Human muscle	0.35	20	30	10	0.035	0.35
Hydraulic	20	50	80	4	5	20
Pneumatic	0.7	50	90	20	0.175	3.5

of SMAs that make them a strong candidate for actuator applications are low power requirement, quiet operating ability, ease of application, biocompatibility and high corrosion resistance due to phase transformation [171,172,190,191].

Utilizing the aforementioned properties of SMAs, various actuators are designed, patented and presented in offshore oil and gas area. Hydraulic accumulators [171], blowout preventers (BOPs) [190] and downhole tool actuators [170,172] are some of these actuator applications designed based on SMAs.

In oil and gas industries, many types of downhole tools are used to carry sensors or collect data to obtain geological information inside the well. These devices are mostly utilized to get information from the deep water. Thus, it is crucial to keep them non-damaged or disturbed during and after the operations.

While the process is occurring, these tools need to stay stable to obtain reliable data. For the current devices, this goal is achieved by using motors which are heavy and powerful so that they create disturbances and safety concerns in the system [172]. Eliminating these undesired behaviors are important for better performance and high yield in oil and gas industry. Current downhole devices employ various mechanisms for actuation purposes. SMAs have emerged as a promising candidate due to its inherent abilities such as SME, superelasticity, and high force output.

One of the SMA used device application is a downhole actuator supported with SMA springs [192]. This design is used to stabilize downhole tools and comprised of pistons, expandable arms and SMA springs. In Fig. 1.7, it is depicted that the downhole tool actuator has two states; the first one is passive and the other one is the activated situation. At some specified depth of the deep water, the SMA springs are heated by the electrical current which triggers the piston move down and expand the movable arms.

A reaction force occurs on the arms as they contact the interior wall of the tool. This force helps the tool stabilize during the process. Then, when the current flow to the SMA springs are terminated, the bias springs force the arms turn back to its initial position [172]. For some of the SMA actuators, bias mechanisms are used to return the

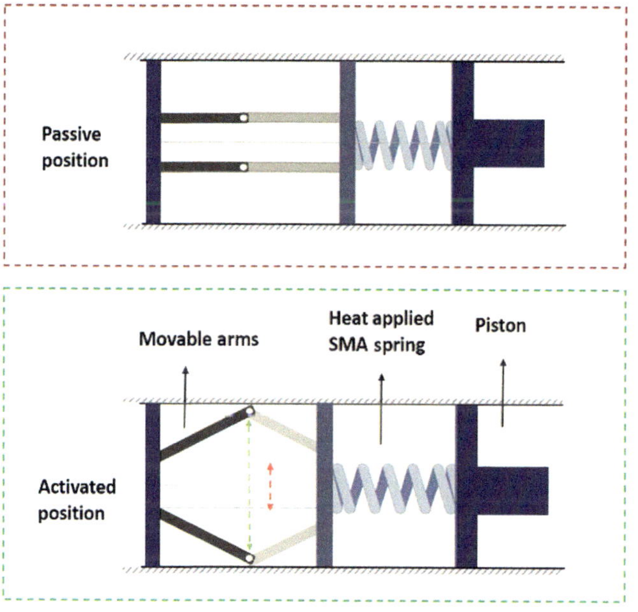

Figure 1.7 SMA downhole actuator using expanding arms. *Adapted from D. Patil, G. Song, A review of shape memory material's applications in the offshore oil and gas industry, Smart Mater. Struct. 26 (9) (2017) 93002.*

device to its original position after exposed to heating. There are couple bias mechanisms such as spring bias, gravitational bias and antagonistic arrangement of two SMA actuators [189], see Fig. 1.8.

Another SMA-based downhole tool is a SMA trigger valve which is prototyped to provide safe operation of the well [170]. Some of the desired performance characteristics of this design is having at least 5-years of installation lifetime, remote operation capability using autonomous energy supply and around 1 minute of response time. Device is composed of SMA rod, heating element, thermal insulators, compression springs and seals as seen in Fig. 1.9. Basically, the SMA rod is heated by using electrical current and as a response to this heat, the rod provides force and lifts the valve seat which opens space for the fluid that enters the hydraulic chamber. Thermal insulation elements are used around the internal components such as the SMA rod and the heating elements to increase the effectiveness of the device.

SMAs can also be used to overcome issues of traditional BOP. BOPs are extensively used as safety equipment in oil and gas industries and they are designed to prevent the well to blowout in case of high-pressure fluid flows. BOPs are mechanically complex structures. Main components of these preventers are seals, pistons, valves, rams and shuttles, see Fig. 1.10. Because these devices are safety equipment, they must

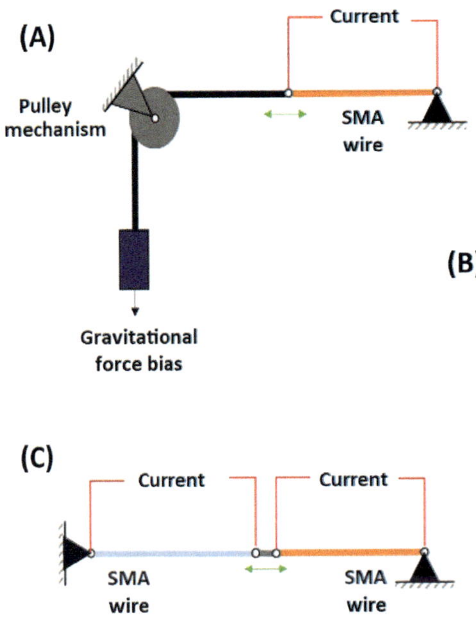

Figure 1.8 Biasing methods in SMA utilized actuators (A) gravitational bias mechanism, (B) Spring bias mechanism, (C) antagonistic arrangement. *Adapted from L. Mikova, S. Medvecka, B.M. Kelemen, F. Trebuna, I. Virgala, Application of shape memory alloy (SMA) as actuator, Metalurgija 54 (2015) 169–172, ISSN 0543-5846.*

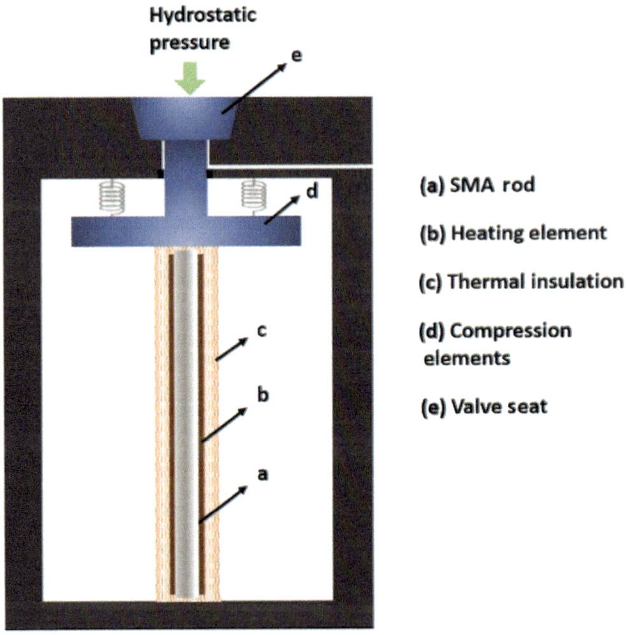

Figure 1.9 SMA-based triggering valve. *Adapted from J. Gore, A. Bowles, M. Maylin, L. Chandrasekaran, M. Buyers, High temperature shape memory alloy actuators through mechanical treatments for an oil & gas down-hole valve, Ind. Commer. Appl. Smart Struct. Technol. 6930 (2008) 69300R.*

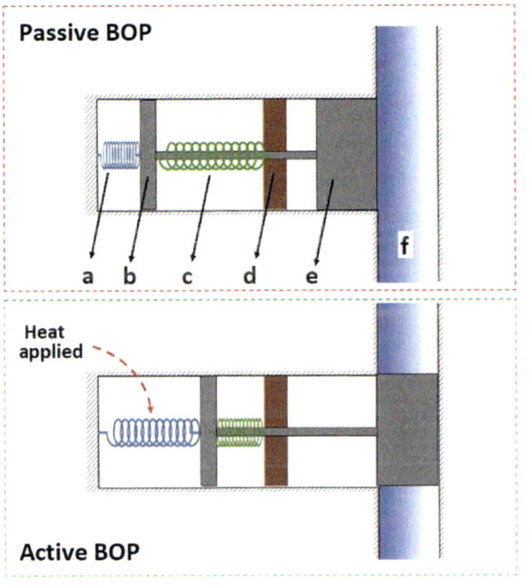

Figure 1.10 SMA actuator using bias spring. *Adapted from G.S. Bigelow, S.A. Padula, A. Garg, D. Gaydosh, R.D. Noebe, Characterization of ternary NiTiPd high-temperature shape-memory alloys under load-biased thermal cycling, Metall. Mater. Trans. A, 41 (12) (2010) 3065–3079.*

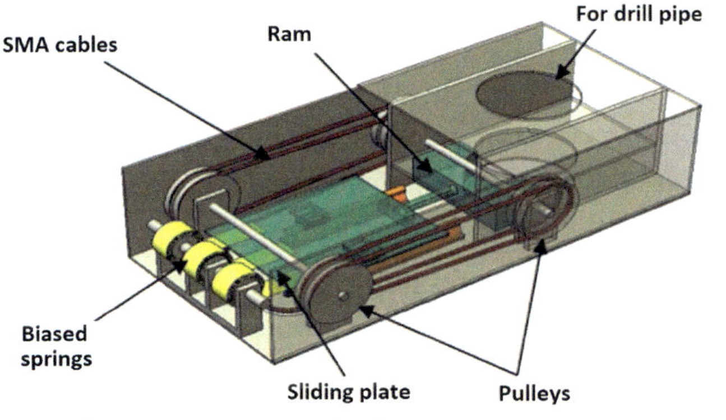

Figure 1.11 Schematic of BOP actuator system [190].

be highly reliable and sustainable. In general, smart materials supported BOPs are consisted of SMA springs (cables), biased spring, moving and blocking pistons.

The device enables the well to be open in its rest position with the help of biased and compressed SMA spring. When an electric current heat the SMA spring, the moving piston acts and moves the biased spring along with the blocking piston which causes the well to be shut-off and cuts the fluid flow. When the spring cools down the process happens in the reverse direction and opens the well. Two methods are described in literature. In the first one (Fig. 1.10), activated SMA spring's force is utilized to push a blocking piston to shutoff well. While in the second method (Fig. 1.11), SMA cables are used to move the ram into shut off position [190]. In both ways, the control is achieved by regulating the electric current provided to SMA. However, with his two methods, there is a risk of using electrical current carrying SMA cables near oil wells, which limits the applicability in practice. To overcome this challenge, an SMA actuated hydraulic accumulator is proposed by the authors of Refs. [193,194]. The proposed accumulator can be located at a safe distance to prevent issues of current carrying SMA.

Patil and Song et al. [171] presented a Nitinol SMA based hydraulic accumulator which is designed to supply high pressure fluid for various tools used in subsea oil and gas exploration phases, the device is presented in Fig. 1.12. These accumulators are also utilized to reduce the effect of hydrostatic head loss regardless of water depth which is a serious issue in offshore oil and gas industry due to the inherent nature of the deep-water environment. The accumulators are expected to enable more efficient and cost friendly deep-water oil extraction systems. They are simply comprised of SMA cables (Nitinol), pistons and cylinder.

The device is triggered by heating the SMA cables which causes the piston to squeeze the fluid and eject it from the cylinder with high pressure. The fluid will be used to activate the blow-out preventer system. In addition to BOPs, these actuators might be also used to activate other actuation mechanisms in subsea oil and gas [171], explorations [172].

Offshore structures are subject to disturbances originating from the sea and the wind. To damp out the disturbances, SMA based dampers can be used to resist vibrations stem from the sea and the wind [195,196].

Vibrations sometimes can be because of equipment changes and building modifications. Additional on-site reinforcement is required to prevent from fractures in the main body of the structure. Structures are traditionally reinforced by introducing additional load sharing structures using welding. However, these processes require to keep the load within certain safety limits. To do this, a common way is to isolate the loads during the addition of structural members. This way of adding structural members requires to recess the system, which is costly. Furthermore, usage of welding cause

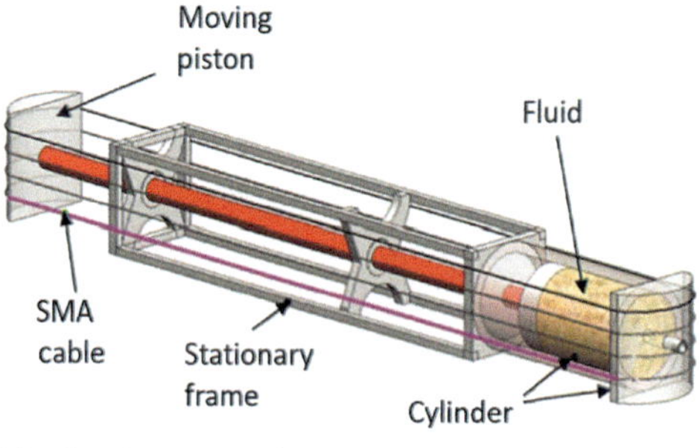

Figure 1.12 SMA-actuated hydraulic accumulator [171].

safety concerns within the platform. Cook et al. used preheated and precooled shims of SMAs to connect additional reinforcement structures [197].

Cementing is an important part of the structural reinforcement. To provide uniform thickness during the cementing process, centralizers are used, which improves the sealing capability of the concrete. Traditionally, centralizers are consisted of moving mechanical parts, they are subject to damage as they get struck by the wall of the well. An SMA based centralizer is suggested by the authors of [198]. The proposed centralizer has radial blades. Radial blades are compressed at low temperature before putting into the well. Once the centralizer placed into the desired location, SMAs are activated through heat and switched to the austenite phase. To minimize the flow space occurring on radial blades during cementing, Mathew et al. tried to minimize gas migration during the cementing process by using SMM based device [199].

Fasteners are used to connect massive structures in oil and gas industries. Torqueing keeps structure connected by applying loads and provides friction between fastener thread and cavity thread, thus unscrewing of the nut and bolt is avoided arising from the vibration of the fastener structure. Traditional solutions to fasteners are expensive, time-consuming and require diligent attention from the workers, yet still, they may not be consistent all the time. Scaling to the large system is also another challenge. A self-torqueing fastener is reported in Ref. [200] using SMA, by which consistent and easy torque is achieved. Such type of selftorqueing fastener do not require additional attention from workers and can still provide consistent torques on connections.

Two type of connectors in underwater connection are hydraulic and electrical connectors, which are used commonly used in offshore oil industry. Efficient functioning of the platform depends greatly on these connectors because these connectors carry crucial control and sensor signals during the operation. A leaking hydraulic connector or coupling can both reduce the efficiency and damage the line. Non-SMA based connector and coupler are also available, however, SMA based counterparts are much preferred due to their long-lasting life, resistance to corrosion and erosion, perfect sealing and so on. SMA based connectors are easy to install and can help to get rid of bulky connector housing [201]. SMA based connectors has also shorter connection time, which is both time-efficient and cost-effective [202].

In oil and gas industry, expansion mandrels are used for sealing purposes in small boreholes. The major effort involves the sealing. Insertion of the mandrel is carried out first into the borehole and it is activated inside. SMA-based ring can circumvent this cumbersome process. In addition, damages within boreholes can also be repaired. Several methods are produced in the literature initiated by Rogen and Adnyana [203]. In a similar vein, an SMP-based approach is also reported in Ref. [204]. Shape memory materials are attached onto the expanded packer such that it covers packer. It is activated using heat or hot wellbore fluid and transformed to the austenite phase.

High pressure fluids can also be used for improved sealing [205]. In addition, SMM also used to improve the efficiency of the some of the downhole tools. Ingram et al. proposed a few methods to improve the sealing system [206]. SMA ring was proposed by the authors of Ref. [207] to increase the bearing performance.

1.3.2 Piezoelectric materials

Piezoelectric materials can generate electricity in the application of a mechanical stress. They will also work in reverse, exhibiting a strain when subjected to an electric field. Pierre and Jacques have discovered piezoelectricity in 1880 [208].

Piezoceramic ceramic materials consist of atoms with negative and positive charges and ionically bonded. Due to lacking a center of symmetry in a unit cell, application of a mechanical stress generates net movement of negative and positive ions, leading to a polarization or electrical dipole [209].

1.3.2.1 Applications of piezoelectric materials in oil and gas industries

With increasing demand for energy, search for hydrocarbon resources into hotter and deeper well is increasing [210]. Piezoelectric materials find variety of applications in the oil and gas industries, particularly in ultrasonic bore-hole imaging, sonic logging-while-drilling, precision pressure measurement and marine seismic survey [210]. Piezoelectric hydrophones are deployed in a benign environment on a massive scale. Thus, device reliability and unit cost are significant factors. Other three applications occur downhole in a severe environment with high pressure and elevated temperature. Small but highly valuable piezoelectric devices are involved in the remaining three applications [210]. Piezoelectric materials are extensively used in measurement on different length scale to characterize, quantity, and locate hydrocarbon deposits. In ocean seismic surveys, large numbers of piezoelectric hydrophones are employed to generate imaging arrays. For oil—well applications, piezoelectric devices are used to perform acoustic and hydrostatic pressure measurements in the ultrasonic and sonic regimes [210].

1.3.2.1.1 Marine seismic survey

Marine seismic surveys are utilized in the exploration phase to discover hydrocarbon reservoirs [211,212]. Aim of seismic surveys is to obtain a high-resolution image of subsurface geology to delineate and identify hydrocarbon deposits.

Piezoelectric material, lead zirconate titanate (PZT) are most widely used in hydrophone applications [213,214]. While air guns send signals, piezoelectric hydrophones receive signals as shown in Fig. 1.13.

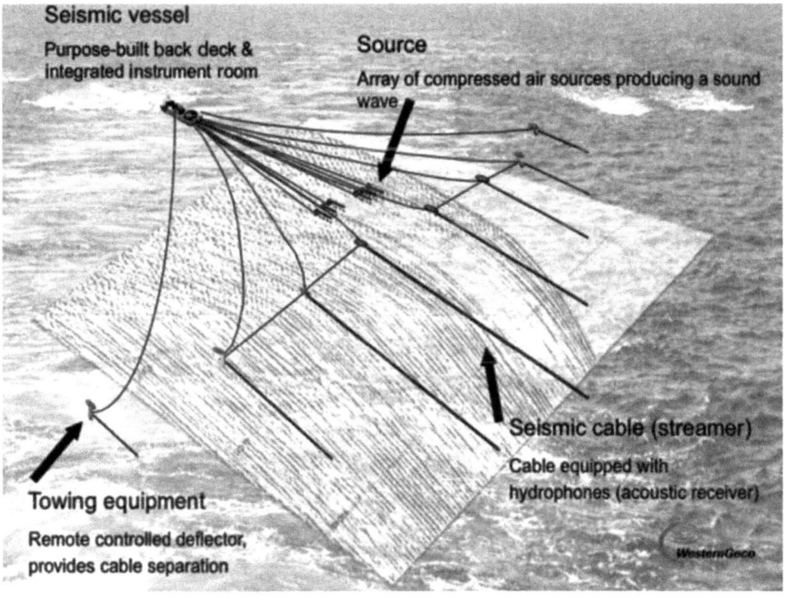

Figure 1.13 Marine seismic survey is performed by expanded tow array made of many hydrophones [210].

1.3.2.1.2 High temperature and pressure operation

Piezoelectric materials can be employed in harsh environments in a wellbore [210]. Fig. 1.14 indicates various pressure and temperature regimes where borehole logging equipment is exposed. For instance, degradation of elastomeric seals occurs when operating temperature and pressure reach to 150°C and 68.9 MPa (10,000 psi) [215,216] of logging tools operates up to 137. 9 MPa (20,000 psi) and 175°C, hence occupying majority of high temperature, high pressure (HPHT) zone [215,216].

Ultra HPHT is new frontier yet undiscovered has challenges. While new wells need to reach even more depths, operating borehole logging tools are subjected to corrosion because of downhole chemistry, erosion as a result of debris and high-G shocks [210]. Piezoelectric devices with low volume, high value components appear promising in the borehole applications.

1.3.2.1.3 Pressure gages

Capacity of hydrocarbon reservoir is estimated using pressure measurement by determining contact of water/gas/oil. Pressure-logging job is showed in Fig. 1.15 [210]. The device is deployed to the depth zone of reservoir and existing hydraulic pressure in the rock pores is determined.

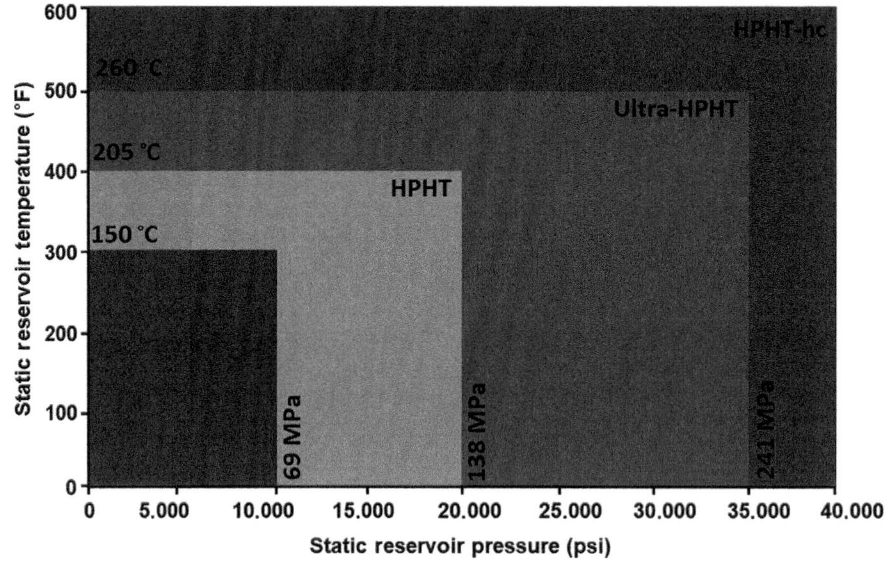

Figure 1.14 Pressure/temperature operating regimes [210].

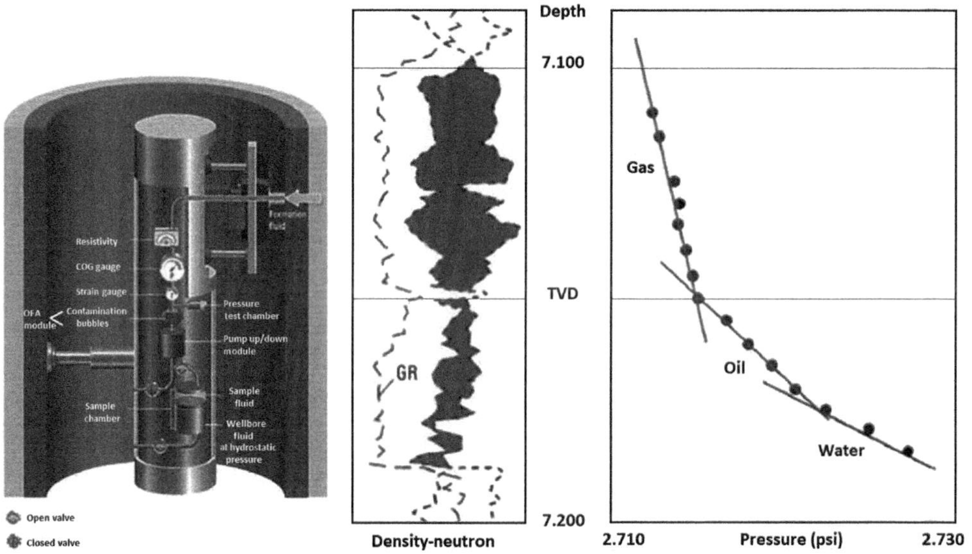

Figure 1.15 Measurement of hydrostatic pressure vs. depth (right) by pressure gage in borehole (left). Precise pressure gage accurately delineates oil, water and gas layers. The middle plot represents a nuclear log indicating availability of water and hydrocarbon reservoir (shaded area) [210].

Quartz technology is used for pressure sensor. Measurement mechanism depends on changes in properties of materials such as geometry, mass density and elastic stiffness with respect to temperature and pressure [217]. Adding piezoelectric into an electrical oscillator with high resonator improves precise measurement of device under harsh temperature and pressure environments [218].

1.3.2.1.4 Ultrasonic imaging

Ultrasonic techniques image the open borehole wall and evaluate integrity of cement and changes in steel casting [219,220]. It is a nondestructive testing in the oil wells. Due to attenuation in borehole well, operating frequency is restricted to below 1 MHz. Available Ultrasonic Imager (USI, Schlumberger) uses a single transducer which rotates at speed to give azimuthal scanning. Fig. 1.16 presents production of 500-kHz transducer array which contains active layer made of piezocomposite material [210]. Thanks to piezocomposite technology—it is possible to modify electromechanical behaviors of composite piezoelectric to certain applications. It is possible to reduce considerably acoustic impedance of the transducer by changing the volume fraction of a high-coupling piezoceramic (PZT-5A) [210,221].

Operation of an ultrasonic transducer takes place in the borehole within a low-impedance liquid environment [221]. Broader bandwidth of operation and better coupling of acoustic power are obtained when transducer impedance is brought closer to those of borehole liquid [210].

1.3.2.1.5 Sonic logging-while-drilling

In the liquid-filled wellbore, sonic logging is used to measure elastic wave speeds of rock formation [210]. Primary interest of sonic logging is shear and compressional velocities (Vs and Vp). Sound waves propagate through rock formation. It is affected

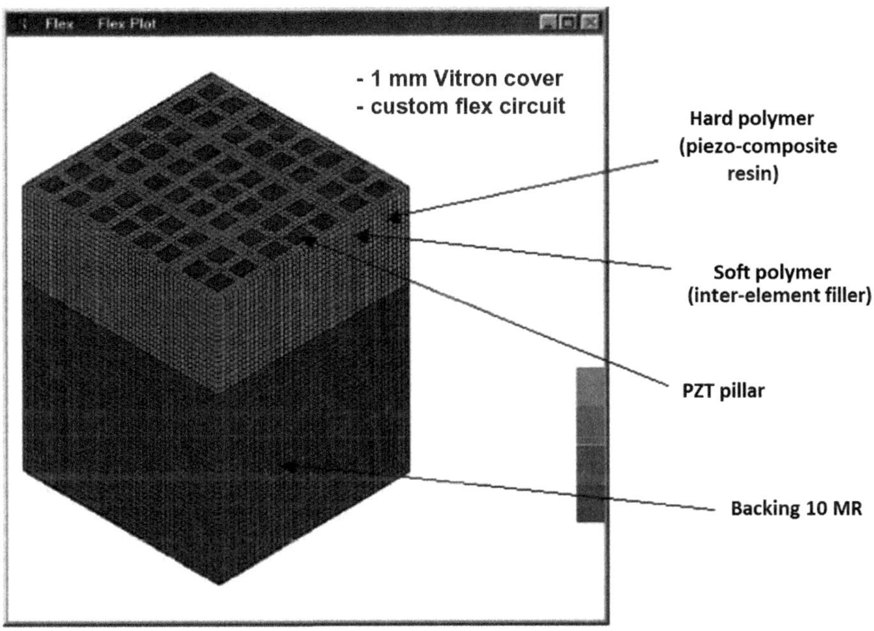

Figure 1.16 High temperature, high pressure (HTHP) 1–3 piezocomposite transducer array production [210].

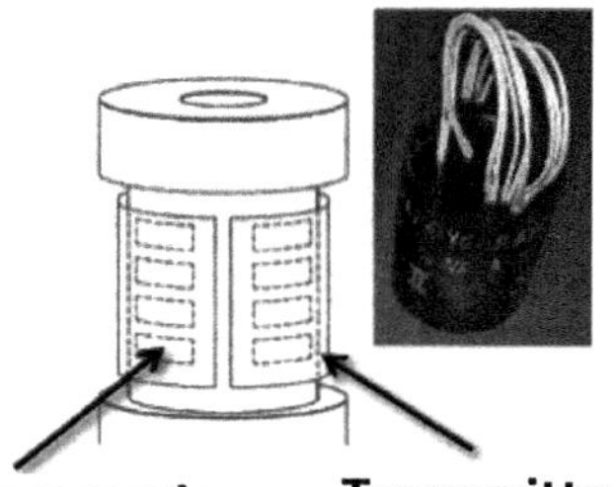

Figure 1.17 Design of sonic multipored transmitter containing four sectors [210].

by mechanical properties of rock formation. Sonic tools infer properties and factors affecting borehole stability such as tectonic stresses, formation pore pressure [222].

To improve resolution and acquire geometrical model of the subsurface geology, sonic velocity method has been widely employed. Thus, wellbore trajectory can be accurately deployed for optimum production of hydrocarbon deposits. Operation of sonic logging occurs within the range of 1–20 kHz [210].

Logging while drilling (LWD) sonic transmitter consists of piezoceramics (Fig. 1.17) [210]. There are four sectors which contain several piezoceramic plates connected in parallel. Electrical excitation (wide-band pulse) has amplitude of 3 kV [210].

1.3.3 Supramolecular assembly solutions in unconventional oil and gas recovery

As a source of abundant, high-quality underground resources, oil and natural gas extracted from shale formations are considered to be significant in meeting the future global energy demand. Production of shale oil and gas has increased substantially in recent years due to advances in methods of shale gas extraction technology including horizontal drilling and hydraulic fracturing [223,224]. Hydraulic fracturing (fracking) fluids, which mainly consist of water, proppant and various chemicals, are used in these operations [225]. However, recovery factors using these fluids are still considered to be low, and the industry has recently been seeking to employ methods and injection fluids including those used in conventional EOR to enhance the efficiency and amount of recovery. For instance, feasibility of various EOR methods such as water and chemical flooding and gas injection in shale oil/gas recovery has been reported [226,227]. Also CO_2 injection efforts in Eagle Ford Shale has been shown to enhance the crude oil production [228].

Characteristics and properties of injected fluids play a major role in efficiency and amount of recovered oil and gas. Shales are unconventionally tight hydrocarbon reservoirs, that is, very confined and low-permeability environments, in which the injected fluids must withstand extreme pressures and perform the required function. Hence, two main factors are significant in design and development of high-performance injection fluids: (1) flow characteristics under confinement, elevated temperatures and high pressures and (2) effect of the flow characteristics on cumulative oil and gas recovery. Smart injection fluids involving stimuli-responsive supramolecular assemblies have recently been developed and show promising potential to enhance hydrocarbon recovery due to their unique and superior rheological properties and endurance against extreme conditions.

Supramolecular are well-defined species containing discrete number of chemicals or building blocks [229] held together by molecular interactions such as electrostatic forces, hydrogen bonding, van der Waals, and steric interactions [230,231]. Balance of these interactions between the building blocks defines the spatial configuration and orientation of the building blocks and their physicochemical properties. Hence, materials properties of supramolecular assemblies can be adjusted by manipulation of the noncovalent forces that hold the building blocks together. In addition, supramolecular materials are, by nature, dynamic materials whose building blocks are linked together via reversible connections; and, therefore, may undergo assembly-disassembly processes under specific conditions [232]. In other words, they are dynamic materials and can generally select their constituents in response to external stimuli or environmental factors. Because of their tunable properties, ability to form gels at extremely low concentrations, and stimuli-responsive properties; supramolecular assemblies have recently received an increased attention as fracturing fluids.

1.3.3.1 Currently available viscosity modifiers and challenges

Supramolecular assemblies can mainly be used as additives in fracturing fluids to control viscoelastic properties of the fluid and enhance recovery by delivering proppant packs to low-permeability formations in shales, that is, fissures. There are various types of viscosity modifiers that are usually added to the injection and fracturing fluids to increase the viscosity [233]. In conventional oil recovery, addition of viscosity modifiers (mainly polymer-based) is used to obtain a lower mobility ratio and reduce viscosity mismatch at the fluid–oil interface, to prevent fingering and thus attain a higher sweep efficiency [234]. In hydraulic fracturing, viscosifying the fluid is to increase fracture width so it can accept higher concentrations of proppant, reduce fluid loss to improve fluid efficiency, improve proppant transport and reduce friction pressure [235]. First of all, polyacrylamide-based solutions and guar derivatives [hydroxypropyl guar (HPG) and carboxymethyl guar (CMG)] are the most commonly gelling agents for fracturing [236,237]. Hydroxyethyl cellulose (HEC) and carboxymethylhydroxyethyl cellulose (CMHEC) are other common types of natural polymers used as viscosifying agents [236,238]. The main limitation of the polyacrylamide-, guar- and cellulose-based fracturing fluids is poor fluid conductivity, which is defined as the width of the fracture multiplied by the apparent permeability of the proppant pack. Another class of viscosifying agents is viscoelastic surfactants, which are aqueous, polymer-free fluids whose viscosity originates from the assembly of surfactant molecules. Assembly and aggregation of the surfactants initially begin with formation of micelles, then turning into bilayers, and ending with rod- or worm-like micelles, which are highly networked and entangled structures. Polymer-free formulation of viscoelastic surfactants enable highly conductive proppant packs with no polymer damage [239]. In addition, minimizing fracture height growth and increasing effective fracture length are a few other advantages of using viscoelastic surfactant fluids [239].

Supramolecular gels is also another type of gelling agents for potential use in fracturing fluids [240], which are formed via self-assembly of molecules through noncovalent interactions such as hydrogen bonding, π—π stacking, van der Waals forces, and hydrophobic and electrostatic interactions. Recently, various examples of supramolecular gels have been reported, most of which involve one of the following building blocks: amino acids, carbohydrates, and bola amphiphiles. *Amino acid* molecules involve concurrent actions of hydrophobic effects and hydrogen bonding, and constitute one of the largest groups of hydrogelators [241,242]. *Carbohydrates* are hydrophilic building blocks that are able to form multiple H-bonds due to the presence of hydroxyl groups, which also favor solubilization in water [243]. Previous studies have shown that carbohydrate-based molecules facilitate supramolecular gelation behavior [243,244]. Furthermore, *bolaamphiphiles* have recently received an increasing attention as supramolecular gelators due to their thermoreversible and rapid gelation properties. For instance, it has been shown that they form gels at concentrations of 2—10 wt.% over a wide range of pH (2—12) in buffered aqueous solutions of varying ionic strengths [245].

1.3.3.2 Structure and dynamics of supramolecular gels

Structure and dynamics determine the physicochemical properties of supramolecular gels. Therefore, to better understand their gelation properties, their structure and dynamics need to be investigated. Various spectroscopic techniques can be applied to supramolecular gel systems, including NMR, UV, and IR spectroscopies, all of which can provide information on the molecular structure of the gels [246]. For example, comparative IR and NMR data in the gel and sol phases have indicated the extent of hydrogen bonding in the assemblies [247]. On the other hand, morphology and geometry of the supramolecular assemblies can be observed via microscopy techniques, including scanning (SEM) and transmission (TEM) electron microscopies [248,249]. While there are a significant number of studies focusing on structure and dynamics of supramolecular gels under bulk conditions and mild pressures, we do not have much information about the nature of supramolecular assemblies under confinement environments and extreme pressures.

1.3.3.3 Rheological properties of supramolecular gels

Wormlike micelles, which are elongated and semi-flexible aggregate structures resulting from the supramolecular assembly of surfactant molecules in aqueous solutions, are rheologically most common types of supramolecular gels [250]. Because wormlike micelles entangle above a critical concentration as polymers, their aqueous solutions then become viscoelastic. Previous studies have shown that, in the entangled regime, the properties of wormlike micelles depend on the micellar volume fraction (i.e., the surfactant concentration) and not on the properties of individual micelles such as their average length or their length distribution. Therefore, their rheological properties such as the elastic modulus G_0, and the static viscosity η_0, could be expressed using scaling laws [251,252]. For other type of supramolecular assembly systems, most rheological studies primarily generate experimental data relating shear rate and the storage modulus, G'; shear rate and the loss modulus, G"; zero frequency viscosity and temperature; and zero frequency viscosity and concentration. Overall, most of the work focusing on rheological properties of supramolecular assemblies have generally been conducted under bulk condition and at pressures near atmospheric conditions. However, potential use of supramolecular gels in oil/gas recovery requires determination of their rheological properties under confined environments and high pressures. Investigation of properties of supramolecular assembly solutions under such extreme conditions will provide insights to their use in oil and gas industry.

1.3.3.4 Recent research efforts and selection criteria

Structure, dynamics, and interfacial properties of supramolecular assemblies are related to one another under confinement and extreme pressures. For this purpose, some types of supramolecular systems have been investigated. Since there are numerous types of supramolecular assembly systems, the selection should be based on the potential of the selected system to the area of fracturing fluids and capable of enhancing the hydrocarbon recovery from tight formations. Therefore, it is desirable that the supramolecular system is stable at wide range of temperatures. The previous research have shown that supramolecular systems with metal-coordination typically have much higher activation energy of disassembly compared to the supramolecular systems without metal-coordination [253]. In addition, for most materials, an increase in temperature leads to a decrease in viscosity. This is a big problem because the fracturing fluids need to have relatively large viscosities at the ground level to account for the loss of viscosity due to the geothermal gradient-induced heating and to maintain high viscosity values required near fissures to effectively transport proppants. Thus, a supramolecular system having an increased viscosity with increasing temperatures can lead to significant advances in the fracturing fluid designs. Recent studies have shown that supramolecular system having twisted nanoribbons structures obtained through metal-coordination and hydrogen bonding of sodium cholate and lanthanum or calcium ions have such a viscosity behavior [254]. Alternatively, adjustable (tunable) viscosity with respect to another external stimulus can be used: For instance, the viscosity can be maintained at low values during injection while the viscosity can later be increased with an external stimuli such as pH or salinity [255]. It has recently been shown that self-assembly of long-chain amino amides with maleic acid through electrostatic and hydrophobic interactions leads to supramolecular materials with pH-sensitive viscosities. These supramolecular systems show gelation behavior at concentrations as low as 0.5% weight. On the other hand, a contemporary key issue is the environmental concerns associated with hydraulic fracturing fluids such as potential contamination of ground water, potential migration of hydraulic fracturing chemicals to the surface, and potential mishandling of waste. Therefore, designing supramolecular systems based on natural products, which are environmentally friendly and can be obtained in a sustainable fashion, can be beneficial for the field of fracturing fluids. The self-assembly of β-cyclodextrin (which can be obtained from starch) and stearic acid (which can be obtained from coconuts) through hydrogen bonding and hydrophobic effect gives rise to supramolecular assemblies with multilayer tubular structures. Such assemblies are shown to display ultralong persistence lengths and high rigidity, which may also persevere against high pressures. Based on these points, the selection can be shortlisted to three classes of supramolecular systems as shown in Fig. 1.18.

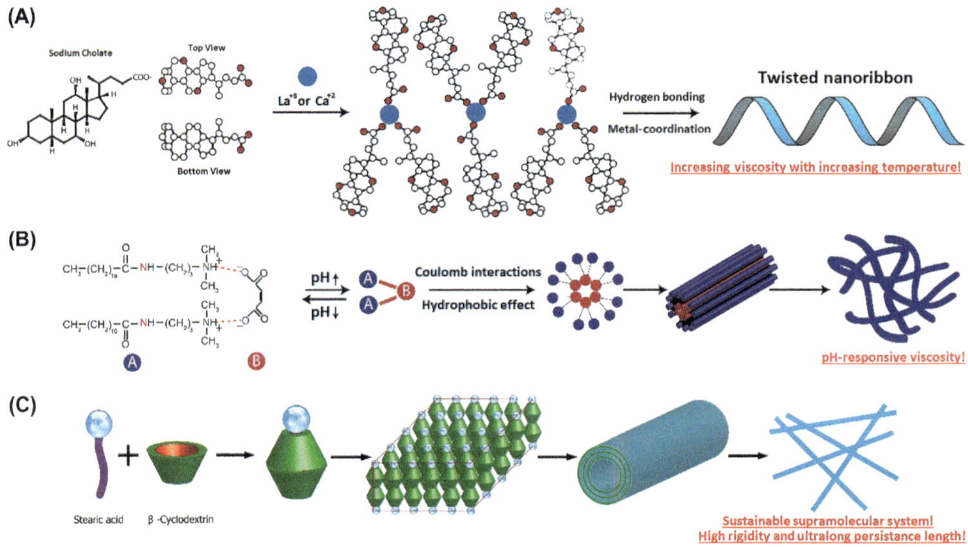

Figure 1.18 Possible types of supramolecular assemblies to be used in oil/gas recovery. (A) The first type involves the formation of twisted nanoribbons through metal-coordination and hydrogen bonding of sodium cholate and lanthanum or calcium ions. The unique feature of such assemblies is that their viscosity increases with increasing temperature. (B) The second type involves the assembly of a long-chain amino amide with maleic acid through electrostatic and hydrophobic interactions. The viscosity of such a supramolecular assembly can significantly and reversibly adjusted via pH-stimulus. (C) The third type involves the assembly of β-cyclodextrin and stearic acid to produce multilayer tubular structures through hydrogen bonding and hydrophobic effect. Such assemblies display high rigidity and ultralong persistence lengths. In addition, the starting materials can be obtained from natural products and are environmentally benign [254,255].

A recent discovery was a new type of temperature (T)-responsive supramolecular system involving the assembly of the long-chain amino amide compound described in Fig. 1.18B with citric acid through electrostatic interactions and hydrogen bonding. This system is chemical analog of the supramolecular system in Fig. 1.18B. The new temperature-responsive system shows high viscosity at elevated temperatures, the same outstanding property offered by the supramolecular system in Fig. 1.18A and provides a huge benefit for fracking applications that require high operational temperatures.

1.3.3.5 Visual observation of pH- and T-responsive gelation behavior

pH and temperature-responsive supramolecular solutions were visually observed. Initially, the pH sensitive solutions with a single-phase milky color at a pH around 6.5 (two identical solutions in Vials A and B as shown in Fig. 1.19A and B). Increasing pH of the solution in Vial B via base addition resulted in an abrupt increase in viscosity as indicated in Fig. 1.19C. Then, the pH of the same solution was reduced by acid addition, which led to the recovery of this solution back to its original viscosity. It was shown that the process was fully reversible as the pH of the solution was increased and decreased multiple times. Following the confirmation of the adjustable viscosity behavior, the solution in Vial B was set to a high pH while that in Vial A remained in its original pH, both solutions were added sand pieces in order to observe the microstructural changes occurred as a result of pH increase. The SEM images obtained from each solution are shown at the right end of Fig. 1.19. Network formation and entanglement, which are the main driving mechanisms of gelation and high viscosity, was observed from the high-pH solution while such networks and entanglements were not observed in the low-pH solution.

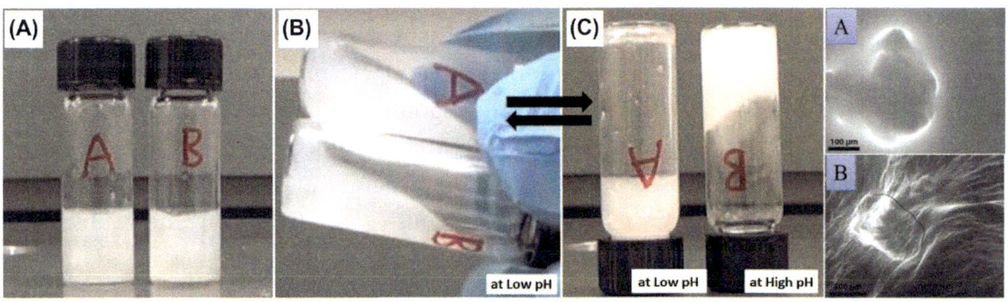

Figure 1.19 (A,B) Identical pH-responsive supramolecular assembly solutions in Vial A and Vial B. (C) Vial B is added NaOH and pH is increased, while Vial A remains the same. Increased pH increases the viscosity and gelation. When the pH is reduced the viscosity goes back to its original level (reversible viscosity). Sand pieces were added to both vials and then SEM images were obtained from the solutions (right end). Entanglement and network formation are clearly seen from the solution in Vial B at a higher pH value.

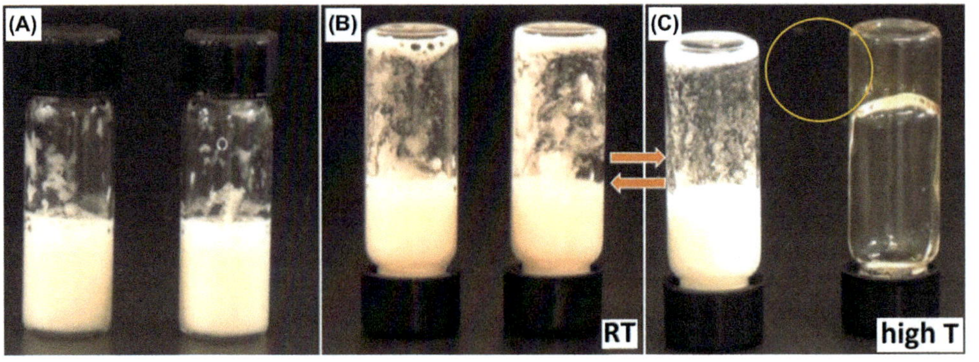

Figure 1.20 (A,B) Two identical temperature-responsive supramolecular assembly solutions. Same viscosity level is observed when the vials are shaken and turned upside down at room temperature. (C) The vial in the right is heated up to 90 °C and the solution changes color and turns into a highly viscous gel as indicated in the circle [256].

Then, similar visual tests were made for the T-responsive supramolecular solution. The original T-responsive solutions had a milky meta-stable phase (Fig. 1.20A and B). Heating of the suspension to 90°C led to a staggering result as the solution turned to be a single-phase transparent solution with a very high viscosity (Fig. 1.20C). The solution was subjected to heating-cooling cycles and a fully-reversible transition was observed similar to the case of the pH-responsive solutions. The distinguishing point of this solution type is the ability of gelation at higher temperatures while the viscosity of most fluids descends with increasing temperatures.

To better evaluate the gelation behavior and rheological properties of the supramolecular assembly solutions, flow-sweep experiments were conducted to determine the viscosity trends. For comparison, the same experiments were performed for polyacrylamide solutions, which are known to be commonly used displacement fluids [257]. Fig. 1.21 depicts the viscosity trends of the pH- and T-responsive supramolecular solutions, and corresponding results obtained from the polyacrylamide solutions. In both cases, the viscosity of the polyacrylamide solutions did not show a significant difference in viscosity as a function of pH and temperature. The pH-responsive solution showed a substantial increase in viscosity from pH 4 to pH 8, and reached 450 Pa.s that is 4.5×10^5 times more viscous than water (1 mPa.s) [255]. It can be attributed from the rheological data that the network structure responsible for the viscoelastic behavior changes with respect to pH. On the other hand, the viscosity variation of the temperature-responsive solution was much higher as it increased by eight orders of magnitude from around 10 mPa.s at RT to 10^9 mPa.s at 90°C. Like the pH-responsive system, the viscosity variation in the T-responsive system was reversible.

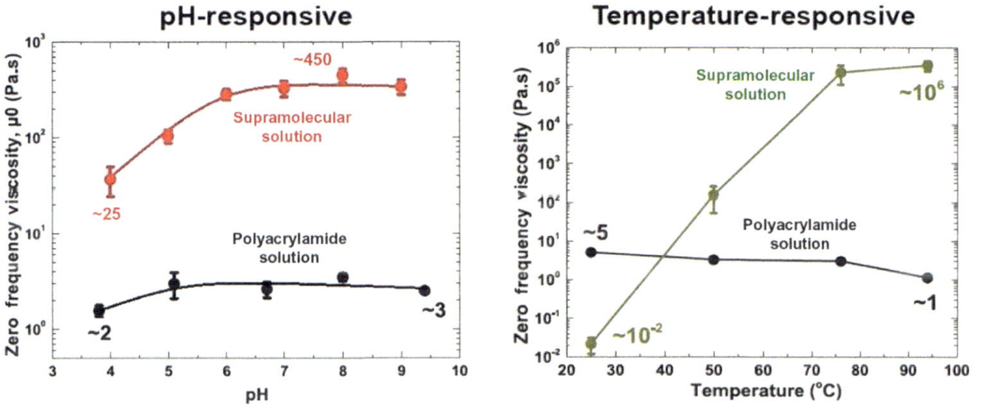

Figure 1.21 (Left) Zero frequency viscosity of pH-responsive solution and polyacrylamide solution at several pH values obtained at room temperature, and (Right) Zero frequency viscosity of temperature-responsive solution and polyacrylamide solution at several temperatures. Data were obtained from solutions with 2% wt. concentration [255,256].

Overall, pH- and temperature-responsive supramolecular systems reported in section have great potential to be used in conventional oil recovery and fracking fluids with their outstanding features including: (1) adjustable viscosity behavior which enable high efficiency in the recovery by setting the viscosity of the injected fluid to a desired value at different stages of oil/gas recovery. (2) self-healing property of these solutions that will help them repair themselves upon breakage when subjected to high temperatures and sudden extreme shear stresses during the operation. (3) sustainability and environmentally-friendly nature of these solutions that will eliminate the harmful effects of polymers and other types of gelling agents in injection fluids. Especially, the thermo-responsive system with its unique property of increased gelling strength at elevated temperatures has a great potential to maximize the oil/gas production at deep and tight shale formations where the injection fluids suffer from shear degradation and geothermal gradient.

1.4 Conclusion

The world of materials is an exciting and challenging field of research since it has always played a dominant role in the evolution of human civilization. The demands from aerospace, defense, automotive, oil and gas, and industrial branches on more advanced and innovative materials has led to the development of a new generation of materials with much better performance and capabilities than the existing conventional structural and functional materials. "Smart materials" refer to those materials which can change their composition or structure, their electrical and/or mechanical properties, or even their functions in respond to some environmental stimuli like temperature, pressure, magnetic and electric fields, chemical, or nuclear radiation. Over the last two decades, smart materials have received significant attention over a broad range of engineering applications because of their unique and inherent characteristics for actuating and sensing aspects.

Furthermore, most widely known SMMs are SMAs, and SMP. SMA can endure erosive stresses due to martensitic phase transformation behavior, which in turn may minimize the downhole well bores erosion. Numerous limitations and defects exist for method of traditional passive damping, while superelastic properties of the SMA can enable enough energy dissipation. Even though, SMA is not suited for high frequency required applications, it is a viable candidate in low frequency required applications owing to high number of cycles to failure, adjustable actuation strain and high energy density compared to its conventional counterparts. One another significant advantages of these materials is to offer reduction in maintenance time and human labor under the ocean as well as cost-effective deep-water oil extraction. As a consequence of growing interest in this field, extensive research has been performed on Smart and State-of-the-art materials.

This chapter describes recent research and technical achievements in two sections prepared by different researchers. Coverage is equally divided among state-of-the-art materials and smart materials. The topics are discussed in a concise, comprehensive way that will give the chapter lasting value. As a result, the era of smart and state-of-the-art materials has started. The objective of this chapter is to define the field of smart and state-of-the-art materials, together with its current status and potential benefits. However, more focus will be devoted to additives, NPs, shape memory materials and piezoelectric materials are discussed. Finally, supramolecular assembly solutions are reviewed. In the literature review, advantages and disadvantages of each control scheme are discussed so that potential researchers can develop more effective strategies to achieve higher control performance of many application systems utilizing smart and state-of-the-art materials.

References

[1] J.K. Cowan, Rapid enumeration of sulfate-reducing bacteria, Corros. Nace Int. 05485 (2005).

[2] E.A. Morris III, D.M. Dziewulski, D.H. Pope, S.T. Paakkonen, Field and laboratory studies into the detection and treatment of microbiologically influenced souring (MIS) in natural gas storage facilities, in: Field and laboratory studies into the detection and treatment of microbiologically influenced souring (MIS) in natural gas storage facilities, U.S. Department of Energy Office of Scientific and Technical Information.

[3] R. Prasad, Pros and Cons of ATP (adenosine triphosphate) measurement in oil field waters, in: Proceedings Volume. NACE Corrosion 88 (St Louis, MO, 3/21–3/25), 1988, pp. 571–579.

[4] L. Barton, W.A. Hamilton (Eds.), Sulphate-Reducing Bacteria: Environmental and Engineered Systems, Cambridge University Press, Cambridge, 2007.

[5] A. Rajasekar, B. Anandkumar, S. Maruthamuthu, Y.P. Ting, P.K.S.M. Rahman, Characterization of corrosive bacterial consortia isolated from petroleum-product-transporting pipelines, Appl. Microbiol. Biotechnol. 85 (4) (2010) 1175–1188.

[6] N.A.F. Sarioglu, R. Javaherdashti, Sarioğlu_1997_International-Journal-of-Pressure-Vessels-and-Piping.pdf, Int. J. Pres. Ves. Pip. 73 (2) (1997) 127–131.

[7] C.W.S. Cheung, I.B. Beech, S.A. Campbell, J. Satherley, D.J. Schiffrin, The effect of industrial biocides on sulphate-reducing bacteria under high pressure, Int. Biodeterior. Biodegrad. 33 (4) (1994) 299–310.

[8] J. Boivin, Oil industry biocides, Mater. Perform. 34 (2) (1994) 65–68.

[9] Y. Zhou, Bactericide for drilling fluid, Fluid Complet. Fluid 7 (3) (1990) 10–12.

[10] M. Smith, K., Persinski, L.J. Wanner, "Effervescent biocide compositions for oilfield applications," 20080004189, 2008.

[11] D. Oppong, V.M. King, Synergistic antimicrobial compositions containing a halogenated acetophenone and an organic acid, WO Patent 9 520 319, 1995.

[12] D.O. Kriel, B.G. Crews, A.B. Burger, E.D. Vanderwende, E. Hitzman, The efficacy of formaldehyde for the control of biogenic sulfide production in porous media, in: Proceedings Volume, 1993, pp. 441–448.

[13] M.A. Sears, J.T., Mueller, R., Reinsel, Inhibition of sulfate-reducing bacteria via nitrite production, 9 (1996) 612–867.

[14] D. Oppong, C.G. Hollis, Synergistic antimicrobial compositions containing (thiocyanomethylthio) benzothiazole and an organic acid, 9 (508) (1995) 267.

[15] A.A. Karaseva, E.V. Dedyukhina, S.N. Dedyukhin, Treatment of water-based drilling solution to prevent microbial attack – by addition of dimethyl-tetrahydro-thiadiazine-thione bactericide, 2 (036) (1995) 216.

[16] L.I. Khanlarova, A.G. Musaev, M.R. Samedov, A.M. Kandinskaya, L.I. Gasanov, A.G. Alieva, Inhibiting growth of sulphate-reducing bacteria-involves introducing diammonium salts of tetrahydrophthalic acid or methyl-tetrahydrophthalic acid into bacteria-containing circulating water, 1 (828) 917, 1993.

[17] J.W. Boivin, J.W. Shapka, R. Khoury, A.E. Blenkinsopp, S. Costerton, An old and a new method of control for biofilm bacteria, in: Annual NACE Corrosion Conference (Corrosion 92), 1992.

[18] V.K.U. Muganlinskij, F.F. Lyushin, M.M. Samedov, A.M.O. Akosta, Suppressing activity of sulphate-reducing bacteria on petroleum extraction – by treating flooding water with di-(tri-n-butyl)-(1,4-benzodioxan-6, 7-dimethyl) diammonium dichloride, 2 (033) (1995) 393.

[19] J.B. Martin, R.L., Brock, G.F., Dobbs, Corrosion inhibitors and methods of use, 6 (866) (2005) 797.

[20] A. Dietsche, F. Essig, M. Friedrich, R. Kutschera, M. Schrepp, W. Witteler, M.J. Friedrich, Organic corrosion inhibitors for interim corrosion protection, in: CORROSION 2007, 2007, p. 12.

[21] N. Obeyesekere, A. Naraghi, J.S. McMurray, Synthesis and evaluation of biopolymers as low toxicity corrosion inhibitors for north sea oil fields, in: Corrosion 2001, 2001, pp. 1–14.

[22] A.J. McMahon, D. Harrop, Green corrosion inhibitors: an oil company perspective, in: Corrosion 95, 1995.

[23] D.C. Silverman, D.J. Kalota, F.S. Stover, Effect of pH on corrosion inhibition of steel by polyaspartic acid, in: Corrosion 95, 1995, pp. 1–8.

[24] R.F.F. Mainier, F.B. Lazaro, W. Do, Silicate-based corrosion-inhibitor in drilling fluids: An environmentally-friendly option (inibidor de corrosao a base de silicato em fluidos de perfuracao: Uma opcao nao agressiva ao meio ambiente), in: 8th Petrobras et al Latin Amer. Drilling Congr, 1992, pp. 467–475.

[25] J.R. Ohlsen, J.M. Brown, G.F. Brock, V.K. Mandlay, Corrosion inhibitor composition and method of use, 5 (459) (1995) 125.

[26] F.B. Growcock, V.R. Lopp, The inhibition of steel corrosion in hydrochloric acid with 3-phenyl-2-propyn-1-ol oil-bearit ~ 6 rock formations are commonly treated with mineral acids to increase permeability and, thus, enhance oil production. As a result, the well tubing throug, 28 (4) (1988) 397–410.

[27] D.W. Fong, B.S. Shambatta, Hydroxamic acid containing polymers used as corrosion inhibitors, 2 (074) (1993) 535.
[28] Y.S. Ahn, V. Jovancicevic, Mercaptoalcohol corrosion inhibitors, 112 (2001) 878.
[29] S. Teeters, Corrosion inhibitor, 5 (084) (1992) 210.
[30] F. Bentiss, M. Lagrenée, M. Traisnel, Corrosion inhibitors for mild steel in acidic media, 56 (7) (2000) 733–742.
[31] E.R. Fischer, J.E. Parker, Technical note: tall oil fatty acid anhydrides as corrosion inhibitor intermediates, Corrosion 53 (1) (1997) 62–64.
[32] H. Wirtz, H. Hoffmann, W. Ritschel, M. Hofinger, M. Mitzlaff, D. Wolter, Optionally quaternized fatty esters of alkoxylated alkyl-alkylene diamines (gegebenenfalls quaternierte fettsaeureester von oxyalkylierten alkyl-alkylendiaminen), 320 (1989) 769.
[33] E. Babaian-Kibala, Naphthenic acid corrosion inhibitor, 5 (1993) 252–254.
[34] B.A.M.O. Alink, B.T. Outlaw, Thiazolidines and use thereof for corrosion inhibition, 140 (2001) 205.
[35] I. Sekine, M. Yuasa, T. Shimode, K. Takaoka, Inhibition of corrosion, 2 (234) (1991) 501.
[36] S.J. Weghorn, C.W. Reese, B. Oliver, Field evaluation of an encapsulated time-release corrosion inhibitor, in: NACE International, 2007.
[37] C.P. Anderson, S.A. Blenkinsopp, F.M. Cusack, J.W. Costerton, Drilling mud fluid loss – an alternative to expensive bulk polymers, in: 4th Inst. Gas Technol. Gas, Oil, & Environ. Biotechnol. Int. Symp., 1991, pp. 481–489.
[38] P. Sullivan, Y. Christanti, I. Couillet, S. Davies, T. Hughes, A. Wilson, Methods for controlling the fluid loss properties of viscoelastic surfactant based fluids, 7 (081) (2006) 439.
[39] T. Huang, J.B. Crews, Use of mineral oils to reduce fluid loss for viscoelastic surfactant gelled fluids, 7 (615) (2009) 517.
[40] T. Huang, J.B. Crews, J.H. Treadway Jr., Fluid loss control agents for viscoelastic surfactant fluids, 7 (550) (2009) 413.
[41] T.F. Willey, R.J. Willey, S.T. Willey, Rock bit grease composition, 7 (312) (2007) 185.
[42] D.A. Oldiges Jr., H. McDonald, T. Blake, K. Stroup, D.A. Oldiges III, Non-metallic thread sealant and anti-seize compound having improved anti-galling properties for metal alloys, 7 (091) (2006) 161.
[43] J.K. Fink, Petroleum Engineer's Guide to Oil Field Chemicals and Fluids, Gulf Professional Publishing, 2012.
[44] C. Stowe, R.G. Bland, D. Clapper, T. Xiang, S. Benaissa, Water-based drilling fluids using latex additives, 2002.
[45] A. Zaitoun, N. Berton, Stabilization of montmorillonite clay in porous media by high-molecular-weight polymers, in: 9th SPE Formation Damage Control Symposium, 1990, pp. 155–164.
[46] C. Aviles-Alcantara, C.C. Guzman, M.A. Rodriguez, Characterization and synthesis of synthetic drilling fluid shale stabilizer, in SPE International Petroleum Conference in Mexico, 2000.
[47] A.H. Hale, E. Van Oort, Efficiency of ethoxylated/propoxylated polyols with other additives to remove water from shale, 1997.
[48] A. Westerkamp, C. Wegner, H.E. Miiller, Borehole treatment fluids with clay swelling-inhibiting properties (II), 1991.
[49] R.E. Himes, Method for clay stabilization with quaternary amines, 1992.
[50] S.E. Alford, North Sea field application of an environmentally responsible water-base shale stabilizing system, in: SPE/IADC Drilling Conference, 1991, pp. 341–355.
[51] S. Palumbo, D. Giacca, M. Ferrari, P. Pirovano, The development of potassium cellulosic polymers and their contribution to the inhibition of hydratable clays, in SPE Oilfield Chemicals International Symposium, 1989, pp. 173–182.
[52] V.L. Yashchenko, et al., Odorimparting composition for natural gas—consists of mixture of mercaptan(s) and additionally contains pyridine and o ∼ -picoline, 1997.
[53] M. Fakhriev, F.R. Ismagilov, M.M. Latypova, Odorant for imparting smell to natural gas—contains a mixture of ethyl-, propyl-, butyl- and amyl-mercaptans, and additionally prescribed dialkyldisulphides, 1995.
[54] E.E. Fant, Odorization-a regulatory perspective, Chicago, IL, 1993.

[55] J.B. Lawson, J. Reisberg, Alternate slugs of gas and dilute surfactant for mobility control during chemical flooding, in: SPE/DOE Enhanced Oil Recovery Symposium, 1980.
[56] W.R. Rossen, Foams in enhanced oil recovery, Foam. Theory Meas. Appl. 57 (1996) 413–464.
[57] F.A. Hamed, M. Zoveidavianpoor, M. Jalilavi, The incorporation of silica nanoparticle and alpha olefin sulphonate in aqueous CO_2 foam: investigation of foaming behavior and synergistic effect, Pet. Sci. Technol. 32 (21) (2014) 2549–2558.
[58] A.S. Emrani, A.F. Ibrahim, H.A. Nasr-El-Din, Evaluation of mobility control with nanoparticle-stabilized CO_2 foam, in: SPE Latin America and Caribbean Petroleum Engineering Conference, 2017, January.
[59] I. Kim, A.J. Worthen, K.P. Johnston, D.A. DiCarlo, C. Huh, Size-dependent properties of silica nanoparticles for Pickering stabilization of emulsions and foams, J. Nanopart. Res. 18 (4) (2016) 82.
[60] T. Lu, Z. Li, Y. Zhou, Flow behavior and displacement mechanisms of nanoparticle stabilized foam flooding for enhanced heavy oil recovery, Energies 10 (4) (2017).
[61] P. Nguyen, H. Fadaei, D. Sinton, Pore-scale assessment of nanoparticle-stabilized CO_2 foam for enhanced oil recovery, Energy Fuels 28 (10) (2014) 6221–6227. Oct.
[62] J. Yu, S. Wang, N. Liu, R. Lee, Study of particle structure and hydrophobicity effects on the flow behavior of nanoparticle-stabilized CO_2 foam in porous media, in: SPE Improved Oil Recovery Symposium, 2014.
[63] D.A. Espinoza, F.M. Caldelas, K.P. Johnston, S.L. Bryant, C. Huh, "Nanoparticle-stabilized supercritical CO_2 foams for potential mobility control applications," in SPE Improved Oil Recovery Symposium, 2010.
[64] A. Worthen, et al., Multi-scale evaluation of nanoparticle-stabilized CO_2 in water foams: from the benchtop to the field, in: SPE Annual Technical Conference and Exhibition, 2015.
[65] S. Siripurapu, J.M. DeSimone, S.A. Khan, R.J. Spontak, Controlled foaming of polymer films through restricted surface diffusion and the addition of nanosilica particles or CO_2-philic surfactants, Macromolecules 38 (6) (2005) 2271–2280.
[66] X. Dong, J. Xu, C. Cao, D. Sun, X. Jiang, Aqueous foam stabilized by hydrophobically modified silica particles and liquid paraffin droplets, Colloids Surf. A Physicochem. Eng. Asp. 353 (2–3) (2010) 181–188. Jan.
[67] A.J. Worthen, S.L. Bryant, C. Huh, K.P. Johnston, Carbon dioxide-in-water foams stabilized with nanoparticles and surfactant acting in synergy, AIChE J. 59 (9) (2013) 3490–3501.
[68] S. Limage, J. Krägel, M. Schmitt, C. Dominici, R. Miller, M. Antoni, Rheology and structure formation in diluted mixed particle–surfactant systems, Langmuir 26 (22) (2010) 16754–16761.
[69] M.S. Wong, R. Verduzco, H.S. Jazeyi, C.A. Miller, J.M. Tour, Polymer-coated nanoparticles for enhanced oil recovery, J. Appl. Polym. Sci. 131 (15) (2014). p. n/a-n/a.
[70] R. Singh, A. Gupta, K.K. Mohanty, C. Huh, D. Lee, Fly ash nanoparticle-stabilized CO_2-in-water foams for gas mobility, in: SPE Annu. Tech. Conf. Exhib., pp. 1–13, 2015.
[71] S. Ghosal, J.L. Ebert, S.A. Self, Chemical composition and size distributions for fly ashes, Fuel Process. Technol. 44 (1–3) (1995) 81–94.
[72] A.A. Eftekhari, R. Krastev, R. Farajzadeh, Foam stabilized by fly ash nanoparticles for enhancing oil recovery, Ind. Eng. Chem. Res. 54 (50) (2015) 12482–12491.
[73] L. Hendraningrat, O. Torsaeter, Unlocking the potential of metal oxides nanoparticles to enhance the oil recovery, in: Offshore Technology Conference-Asia, 2014.
[74] M.O. Onyekonwu, N.A. Ogolo, Investigating the use of nanoparticles in enhancing oil recovery, in: *Nigeria Annual International Conference and Exhibition*, 2010.
[75] R. Nazari Moghaddam, A. Bahramian, Z. Fakhroueian, A. Karimi, S. Arya, Comparative study of using nanoparticles for enhanced oil recovery: wettability alteration of carbonate rocks, Energy Fuels 29 (4) (2015) 2111–2119.
[76] D.S. Shekhawat, A. Aggarwal, S. Agarwal, M. Imtiaz, Magnetic recovery-injecting newly designed magnetic fracturing fluid with applied magnetic field for EOR, in: SPE Asia Pacific Hydraulic Fracturing Conference, 2016.
[77] N.A. Ogolo, O.A. Olafuyi, M.O. Onyekonwu, Enhanced oil recovery using nanoparticles, in: SPE Saudi Arabia Section Technical Symposium and Exhibition, 2012.

[78] E. Joonaki, S. Ghanaatian, The application of nanofluids for enhanced oil recovery: effects on interfacial tension and coreflooding process, Pet. Sci. Technol. 32 (21) (2014) 2599–2607.

[79] M.R. Haroun, et al., Smart nano-EOR process for Abu Dhabi carbonate reservoirs, in: Abu Dhabi International Petroleum Conference and Exhibition, 2012.

[80] L.N. Nwidee, S. Al-Anssari, A. Barifcani, M. Sarmadivaleh, S. Iglauer, Nanofluids for enhanced oil recovery processes: wettability alteration using zirconium oxide, Offshore Technology Conference Asia, 2016.

[81] M. Al-Jabari, N.N. Nassar, M.M. Husein, M.E. Al-Jabari, Removal of asphaltenes from heavy oil by nickel nano and micro particle adsorbents, in: Proceedings of the IASTED International Conference, 2018, pp. 171–175.

[82] H. Ehtesabi, M.M. Ahadian, V. Taghikhani, M.H. Ghazanfari, Enhanced Heavy Oil Recovery in Sandstone Cores Using TiO_2 Nanofluids, Energy Fuels 28 (1) (2014) 423–430.

[83] G. Cheraghian, Effects of titanium dioxide nanoparticles on the efficiency of surfactant flooding of heavy oil in a glass micromodel, Pet. Sci. Technol. 34 (3) (2016) 260–267.

[84] M.H. Sedaghat, H. Mohammadi, R. Razmi, Application of SiO_2 and TiO_2 nano particles to enhance the efficiency of polymer-surfactant floods, Energy Source 38 (1) (2016) 22–28.

[85] H. Ehtesabi, M.M. Ahadian, V. Taghikhani, Enhanced heavy oil recovery using TiO_2 nanoparticles: investigation of deposition during transport in core plug, Energy Fuels 29 (1) (2015) 1–8.

[86] H.M. Zaid, N. Yahya, N.R.A. Latiff, The effect of nanoparticles crystallite size on the recovery efficiency in dielectric nanofluid flooding, J. Nano Res. 21 (2012) 103–108.

[87] M. Tajmiri, S.M. Mousavi, M.R. Ehsani, E. Roayaei, A. Emadi, Wettability alteration of sandstone and carbonate rocks by using ZnO nanoparticles in heavy oil reservoirs, Iran. J. Oil Gas. Sci. Technol. 4 (4) (2015) 50–66.

[88] N.R.A. Latiff, N. Yahya, H.M. Zaid, B. Demiral, Novel enhanced oil recovery method using dielectric zinc oxide nanoparticles activated by electromagnetic waves, in: 2011 National Postgraduate Conference, 2011, pp. 1–7.

[89] M. Adil, H.M. Zaid, L.K. Chuan, N.R.A. Latiff, Effect of dispersion stability on electrorheology of water-based ZnO nanofluids, Energy Fuels 30 (7) (2016) 6169–6177. Jul.

[90] C. Negin, S. Ali, Q. Xie, Application of nanotechnology for enhancing oil recovery – a review, Petroleum 2 (4) (2016) 324–333.

[91] A. Karimi, et al., Wettability alteration in carbonates using zirconium oxide nanofluids: EOR implications, Energy Fuels 26 (2) (2012) 1028–1036.

[92] M.S. Moslan, W.R. Wan Sulaiman, A.R. Ismail, M.Z. Jaafar, I. Ismail, Wettability alteration of dolomite rock using nanofluids for enhanced oil recovery, Mater. Sci. Forum 864 (2016) 194–198. Aug.

[93] P. Esmaeilzadeh, N. Hosseinpour, A. Bahramian, Z. Fakhroueian, S. Arya, Effect of ZrO2 nanoparticles on the interfacial behavior of surfactant solutions at air–water and n-heptane–water interfaces, Fluid Phase Equilib. 361 (2014) 289–295. Jan.

[94] M.S. Alnarabiji, et al., The influence of hydrophobic multiwall carbon nanotubes concentration on enhanced oil recovery, Procedia Eng. 148 (2016) 1137–1140. Jan.

[95] S. Horikoshi, N. Serpone, Microwaves in Nanoparticle Synthesis: Fundamentals and Applications, Wiley-VCH, 2013.

[96] M. Shen, D.E. Resasco, Emulsions stabilized by carbon nanotube – silica nanohybrids, Langmuir 25 (18) (2009) 10843–10851.

[97] L.C. Villamizar, P. Lohateeraparp, J.H. Harwell, D.E. Resasco, B.J. Ben Shiau, Interfacially active SWNT/Silica nanohybrid used in enhanced oil recovery, in: SPE Improved Oil Recovery Symposium, 2010.

[98] M.J. Kadhum, D.P. Swatske, C. Chen, D.E. Resasco, J.H. Harwell, B. Shiau, Propagation of carbon nanotube hybrids through porous media for advancing oilfield technology, in: SPE International Symposium on Oilfield Chemistry, 2015.

[99] M. AfzaliTabar, M. Alaei, R. Ranjineh Khojasteh, F. Motiee, A.M. Rashidi, Preference of multi-walled carbon nanotube (MWCNT) to single-walled carbon nanotube (SWCNT) and activated carbon for preparing silica nanohybrid pickering emulsion for chemical enhanced oil recovery (C-EOR), J. Solid. State Chem. 245 (2017) 164–173.

[100] D. Zhu, et al., Aqueous hybrids of silica nanoparticles and hydrophobically associating hydrolyzed polyacrylamide used for EOR in high-temperature and high-salinity reservoirs, Energies 7 (6) (2014) 3858–3871.
[101] Z. Hu, M. Haruna, H. Gao, E. Nourafkan, D. Wen, Rheological properties of partially hydrolyzed polyacrylamide seeded by nanoparticles, Ind. Eng. Chem. Res. 56 (12) (2017) 3456–3463.
[102] A. Maghzi, R. Kharrat, A. Mohebbi, M.H. Ghazanfari, The impact of silica nanoparticles on the performance of polymer solution in presence of salts in polymer flooding for heavy oil recovery, Fuel 123 (2014) 123–132.
[103] T. Sharma, G.S. Kumar, J.S. Sangwai, Comparative effectiveness of production performance of Pickering emulsion stabilized by nanoparticle–surfactant–polymerover surfactant–polymer (SP) flooding for enhanced oil recoveryfor Brownfield reservoir, J. Pet. Sci. Eng. 129 (2015) 221–232.
[104] A.I. El-Diasty, A.M. Aly, Understanding the mechanism of nanoparticles applications in enhanced oil recovery, in: SPE North Africa Technical Conference and Exhibition, 2015.
[105] M. Jafarnezhad, M.S. Giri, M. Alizadeh, Impact of SnO_2 nanoparticles on enhanced oil recovery from carbonate media, Energy Sources A 39 (1) (2017) 121–128.
[106] B.F. Towler, et al., Spontaneous imbibition experiments of enhanced oil recovery with surfactants and complex nano-fluids, J. Surfactants Deterg. 20 (2) (2017) 367–377.
[107] A. Shahrabadi, H. Bagherzadeh, A. Roostaie, H. Golghanddashti, Experimental investigation of HLP nanofluid potential to enhance oil recovery: a mechanistic approach, in: SPE International Oilfield Nanotechnology Conference and Exhibition, 2012.
[108] L. Hendraningrat, S. Li, O. Torsaeter, Enhancing oil recovery of low-permeability berea sandstone through optimised nanofluids concentration, in SPE Enhanced Oil Recovery Conference, 2013.
[109] B. Najafiazar, J. Yang, C.R. Simon, F. Karimov, O. Torsæter, T. Holt, Transport properties of functionalised silica nanoparticles in porous media, in SPE Bergen One Day Seminar, 2016.
[110] F. Mannale, C. Simon, J. Beylich, K. Redford, Polybranched, organic/inorganic hybrid polymer and method for its manufacture, US20080039607A1, 2005.
[111] J. Huang, A. Al-Mohsin, M. Bataweel, P. Karadkar, W. Li, A. Shaikh, Systematic approach to develop a colloidal silica based gel system for water shut-off, in: SPE Middle East Oil & Gas Show and Conference, 2017.
[112] B.A. Suleimanov, E.F. Veliyev, Nanogels for deep reservoir conformance control, in: SPE Annual Caspian Technical Conference & Exhibition, 2016.
[113] K.M. Rankin, Q.P. Nguyen, S.C. Austin, Conformance control through in-situ gelation of silica nanoparticles, 3, pp. 521–524, 2014.
[114] I.J. Lakatos, A. Vago, T. Bodi, J. Lakatos-Szabo, G. Szentes, Z. Karaffa, New alternatives in conformance control: nanosilica and liquid polymer aided silicate technology, 2015.
[115] B.A. Suleimanov, E.F. Veliyev, Nanogels for deep reservoir conformance control in-depth fluid diversion nanogel strenght evaluation experiments, in: SPE Annu. Casp. Tech. Conf. Exhib., no. November, pp. 1–3, 2016.
[116] V. Goel, P.J. Boul, X. Pang, B.R. Reddy, L. Eoff, A. Ye, Nanosilica-based conformance gels, no. June, pp. 3–5, 2015.
[117] R. Kalgaonkar, V. Wagle, K. Alnoaimi, I. Alhussain, Novel compositions based on nanomaterials designed for use as conformance sealants, 2017.
[118] M.K. Panga, et al., A particulate gel based system for water shut-off applications, pp. 1–12, 2018.
[119] A. Pandey, A. Telmadarreie, M. Trifkovic, S. Bryant, Cellulose nanocrystal stabilized emulsions for conformance control and fluid diversion in porous media, 2018.
[120] K. Panthi, K.K. Mohanty, C. Huh, Precision control of gel formation using superparamagnetic nanoparticle-based heating, in: SPE Annual Technical Conference and Exhibition, 2015.
[121] S. Ko, C. Huh, Use of nanoparticles for oil production applications, J. Pet. Sci. Eng. 172 (2019) 97–114.
[122] R. Kainuma, M. Ise, C.-C. Jia, H. Ohtani, K. Ishida, Phase equilibria and microstructural control in the Ni-Co-Al system, Intermetallics 4 (1996) S151–S158.
[123] J.M. Sater, Smart structures and materials 1997: industrial and commercial applications of smart structures technologies, in: Smart Structures and Materials 1997: Industrial and Commercial Applications of Smart Structures Technologies, 1997, 3044.

[124] J.M. Jani, M. Leary, A. Subic, M.A. Gibson, A review of shape memory alloy research, applications and opportunities, Mater. Des. 56 (2014) 1078–1113.

[125] D.C. Lagoudas (Ed.), Shape Memory Alloys: Modeling and Engineering Applications, Springer Science & Business Media, 2008.

[126] A. Ölander, An electrochemical investigation of solid cadmium-gold alloys, J. Am. Chem. Soc. 54 (10) (1932) 3819–3833.

[127] W.J. Buehler, J.V. Gilfrich, R.C. Wiley, Effect of low-temperature phase changes on the mechanical properties of alloys near composition TiNi, J. Appl. Phys. 34 (5) (1963) 1475–1477.

[128] J. Ma, I. Karaman, R.D. Noebe, High temperature shape memory alloys, Int. Mater. Rev. 55 (5) (2010) 257–315.

[129] G. Kang, D. Song, Review on structural fatigue of NiTi shape memory alloys: Pure mechanical and thermo-mechanical ones, Theor. Appl. Mech. Lett. 5 (6) (2015) 245–254.

[130] T.H. Nam, T. Saburi, K. Shimizu, Cu-content dependence of shape memory characteristics in Ti–Ni–Cu alloys, Mater. Trans. JIM 31 (11) (1990) 959–967.

[131] G.S. Bigelow, S.A. Padula, A. Garg, D. Gaydosh, R.D. Noebe, Characterization of ternary NiTiPd high-temperature shape-memory alloys under load-biased thermal cycling, Metall. Mater. Trans. A 41 (12) (2010) 3065–3079.

[132] K.C. Atli, I. Karaman, R.D. Noebe, Work output of the two-way shape memory effect in Ti50.5 Ni24.5Pd25 high-temperature shape memory alloy, Scr. Mater. 65 (10) (2011) 903–906.

[133] J.A. Monroe, I. Karaman, D.C. Lagoudas, G. Bigelow, R.D. Noebe, S. Padula, Determining recoverable and irrecoverable contributions to accumulated strain in a NiTiPd high-temperature shape memory alloy during thermomechanical cycling, Scr. Mater. 65 (2) (2011) 123–126.

[134] A. Evirgen, Microstructural characterization and shape memory response of Ni-rich NiTiHf and NiTiZr high temperature shape memory alloys, Ph.D. Dissertation, Texas A&M University, College Station, USA, 2014.

[135] A.P. Stebner, et al., Transformation strains and temperatures of a nickel–titanium–hafnium high temperature shape memory alloy, Acta Mater. 76 (2014) 40–53.

[136] K.C. Atli, I. Karaman, R.D. Noebe, G. Bigelow, D. Gaydosh, Work production using the two-way shape memory effect in NiTi and a Ni-rich NiTiHf high-temperature shape memory alloy, Smart Mater. Struct. 24 (12) (2015) 125023.

[137] B.C. Hornbuckle, T.T. Sasaki, G.S. Bigelow, R.D. Noebe, M.L. Weaver, G.B. Thompson, Structure–property relationships in a precipitation strengthened Ni–Ti29.7 –Hf20 (at%) shape memory alloy, Mater. Sci. Eng. A 637 (2015) 63–69.

[138] D. Canadinc, et al., On the deformation response and cyclic stability of Ni50Ti35Hf15 high temperature shape memory alloy wires, Scr. Mater. 135 (2017) 92–96.

[139] A. Evirgen, I. Karaman, R.D. Noebe, R. Santamarta, J. Pons, Effect of precipitation on the microstructure and the shape memory response of the Ni50.3Ti29.7Zr20 high temperature shape memory alloy, Scr. Mater. 69 (5) (2013) 354–357.

[140] A. Evirgen, I. Karaman, R. Santamarta, J. Pons, C. Hayrettin, R.D. Noebe, Relationship between crystallographic compatibility and thermal hysteresis in Ni-rich NiTiHf and NiTiZr high temperature shape memory alloys, Acta Mater. 121 (2016) 374–383.

[141] A. Evirgen, F. Basner, I. Karaman, R.D. Noebe, J. Pons, R. Santamarta, Effect of aging on the martensitic transformation characteristics of a Ni-rich NiTiHf high temperature shape memory alloy, Funct. Mater. Lett. 5 (04) (2012) 1250038.

[142] A. Evirgen, I. Karaman, R. Santamarta, J. Pons, R.D. Noebe, Microstructural characterization and shape memory characteristics of the Ni50.3Ti34.7Hf15 shape memory alloy, Acta Mater. 83 (2015) 48–60.

[143] A. Evirgen, I. Karaman, J. Pons, R. Santamarta, R.D. Noebe, Role of nano-precipitation on the microstructure and shape memory characteristics of a new Ni50.3Ti34.7Zr15 shape memory alloy, Mater. Sci. Eng. A 655 (2016) 193–203.

[144] R. Santamarta, et al., TEM study of structural and microstructural characteristics of a precipitate phase in Ni-rich Ni–Ti–Hf and Ni–Ti–Zr shape memory alloys, Acta Mater. 61 (16) (2013) 6191–6206.

[145] O. Karakoc, et al., Effects of upper cycle temperature on the actuation fatigue response of NiTiHf high temperature shape memory alloys, Acta Mater. 138 (2017) 185–197.
[146] O. Karakoc, C. Hayrettin, D. Canadinc, I. Karaman, Role of applied stress level on the actuation fatigue behavior of NiTiHf high temperature shape memory alloys, Acta Mater. (2018).
[147] P.G. Lindquist, C.M. Wayman, in: T.W. Duerig, et al. (Eds.), Engineering Aspects of Shape Memory Alloys, Butterworth-Heinemann Ltd, London, 1990.
[148] M. Sade, M. Ahlers, Low temperature fatigue in Cu-Zn-Al single crystals, Scr. Metall. 19 (4) (1985) 425–430.
[149] K. Otsuka, C.M. Wayman, Shape Memory Materials, Cambridge University Press, 1999.
[150] H. Funakubo, Shape Memory Alloys, Gordon and Breach Science Publishers, New York, 1987.
[151] D. Schryvers, P. Boullay, P.L. Potapov, R.V. Kohn, J.M. Ball, Microstructures and interfaces in Ni–Al martensite: comparing HRTEM observations with continuum theories, Int. J. Solids Struct. 39 (13–14) (2002) 3543–3554.
[152] H. Morito, A. Fujita, K. Fukamichi, R. Kainuma, K. Ishida, K. Oikawa, Magnetocrystalline anisotropy in single-crystal Co–Ni–Al ferromagnetic shape-memory alloy, Appl. Phys. Lett. 81 (9) (2002) 1657–1659.
[153] P.J. Webster, K.R.A. Ziebeck, S.L. Town, M.S. Peak, Magnetic order and phase transformation in Ni2MnGa, Philos. Mag. B 49 (3) (1984) 295–310.
[154] S.J. Murray, M. Marioni, S.M. Allen, R.C. O'handley, T.A. Lograsso, 6% magnetic-field-induced strain by twin-boundary motion in ferromagnetic Ni–Mn–Ga, Appl. Phys. Lett. 77 (6) (2000) 886–888.
[155] R.C. O'Handley et al., Magnetic-field-induced strain in single-crystal Ni-Mn-Ga, in: Smart Structures and Materials 2003: Active Materials: Behavior and Mechanics, 2003, 5053, pp. 200–207.
[156] A. Sozinov, A.A. Likhachev, N. Lanska, O. Soderberg, K. Ullakko, V.K. Lindroos, Effect of crystal structure on magnetic-field-induced strain in Ni-Mn-Ga, in: Smart Structures and Materials 2003: Active Materials: Behavior and Mechanics, 2003, 5053, pp. 586–595.
[157] H.E. Karaca, I. Karaman, B. Basaran, D.C. Lagoudas, Y.I. Chumlyakov, H.J. Maier, On the stress-assisted magnetic-field-induced phase transformation in Ni2MnGa ferromagnetic shape memory alloys, Acta Mater. 55 (13) (2007) 4253–4269.
[158] S.J. Murray, R. Hayashi, M.A. Marioni, S.M. Allen, R.C. O'Handley, Magnetic and mechanical properties of FeNiCoTi and NiMnGa magnetic shape memory alloys, in: Smart Structures and Materials 1999: Smart Materials Technologies, 1999, 3675, pp. 204–212.
[159] A. Fujita, K. Fukamichi, F. Gejima, R. Kainuma, K. Ishida, Magnetic properties and large magnetic-field-induced strains in off-stoichiometric Ni–Mn–Al Heusler alloys, Appl. Phys. Lett. 77 (19) (2000) 3054–3056.
[160] M. Wuttig, J. Li, C. Craciunescu, A new ferromagnetic shape memory alloy system, Scr. Mater. 44 (10) (2001) 2393–2397.
[161] R.D. James, M. Wuttig, Magnetostriction of martensite, Philos. Mag. A 77 (5) (1998) 1273–1299.
[162] T.W. Shield, Magnetomechanical testing machine for ferromagnetic shape-memory alloys, Rev. Sci. Instrum. 74 (9) (2003) 4077–4088.
[163] J. Cui, T.W. Shield, R.D. James, Phase transformation and magnetic anisotropy of an iron–palladium ferromagnetic shape-memory alloy, Acta Mater. 52 (1) (2004) 35–47.
[164] T. Yamamoto, M. Taya, Y. Sutou, Y. Liang, T. Wada, L. Sorensen, Magnetic field-induced reversible variant rearrangement in Fe–Pd single crystals, Acta Mater. 52 (17) (2004) 5083–5091.
[165] T. Sakamoto, T. Fukuda, T. Kakeshita, T. Takeuchi, K. Kishio, Magnetic field-induced strain in iron-based ferromagnetic shape memory alloys, J. Appl. Phys. 93 (10) (2003) 8647–8649.
[166] O. Benafan, A decade of SMA activities at NASA-GRC, material challenges, and future prospects, 2015.
[167] P.K. Kumar, et al., Experimental investigation of simultaneous creep, plasticity and transformation of Ti50.5Pd30Ni19.5 high temperature shape memory alloy during cyclic actuation, Mater. Sci. Eng. A 530 (2011) 117–127.
[168] B. Kockar, I. Karaman, J.I. Kim, Y. Chumlyakov, A method to enhance cyclic reversibility of NiTiHf high temperature shape memory alloys, Scr. Mater. 54 (12) (2006) 2203–2208.

[169] S. Padula, G. Bigelow, R. Noebe, D. Gaydosh, A. Garg, Challenges and progress in the development of high-temperature shape memory alloys based on NiTiX compositions for high-force actuator applications, 2006.
[170] J. Gore, A. Bowles, M. Maylin, L. Chandrasekaran, F.D., M. Buyers, High temperature shape memory alloy actuators through mechanical treatments for an oil & gas down-hole valve, Ind. Commer. Appl. Smart Struct. Technol. 6930 (2008) 69300R.
[171] P. Devendra, G. Song, Shape memory alloy actuated accumulator for ultra-deepwater oil and gas exploration, Smart Mater. Struct. 25 (4) (2016) 45012.
[172] D. Patil, G. Song, A review of shape memory material's applications in the offshore oil and gas industry, Smart Mater. Struct. 26 (9) (2017) 93002.
[173] L.B. Vernon, H.M. Vernon, Process of manufacturing articles of thermoplastic synthetic resins, 1941.
[174] W.C. Rainer, E.M. Redding, J.J. Hitov, A.W. Sloan, W.D. Stewart, Polyethylene product and process, 1964.
[175] F. Pilate, A. Toncheva, P. Dubois, J.-M. Raquez, Shape-memory polymers for multiple applications in the materials world, Eur. Polym. J. 80 (2016) 268–294.
[176] F. Zhang, T. Zhou, Y. Liu, J. Leng, Microwave synthesis and actuation of shape memory polycaprolactone foams with high speed, Sci. Rep. (2015).
[177] D. Haiyan, Z. Song, J. Wang, Z. Liang, Y. Shen, F. You, Microwave-induced shape-memory effect of silicon carbide/poly (vinyl alcohol) composite, Sens. Actuators A Phys. 228 (2015) 1–8.
[178] F. Yin, Y. Ni, S.-Y. Leo, C. Taylor, V. Basile, P. Jiang, Reconfigurable photonic crystals enabled by pressure-responsive shape-memory polymers, Nat. Commun. 6 (2015).
[179] G. Weiwei, et al., pH-Stimulated DNA Hydrogels Exhibiting Shape-Memory Properties, Adv. Mater. 27 (1) (2015) 73–78.
[180] J. Hu, Y. Zhu, H. Huang, J. Lu, Recent advances in shape–memory polymers: Structure, mechanism, functionality, modeling and applications, Prog. Polym. Sci. 37 (12) (2012) 1720–1763.
[181] W. Sokolowski, S. Tan, M. Pryor, Lightweight shape memory self-deployable structures for gossamer applications, in: 45th AIAA/ASME/ASCE/AHS/ASC Structures, Structural Dynamics & Materials Conference, 2004.
[182] Seward, K.P., P.A. Krulevitch. Shape memory alloy/shape memory polymer tools, U.S. Patent No. 6,872,433. 29 Mar. 2005.
[183] A. Lendlein, R. Langer, Biodegradable, elastic shape-memory polymers for potential biomedical applications, Science (2002).
[184] H. Jiang, S. Kelch, A. Lendlein, Polymers move in response to light, Adv. Mater. 18 (2006) 1471–1475.
[185] I. Bellin, S. Kelch, R. Langer, A. Lendlein, Polymeric triple-shape materials, Proc. Natl. Acad. Sci. 103 (48) (2006) 18043–18047.
[186] J.M. Hollerback, I.W. Hunter, J. Ballantyne, A comparative analysis of actuator technologies for robotics, The Robotics Review 2, MIT Press, Cambridge, MA, 1992, pp. 299–342.
[187] S. Lederle, Issues in the Design of Shape Memory Alloy Actuators, Massachusetts Institute of Technology, Boston, 2002.
[188] P.C. da C. M. Jr, T.A. Netto, L.L. da S. Monteiro, M.A. Savi, Design and Modeling of Shape Memory Actuator for Offshore Applications, in: ASME. International Conference on Offshore Mechanics and Arctic Engineering, Volume 10: Petroleum Technology: V010T11A032, 2015.
[189] L. Mikova, S. Medvecka, B.M. Kelemen, F. Trebuna, I. Virgala, Application of shape memory alloy (SMA) as actuator, Metalurgija 54 (2015) 169–172. no. ISSN 0543-5846.
[190] G. Song, et al., An innovative ultradeepwater subsea blowout preventer control system using shape memory alloy actuator, J. Energy Resour. Technol. 130/033101 (3) (2008) 7.
[191] K. Andrianesis, Y. Koveos, G. Nikolakopoulos, A. Tzes, Experimental study of a shape memory alloy actuation system for a novel prosthetic hand, Shape Memory Alloys, Sciyo, Croatia, 2010, pp. 81–106.
[192] P.A. Reinhardt, Downhole tool actuator. U.S. Patent No. 6,216,779. 17 Apr. 2001.
[193] G. Lewis, G. Song, D. Patil, Shape memory alloy powered hydraulic accumulator having actuation plates, 2015.

[194] D. Patil, G. Song, Shape memory alloy actuated accumulator for ultra-deepwater oil and gas exploration, Smart Mater. Struct. 25 (4) (2016) 45012.
[195] P.W. Clark, I.D. Aiken, J.M. Kelly, M. Higashino, R. Krumme, Experimental and analytical studies of shape-memory alloy dampers for structural control, Smart Struct. Mater. 2445 (1995) 241–252.
[196] D. Mauro, C. Donatello, M. Roberto, Implementation and testing of passive control devices based on shape memory alloys, Earthq. Eng. Struct. Dyn. (2000) 945–968.
[197] J. Cook, L. Chandrasekaran, Method of reinforcing structures, 2003.
[198] J.J. Barnard, D.N. Horner, M.H. Johnson, Wellbore centralizer for tubulars, 2013.
[199] M. Thomas, M.H. Johnson, R. Steve, Shape memory cement annulus gas migration prevention apparatus, 2014.
[200] A.M. Post, Self-torquing fasteners. U.S. Patent No. 6,688,828, 2004.
[201] O.M. Kabir, Methods for testing shape-memory alloy couplers for oil and gas applications, 2016.
[202] L. Peter, M. Vasquez, Shape memory alloy: low-cost manufacturing for the oil and gas industry, in: Offshore Technology Conference, 2013.
[203] N.E. Rogen, D.N. Adnyana, Packing tool apparatus for sealing well bores, U.S. Patent No. 4,515,213. 7 May 1985.
[204] E.J. O'malley, M.R. Bennett, Packer sealing element with shape memory material, U.S. Patent No. 7,743,825, 29 Jun. 2010.
[205] E.J. O'malley, Apparatus and method for completing wells using slurry containing a shape-memory material particles, 2014.
[206] Ingram, G., et al. Wellbore isolation tool using sealing element having shape memory polymer. U.S. Patent No. 8,763,687, 1 Jul. 2014.
[207] N.J. Kar, N.R. Rao, Dual squeeze seal gland, 1984.
[208] A. Manbachi, R.S.C. Cobbold, Development and application of piezoelectric materials for ultrasound generation and detection, Ultrasound 19 (4) (2011) 187–196.
[209] G. Gautschi, Piezoelectric sensors, Piezoelectric Sensorics, Springer, 2002, pp. 73–91.
[210] N. Goujon, H. Hori, K.K. Liang, B.K. Sinha, Applications of piezoelectric materials in oilfield services, IEEE Trans. Ultrason. Ferroelectr. Freq. Control. 59 (9) (2012).
[211] T. Alsos, Seismic applications throughout the life of the reservoir, Oilf. Rev. 14 (2) (2002) 48–65.
[212] J.C. Alfaro, et al., Reducing exploration risk, Oilf. Rev. 19 (1) (2007) 26–43.
[213] A.C. Tims, A new capped-cylinder design for an underwater sound transducer (USRD Type F50), J. Acoust. Soc. Am. 51 (5B) (1972) 1751–1758.
[214] D.F. Jones, S.E. Prasad, S.R. Kavanaugh, An End-Capped Cylindrical Hydrophone for Underwater Sound Detection, Defence Research Establishment Atlantic, 1992.
[215] A. Belani, S. Orr, A systematic approach to hostile environments, J. Pet. Technol. 60 (07) (2008) 34–39.
[216] G. DeBruijn, et al., High-pressure, high-temperature technologies, Oilf. Rev. 20 (3) (2008) 46–60.
[217] B.K. Sinha, Stress compensated orientations for thickness-shear quartz resonators, in: Thirty Fifth Annual Frequency Control Symposium, 1981, pp. 213–221.
[218] B.K. Sinha, Doubly rotated contoured quartz resonators, IEEE Trans. Ultrason. Ferroelectr. Freq. Control. 48 (5) (2001) 1162–1180.
[219] A.J. Hayman, R. Hutin, P.V. Wright, High-resolution cementation and corrosion imaging by ultrasound, in: SPWLA 32nd Annual logging symposium, 1991.
[220] A.J. Hayman, et al., Quantitative corrosion evaluation in wells using a multi-function ultrasonic imager, in: Ultrasonics Symposium, 1995. Proceedings., 1995 IEEE, 1995, vol. 2, pp. 1115–1120.
[221] W.A. Smith, The role of piezocomposites in ultrasonic transducers, in Ultrasonics Symposium, 1989. Proceedings., IEEE 1989, 1989, pp. 755–766.
[222] K. Hsu, M. Hashem, C.L. Bean, R. Plumb, G.N. Minerbo, Interpretation and analysis of sonic-while-drilling data in overpressured formations, in: SPWLA 38th Annual Logging Symposium, 1997.
[223] N. Gaillard, A. Thomas, C. Favero, Novel associative acrylamide-based polymers for proppant transport in hydraulic fracturing fluids, in: SPE International Symposium on Oilfield Chemistry, 2013.
[224] H.B. Jung, et al., Stimuli-responsive/rheoreversible hydraulic fracturing fluids as a greener alternative to support geothermal and fossil energy production, Green. Chem. 17 (2015) 2799–2812.

[225] C. Montgomery, Fracturing fluid components, in: Effective and Sustainable Hydraulic Fracturing, 2013, pp. 25–45.
[226] J.J. Sheng, Enhanced oil recovery in shale reservoirs by gas injection, J. Nat. Gas. Sci. Eng. 22 (2015) 252–259.
[227] Y. Yu, J.J. Sheng, Experimental evaluation of shale oil recovery from eagle ford core samples by nitrogen gas flooding, in: SPE Improved Oil Recovery Conference, 2016, p. SPE-179547-MS.
[228] C. Davis, EOG enhancing eagle ford oil recovery using novel NatGas injection. System, 2016.
[229] M.C. Fyfe, J.F. Stoddart, Synthetic supramolecular chemistry, Acc. Chem. Res. 30 (1997) 393–401.
[230] H.-J.R. Schneider, Interactions in supramolecular complexes involving arenes: experimental studies, Acc. Chem. Res. 46 (2012) 1010–1019.
[231] H.T. Chifotides, K.R. Dunbar, Anion–π interactions in supramolecular architectures, Acc. Chem. Res. 46 (2013) 894–906.
[232] J.-M. Lehn, Toward complex matter: supramolecular chemistry and self-organization, Proc. Natl. Acad. Sci. U. S. A. 99 (8) (2002) 4763–4768.
[233] G.T. Woo, H. Lopez, A.S. Metcalf, J. Boles, Polymer-free fluid for fracturing applications, SPE Drill. Complet. 14 (1999) 240–246.
[234] P. Raffa, A.A. Broekhuis, F. Picchioni, Polymeric surfactants for enhanced oil recovery: a review, J. Pet. Sci. Eng. 145 (2016) 723–733.
[235] R. Veatch, Overview of current hydraulic fracturing design and treatment technology-part 1, J. Pet. Technol. 35 (1983) 677–689.
[236] J. Chatterji, J. Borchardt, Applications of water-soluble polymers in the oil field, J. Pet. Technol. 33 (1981) 2042–2056.
[237] A. Rahy, et al., Enhandced proppant transport for hydraulic fracturing, WO2016201445 A1, 2016.
[238] B. Gall, C. Raible, Molecular size studies of degraded fracturing fluid polymers, in: SPE Oilfield and Geothermal Chemistry Symposium, 1985.
[239] M. Samuel, et al., Viscoelastic surfactant fracturing fluids: applications in low permeability reservoirs, in: SPE Rocky Mountain Regional/Low-Permeability Reservoirs Symposium and Exhibition, 2000.
[240] M.M. Samuel, et al., Polymer-free fluid for fracturing applications, SPE Drill. Complet. 14 (4) (1999) 240–246.
[241] J. Makarević, M. Jokić, L. Frkanec, D. Katalenić, Z. Mladen, Gels with exceptional thermal stability formed by bis (amino acid) oxalamide gelators and solvents of low polarity, Chem. Commun. (2002) 2238–2239.
[242] J. Heeres, A. van der Pol, C. Stuart, M. Friggeri, A. Feringa, B.L. van Esch, Orthogonal self-assembly of low molecular weight hydrogelators and surfactants, J. Am. Chem. Soc. 125 (2003) 14252–14253.
[243] M.J. Clemente, P. Romero, J.L. Serrano, J. Fitremann, L. Oriol, Supramolecular hydrogels based on glycoamphiphiles: effect of the disaccharide polar head, Chem. Mater. 24 (2012) 3847–3858.
[244] Q. Chen, Y. Lv, D. Zhang, G. Zhang, C. Liu, D. Zhu, Cysteine and pH-responsive hydrogel based on a saccharide derivative with an aldehyde group, Langmuir 26 (2009) 3165–3168.
[245] G.R. Newkome, G.R. Baker, S. Arai, M.J. Saunders, P.S. Russo, K.J. Theriot, et al., Cascade molecules. Part 6. Synthesis and characterization of two-directional cascade molecules and formation of aqueous gels, J. Am. Chem. Soc. 112 (1990) 8458–8465.
[246] L.A. Estroff, A.D. Hamilton, Water gelation by small organic molecules, Chem. Rev. 104 (2004) 1201–1218.
[247] J.P. Schneider, D.J. Pochan, B. Ozbas, K. Rajagopal, L. Pakstis, J. Kretsinger, Responsive hydrogels from the intramolecular folding and self-assembly of a designed peptide, J. Am. Chem. Soc. 124 (2002) 15030–15037.
[248] F.M. Menger, H. Zhang, K.L. Caran, V.A. Seredyuk, R.P. Apkarian, Gemini-induced columnar jointing in vitreous ice. Cryo-HRSEM as a tool for discovering new colloidal morphologies, J. Am. Chem. Soc. 124 (2002) 1140–1141.
[249] N.M. Sangeetha, U. Maitra, Supramolecular gels: functions and uses, Chem. Soc. Rev. 34 (10) (2005) 821–836.
[250] S.R. Raghavan, E.W. Kaler, Highly viscoelastic wormlike micellar solutions formed by cationic surfactants with long unsaturated tails, Langmuir 17 (2001) 300–306.

[251] C.A. Dreiss, Wormlike micelles: where do we stand? Recent developments, linear rheology and scattering techniques, Soft Matter 3 (2007) 956–970.
[252] N. Spenley, M. Cates, T. McLeish, Nonlinear rheology of wormlike micelles, Phys. Rev. Lett. 71 (1993) 939.
[253] X. Shi, et al., From a 1-D chain, 2-D layered network to a 3-D supramolecular framework constructed from a metal-organic coordination compound, Cryst. Growth Des. 5 (2005) 207–213.
[254] Y. Yan, Y. Lin, Y. Qiao, J. Huang, Construction and application of tunable one-dimensional soft supramolecular assemblies, Soft Matter 7 (2011) 6385–6398.
[255] I.-C. Chen, C. Yegin, M. Zhang, M. Akbulut, Use of pH-responsive amphiphilic systems as displacement fluids in enhanced oil recovery, SPE J. 19 (06) (2014) 1035–1046.
[256] Y. Min, et al., Thermo-responsive gels based on supramolecular assembly of an amidoamine and citric acid, Soft Matter 14 (3) (2017) 432–439.
[257] D.A.Z. Wever, L.M. Polgar, M.C.A. Stuart, F. Picchioni, A.A. Broekhuis, Polymer molecular architecture as a tool for controlling the rheological properties of aqueous polyacrylamide solutions for enhanced oil recovery, Ind. Eng. Chem. Res. 52 (47) (2013) 16993–17005.

CHAPTER TWO

Advanced materials for geothermal energy applications

Celal Hakan Canbaz[1], Yildiray Palabiyik[2], Mustafa Hakan Ozyurtkan[2], Fatma Bahar Hosgor[3] and Mufrettin Murat Sari[4]

[1]Ege University, Izmir, Turkey
[2]Department of Petroleum and Natural Gas Engineering, Istanbul Technical University, Istanbul, Turkey
[3]Petroleum Software Ltd., London, United Kingdom
[4]Texas A&M University, Commerce, TX, United States

2.1 Introduction

By the development of new technologies, investigation methods used in the exploration, discovery, and monitoring phases of the geothermal systems along with the extraction and utilization of geothermal fluids showed a robust evolution in last two decades. Usage of nano technological materials and advanced tools in the measurements of physical properties of geothermal fluids as well as rocks paved the way of having robust and more accurate results with less uncertainty. New technology materials such as Fiber Optic Sensors, Distributed Temperature Sensing Systems (DTS), Thermal Infrared Remote Sensing (TIR) Tools and Advanced Technology Carriers of these tools such as, Drones, Aircrafts, and Satellites brought a new perspective to geothermal industry and expedited reaching the data effectively.

Advancements in the eophysical tools also ensured more accurate spatial resolution about the delineation of the dimensions of the geothermal systems and to specify the potential locations of the especially subsurface geothermal reservoirs in the deep which does not have any surface manifestation, as well as the shallow systems having upflow zones and their properties prior to the drilling required to ultimately exploit the geothermal fluid. Since a drilling operation is a very costly stage of a geothermal energy application, geophysical methods which are capable of scanning much more extensive area in a much cheaper and faster way compared to the drilling are to become more of an issue to be able to make a cost-effective decision if the potential geothermal zones are worth investing on the drilling [1]. In the geothermal systems, in addition to the parameters like heat source, extension of reservoir area underground, upflow and highly permeable zones, temperature, density, resistivity, porosity, seismic velocity, and magnetic susceptibility can be counted as the most general properties that are

Sustainable Materials for Transitional and Alternative Energy.
DOI: https://doi.org/10.1016/B978-0-12-824379-4.00002-1

aimed to be found by the geophysical techniques. Selection of the appropriate geophysical method(s) depends on numerous factors including the physical and chemical characteristics of the investigation area being explored together with the prior information regarding the geological, geochemical, hydrological, etc. of the hydrothermal system. On the other hand, the geophysical technique that will apply may change based on the current situation of the geothermal field (e.g., if the field is on the exploitation stage with or without re-injection process) [2].

There are also novel materials and techniques used in drilling, coating and measurement for geothermal systems. The main purpose is to provide well integrity and obtain reliable data from the wells. One of the key issues is to deliver zonal isolation. The conventional cement types are not suitable for high-temperature environment. High-temperature environment and CO_2 rich geothermal fluids affects the mechanical and chemical properties of cement slurries. Some newly introduced and conventional materials are added in cement slurries to obtain proper cement properties. These applications are used of Foam Cement, Phosphate Bonded Cement, Self-Healing Cement, Acid and CO_2 Resistant Cement in the fields. Beside cement applications, novel drilling fluids application is also important to provide the integrity during geothermal well drilling. Unlike cement application, conventional clay based and chemical ingredients are not appropriate for such a harsh environment. Functional drilling fluid design is crucial to describe the thermal properties of the drilling mud. The high temperature-resistant drilling fluid structures mainly contain Sulfonated and Polysulfonate Drilling Fluids. There are also some new approaches related to nano-based drilling fluids. Some of these approached were tested in real field applications and some needs to be developed.

The contact with geothermal fluids in the well, the well-head, and the production components distresses their metallurgical strength, since the geothermal fluids are very corrosive. Carbon Steel items coated with anticorrosion materials are generally preferred. There are also other coatings such as Nickel-Based Alloys, Titanium-Based Alloys and Polymeric Coatings used to enhance the strength of components like Heat Exchanger, Casings and Liners in the well, Wellheads and Condensers.

The expected bottomhole temperature fundamentally restricts the logging and measurement methods and tools in geothermal exploration. The conventional Logging Tools generally are not applicable beyond 150°C due to the inadequate temperature tolerance of their electronics and sensors. Different service companies and laboratories generate different application to get measurement from the geothermal environments. The general approach is to use some heat shield around the electronic parts and sensors of the tools. In advanced utilizations, there are novel wireline, high temperature Measurement While Drilling (MWD), and advanced materials used logging tools were proposed.

Furthermore, advanced and nanomaterials also started to be used in geothermal downstream parts such as, heat transfer and energy conversion in thermo electrical power plants. Geothermal heat pumps and heat exchangers are encased assembly units that perform a heating and a cooling process by using geothermal energy. Geothermal Heat Pumps (GHP) are extremely appealing option compared to conventional electric or fossil fuel space conditioning equipment by providing a cost advantage on first cost basis as well as in operation and maintenance. Heat pump concept was revealed by Lord Kelvin in the 19th century but its popularity has begun increasing in the 1960s. Afterwards, GHP becomes one of the rapidly growing renewable energy applications in the world. Ground-Source Heat Pumps (GSHPs) presents energy-efficient and environmental-friendly alternative for heating/cooling system but efforts are still needed to improve their performance since the diffusion of these systems is still limited by their high initial costs. Today, new technologies are searching for system improvements. Nanomaterials and thermoelectric applications may take a vital role in the GSHP system.

For energy conversion and electricity production, the Rankine Cycle is the most dominant technology used in thermal plant energy conversion and electricity generation in the world for high temperature geothermal energy. In recent years, advance technology Hybrid Power Plants that combine two different energy system or sources in a single plant are being constructed for higher resource utilization efficiency and obtain a synergistic outcome.

Conventional power plants require high temperature ranges and they are not convenient in moderate power ranges and for low-temperature application. For lower temperature or low-to-medium grade thermal sources, Organic Rankine Cycle (ORC) power plants are used. ORC technology gives the ability to recover low-grade heat and the possibility to run decentralized lower-capacity power plants. Thermoelectric technology can also be adapted to the ORC to improve, especially the efficiency of waste heat recovery. There has been growing interest in this research area.

2.2 Advanced materials for geothermal energy applications

2.2.1 Geophysical tools

Geophysical tools implement the structural characterization of the geothermal systems. Gravity Survey is one of the most common methods for the characterizations of geological structures of the geothermal resources based on the density variations [3]. Besides, seismic survey which is a more expensive and efficient tool can be used as well to characterize the reservoir structures. Mainly, two different seismic methods such as; induced-seismic and micro-seismic [4,5]. Furthermore, seismic exploration is recently used for the discovery of the geothermal supercritical fluids (SCF) as a new frontier approach [6]. Electrical and electromagnetic methods (Magnetotellurics (MT), a Audio-Magnetotellurics (AMT), Direct Current (DC) Resistivity, Transient Electromagnetic (TEM), etc.) can also be taken into consideration to identify the boundaries and shapes of the geothermal reservoirs in addition to characterization purposes [7]. Different from the electrical methods mentioned above, some new applications such as self-potential (SP), induced polarization (IP), and spectral IP are also available. However, their applications are relatively restricted in comparison with the other electrical approaches to explore the hydrothermal systems owing to the lower depth of investigation they have. In another method, the key parameters Curie point temperature and depth of magnetization along with delineation of the reservoir are determined by the magnetic technique for the geothermal investigations [8]. In last years, the methods such as gravity, time-lapse MT, AMT, and controlled source AMT (CSAMT) are quite popular and useful to monitor and develop the geothermal resources by benefiting from the applications like hydraulic fracturing, re-injection, etc. Nevertheless, it is not be forgotten the fact that, most of the time, making use of only one technique may not be good enough to decide about the destiny of the potential of a geothermal system and a combination of all the data from which different sources might come yields the best result for the evaluation of the whole system [9,10].

Geophysical tools can primarily be divided into two parts as direct and indirect techniques [11].

2.2.1.1 *Indirect surveys*

It is possible to get some crucial properties about the subsurface geological structures of the geothermal systems by means of indirect (structural) geophysical techniques. These parameters help us make the vital and reasonable decisions (drilling, etc.) regarding the resource being investigated [11]. The following subsections are related to those methods.

2.2.1.1.1 Seismic surveys

The seismic tools that make use of the rocks having Low-Frequency Energy Suitable Transmitters have been firstly improved for petroleum industry and then, employed for the needs of geothermal exploration [1]. These tools, that are mainly being classified as active and passive techniques, are usually utilized to determine the sound velocity anomalies for geothermal resource exploration.

In active techniques, the artificial impacts are created externally into the subsurface formations to find out sound velocity distribution or detailed geological structure by recording the sound waves whereas the natural seismic activity can be detected by using the passive techniques to reveal the important properties of hydrothermal system in the earth such as tectonics and tomography of the subsurface structures [11].

Seismic surveys basically benefit from the elastic waves defined as P- and S-Waves to record the sound velocity values. The waves refracted or reflected at the formation boundaries or structural discontinuities propagate through the subsurface formation depending on their velocity values (Fig. 2.1). P-Waves (Pressure Waves) which are one type of the body waves and the fastest ones propagate in the travel direction of the fluids and have the compression and expansion effect on the materials while S-Waves (Shear Waves) which are the other main type of the body waves propagate only through the solid formations. The movements of these waves are perpendicular to the travel direction of the fluids and have the shearing effect on the materials [1,11].

Although some valuable information such as temperature, porosity, rock density, delineation, discontinuities, and the zones that may contain fluid about the geothermal system can be obtained by the active techniques that use the artificial seismic activity with two different refraction and reflection measurements that appropriate method can be chosen depending on the case being evaluated, those methods are more suitable for oil exploration rather than geothermal because of the lower depth of investigation and in their coherence with sedimentary rock environments compared to the passive ones [11,12]. Conversely, passive techniques such as shear-wave splitting can reveal the active faults by finding out the delineation of the reservoir structure and flow paths along with permeable formations. Furthermore, it may be possible to indirectly find the deployment of depth of heat source that is a very important parameter for the

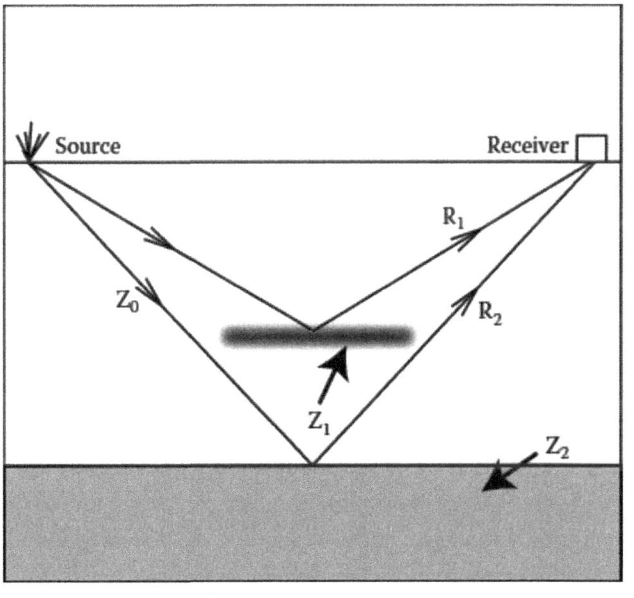

Figure 2.1 Schematic representation of a reflection seismology experiment [1].

geological evolution of the geothermal activity in the related area and may also indicate high temperature zones by using the passive methods [11,13].

On the other hand, there are some geothermal systems that contain SCF in the earth. SCF have remarkable effect particularly for the generation of electricity from a geothermal power plant because these types of fluids have much higher heat capacity values than the fluids at subcritical conditions. At this point, Agostinetti et al. make use of tele-seismic converted waves for the exploration of the deeper parts of the earth for the first time in Larderello geothermal system bearing SCF. Consequently, nominally ductile granitic rocks in the field which contain SCF and fracture permeability are brought a new frontier and promising findings regarding the seismic exploration for the future in respect of new understanding of geothermal production along with the magmatic systems [6].

Another concept about the seismology has been the shear-wave propagation to reveal the seismic anisotropy in the rocks for the decades [14].

2.2.1.1.2 Magnetic surveys

The magnetic tools have been extensively utilized in geothermal investigations from the past to present for mapping of geological structures along with revealing the igneous rocks and relevant magmatic formations such as dykes, veins that may indicate a heat source or individual faults. Magnetic surveys basically rely on the demagnetization process that have been taken place by the hydrothermal alteration due to movement of the thermal fluids through the permeable flow paths in the geothermal system and mainly two kinds of magnetization, induced magnetization in the direction of earth's field and permanent magnetization mostly related to the igneous rocks and their derivatives, can be referred to the rocks of the earth [11–13,15–18]. There are several devices that make use of proton-precession, fluxgate, optical absorption, and overhauser magnetometers to measure the data. The current magnetic field of the earth is disturbed by the magnetic rocks in the earth as a combination of those magnetizations as a function of composition, quantity, and the grain size containing magnetic minerals [15–17]. In those methods, measurements are performed regularly either in a grid structure or through the parallel lines to bring out the igneous and tectonic structures stated above. For example, according to the magnetic mapping studies in the Hengill area, the geothermal activity is inversely proportional to the magnetism [12].

The magnetic techniques can be considered as the quick and cost-effective methods for geothermal exploration. However, they have some limitations such as the lower resolution with increasing depth and the fact that they are not easily applicable to the complex geological structures and not sensitive to the fluid-filled zones. In addition, their usage has been restricted and is not that convenient to apply to geothermal resources as they require the refinement of data. These techniques work better in relatively flat topographies, and uncertainty negatively affects the accuracy of these methods [11].

The mentioned tools are used for the measurements on the ground. If the deeper and more extensive anomalies or distributions are needed to explore, aeromagnetic tools which are very popular and the most commonly used geothermal exploration method in recent years have been utilized [13]. Since this subsection mainly covers the ground-based geophysical tools, aeromagnetic tools will be discussed in detail in the next sections associated with the "Airborne Imaging".

2.2.1.1.3 Gravimetric surveys

The gravity method which is utilized by the gravity concept defined as the attractive force among the bulk structures and employed for the creation of mapping the subsurface geothermal environments by making use of density differences of the topography, latitude, and drift features is a less costly and uncomplicated geophysical tool compared to most of the other ones. Porosity, chemical composition, and saturation of the rocks are the major parameters affecting the density variations. Gravimetric surveys used jointly with the other tools are very powerful techniques particularly for the exploration of the shallow geothermal systems while the fact that their mapping resolutions remarkably go down with increasing depth is the main drawback of the method and their usage in geothermal exploration have been restricted thanks to indefinite numbers of density solutions depending on the complex lithological structures (such as the various combinations of the sedimentary and igneous rock occurrences) and thermal activities [10,18,19]. Árnason exhibited the Hengill Area's Bouguer gravity anomaly map demonstrating the significant monitoring effect of the gravimetric method on the geothermal exploitation realized in the system as can be seen in Fig. 2.2 [12,13].

2.2.1.2 Direct surveys

Direct surveys are utilized for the investigation of the properties affected by the geothermal activities. These techniques are elaborated in the following subsections.

2.2.1.2.1 Electrical surveys

Electrical surveys have been extensively used in the geothermal research and are mostly cost-effective in comparison with the other geophysical methods. The main difference of those methods than the other geophysical methods is that they are utilized in each stage of the geothermal investigations. Among the DC electrical surveys, resistivity methods are employed to reveal the physical parameters of the geothermal systems.

It is possible to benefit from the SP survey to detect fluid and heat flow zones by determining their electrical potentials. Another method, which called as IP method is being employed to explore the alteration mineralogy in time and frequency domain.

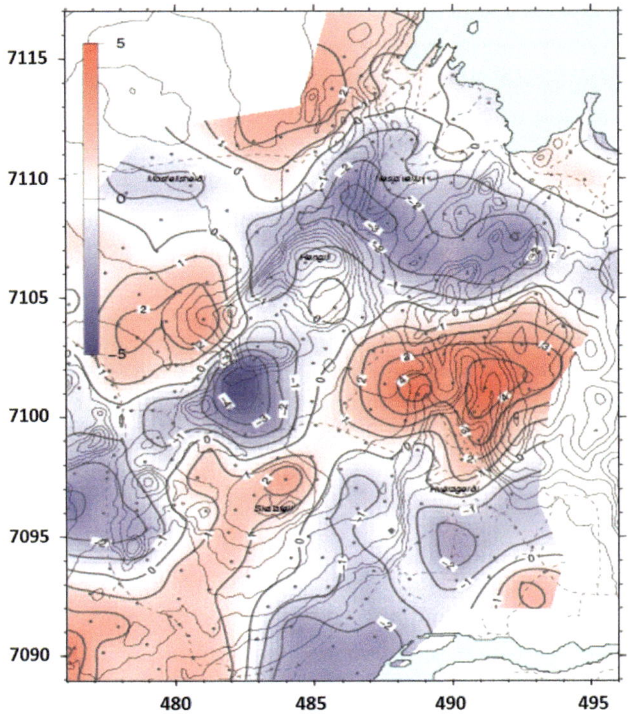

Figure 2.2 Bouger gravity map of the Hengill high-temperature area; areas of high gravity may indicate intrusions at deeper levels [13].

Other electrical methods which cover the electromagnetic surveys such as; MT, AMT, and CSAMT are often used to find out the heat flow paths/zones (thermal waters, hydrothermal alteration zones, features of magma, etc.) along with the structural geology (faults, etc.) up to 10 s of km although their maximum efficiency is valid in depths of 1–3 km [18,20–23].

Direct current surveys DC surveys which are not so popular especially for the geothermal exploration in recent years are regarded as the most conventional geophysical methods which are employed based on measurement of potential difference corresponding to the current passing through the surface to find out the apparent resistivity as is shown in Eq. 2.1 [10].

$$R_a = \frac{K\Delta V}{I} \tag{2.1}$$

where R_a, K, ΔV, and I represent the apparent resistivity, geometrical factor of array, potential difference, and the current, respectively. Generally, resistivity tools are operated depending on two pairs of electrode which the current is transmitted by one and the other stands for the potential difference. It can be stated that mainly three different DC methods are used for the geothermal research [11].

- Schlumberger sounding: This technique is the most commonly employed DC method and the configuration of the electrodes are placed symmetrically based on a reference point in the same line. Electrodes for the potential difference are located near the center while the ones for the current are located progressively going away from the center and this setup increases the sounding with depth.
- Dipole sounding: This method which is used in the past is based on the numerous arrays and is not used anymore for the geothermal exploration.
- Werner technique: In this method, electrodes are placed in the same line and the distance between the electrodes are the same at the beginning of the configuration. Growth in sounding measurements can be achieved as a function of depth by changing the distance between the electrodes.

A regional resistivity map in Husabik, North Iceland at 500 m below the sea level by using the method of Schlumberger is constructed by Georgsson et al. [24]. The resistivity soundings are illustrated by the map showing the low-resistivity anomaly regarding the Hveravellir geothermal area (Fig. 2.3).

Induction surveys The induction surveys depend on the electromagnetic field created by a current transmission. This transmission is based on an audio frequency domain.

Those types of surveys consist of active audiomagnetotellurics, Transient Electro-Magnetic (TEM) sirotem methods, Time Domain Electromagnetic Method (TDEM), Controlled Source electromagnetic (CSEM), and Controlled Source Audiomagnetotellurics (CSAMT) [25].

In the induction methods, measurements of the electromagnetic field are conducted based on the configuration with dipole transmission. The methodology here is to perform measurements regarding the electric field having the components E_x and E_y along with the magnetic field with the components H_x, H_y, and H_z. After measuring those components, the apparent resistivity can be calculated as follows [22]:

$$R_a = 0.2T|Z|^2 \tag{2.2}$$

where R_a and T represent the apparent resistivity and the period in seconds of the spectral component of the electromagnetic wave, respectively while

$$Z = \frac{E_x}{H_y} = \frac{-E_y}{H_x} \tag{2.3}$$

Depth of investigation can be identified as the skin depth in these methods by attenuating the plane wave at 37% of the real amplitude.

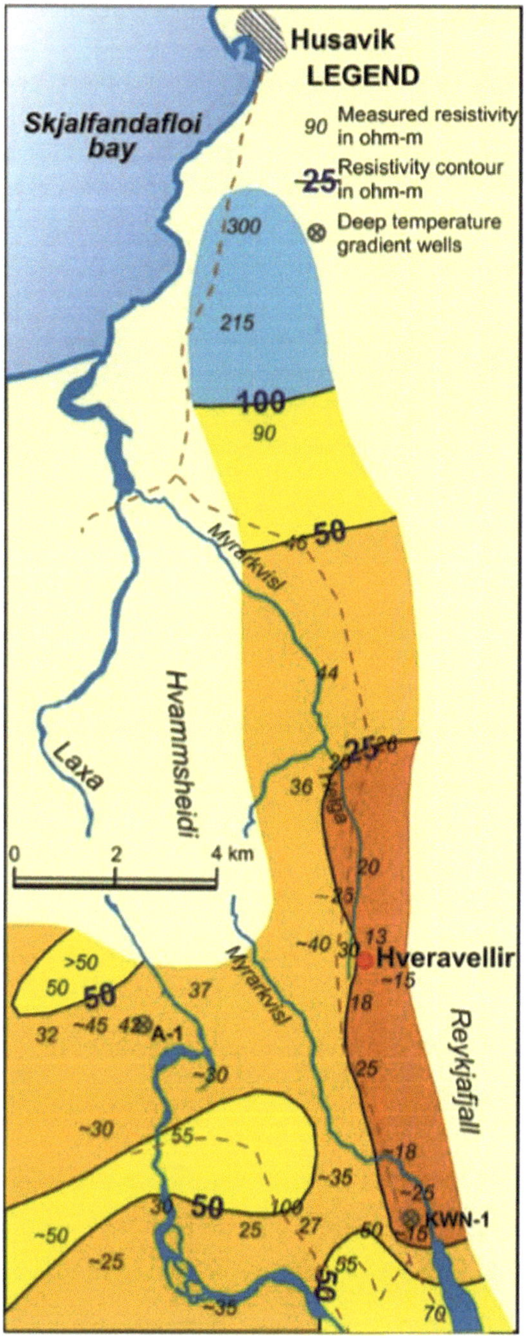

Figure 2.3 Resistivity map of the Hveravellir geothermal area and its surroundings at 500 m below sea level [24].

Frequency domain electromagnetic surveys Telluric, SP, MT, and AMT can be counted among passive electrical methods [11].

Magnetotelluric and Audiomagnetotellurics make use of natural source as the electromagnetic field to perturb the ground. Scalar apparent resistivity concept given to mathematically describe this phenomenon can be expressed in ohm-m as follows [21]:

$$R_a = \frac{|E_x|^2}{5f|H_y|^2} \tag{2.4}$$

where f, E_x, and H_y represent the frequency in hertz, horizontal x-directed electrical field component, and horizontal y-directed magnetic field component in gammas, respectively. Therefore, mutually orthogonal magnetic and electrical fields for the distant lightning can be measured to find the apparent resistivity as a function of the frequency. Skin depth as a result of the depth of penetration can be expressed in meters as follows [21]:

$$\delta = 503\sqrt{\left(\frac{R}{f}\right)} \tag{2.5}$$

As can clearly be seen from the Eq. 2.5, the highest value of the depth of penetration relies on the square root of ratio of the resistivity to the frequency.

Electromagnetic fields having natural source are utilized to explore the resistivity mapping in the earth as the functions of the direction, strength, and frequency [23]. Wayemo has shown the resistivity cross-sections of the Menengai region which is located in Kenya as a high temperature geothermal system (Figs. 2.4–2.7). This is a deep geothermal system up to approximately 8 km depth and indicates low resistivity at shallow depths while it shows high resistivity with increasing depth [26].

Taupo volcanic zone (TVZ) in New Zealand can be regarded as one of the most accomplished applications of the resistivity methods [27]. According to the resistivity distributions, regions having low resistivity indicate the geothermal systems while higher resistivity portions show the cold water saturated rocks as can be seen from Fig. 2.8 [28].

Telluric surveys have come up with the hypothesis that horizontal lateral variations which result from the geological influences affects resistivity and this phenomena also impacts the telluric currents' movement. From this point of view, measurements are performed based on a reference and a remote station regarding the telluric field. Finally, the results of the proportions from two difference locations can be acquired to evaluate the resistivity changes. It should be considered that these findings are qualitative and prior information [11].

In the SP surveys, the electrical polarization of heat flow zones (faults, fractures, etc.) in the geothermal system are perturbed by the elecrokinetic and thermoelectric

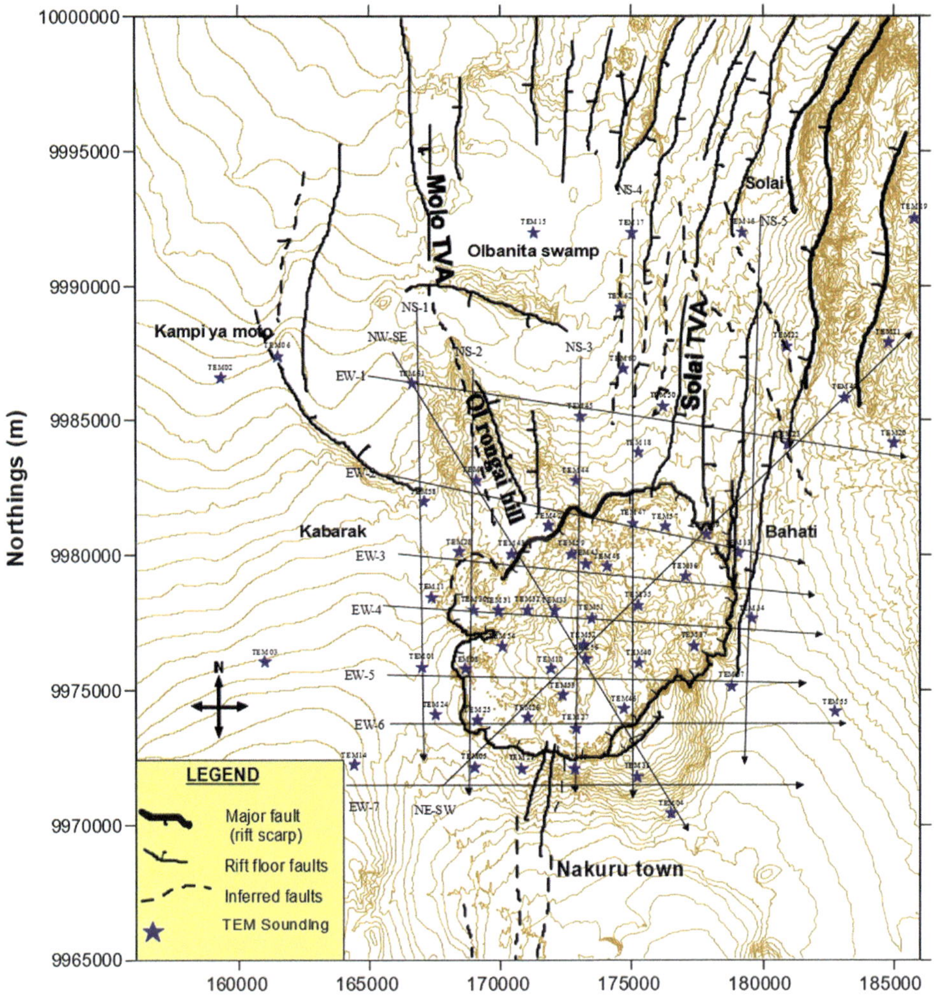

Figure 2.4 Map showing location of Transient Electromagnetic (TEM) soundings and resistivity cross-sections of the Menengai region [26].

effects to obtain the static electrical potential field of the ground by causing the understanding of the exploration area of interest with the SP anomalies such as dipolar anomalies [11]. SP survey was applied to a geothermal system containing single-phase geothermal reservoir in Japan for monitoring purpose. It is come up with this application that the streaming potential is caused by the SP anomaly [33].

A high-temperature geothermal system case from Krafla region of North Iceland is related to the resistivity measurements. It has been seen from this survey that a large low-resistivity region may indicate partially molten magma below the producible geothermal system approximately having a depth of 4 km [17].

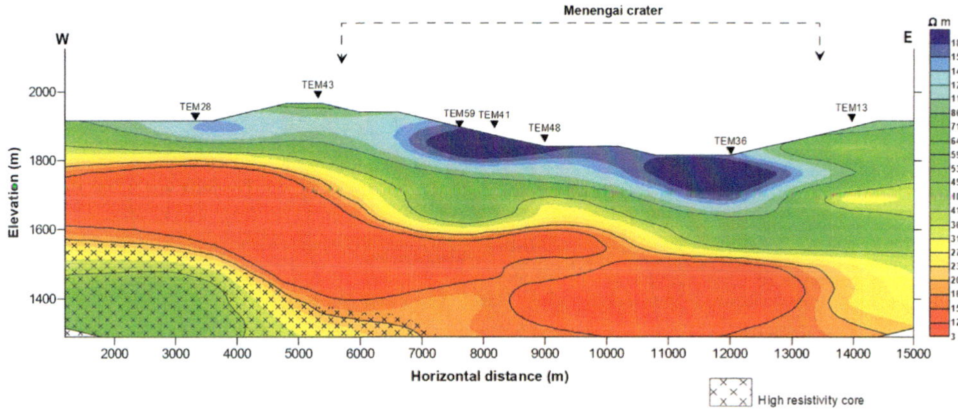

Figure 2.5 TEM resistivity cross-section EW-3 [26].

2.2.1.2.2 Thermal surveys

In those types of surveys, principally two different methods which are shallow and borehole probe techniques can be mentioned. These techniques are usually used to get the thermal gradient and so, the measurements of some parameters such as heat flow (if the thermal conductivity is known as a prior information), surface temperature of the earth that may be obtained from the airborne or satellite imaging like thermal infrared methods that will be discussed in detail in the following sections are also possible [34].

The main distinguished feature of the thermal techniques compared to the other ones is that they can directly measure the temperature and/or heat and, this provides better relationship with the geothermal systems' characteristics. Although they have easy application to the geothermal systems, the primary disadvantage of those methods is that their utilizations are restricted at shallow depths [35].

2.2.2 Advanced well-logging and measurement applications in geothermal fields

The bottomhole temperature depends on the geothermal gradient and the depth. The estimated bottomhole temperature essentially limits the relevant logging and measurement methods and tools in geothermal exploration. In temperatures lower than 150°C, conventional well-logging methods and tools can be efficiently run. However, the conventional logging tools cannot be used beyond 150°C due to the insufficient temperature tolerance of their electronics and sensors. In severe high temperature conditions, memory tools are run. These memory tools do not deliver real-time data. They collect data in an in-built memory. The collected data can be obtained when the tool is pulled to the surface.

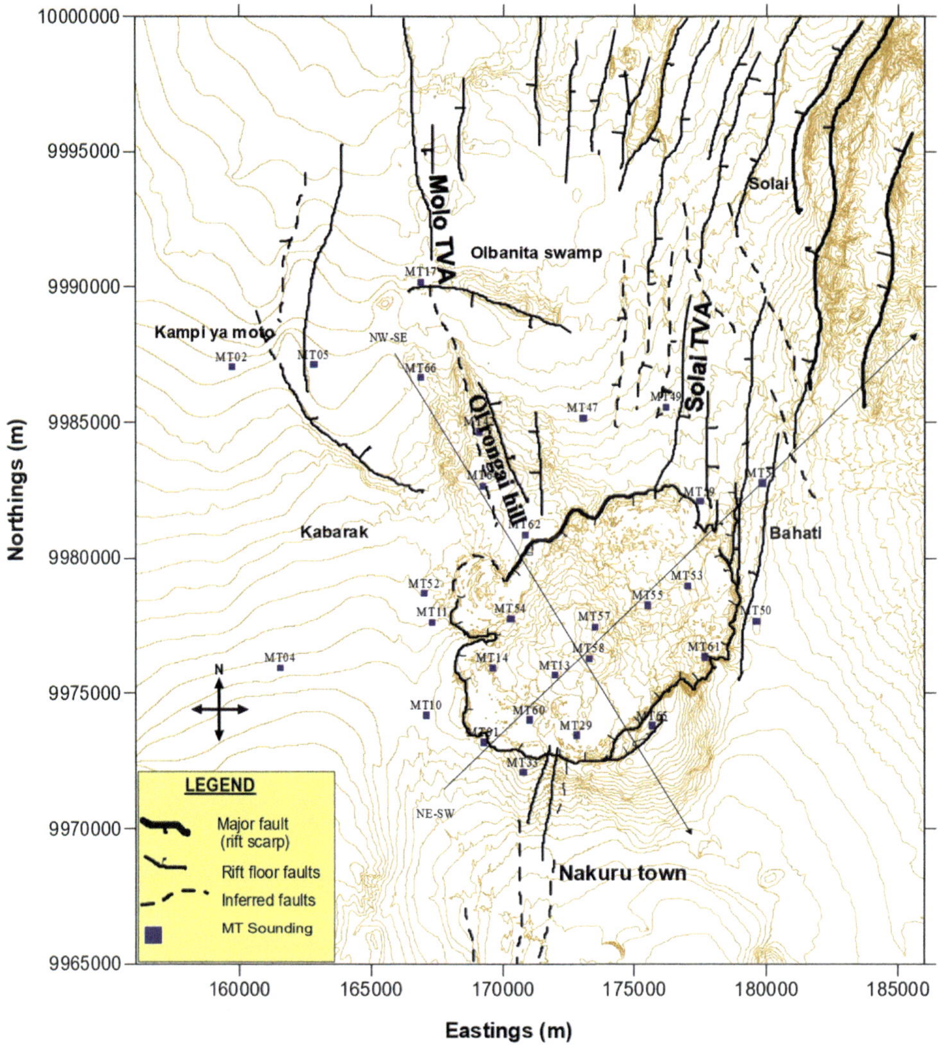

Figure 2.6 Map showing location of magnetotelluric soundings and resistivity cross-sections of the Menengai region [26].

Another approach to battle with high temperature is using shields in tools. These shields are bulky metal thermos with a vacuum located in the middle of the wall pressure covering. The temperature limitation of the shield application is around 400°C. But the heat shields applications are costly. The conventional electronics are placed in a Dewar flask which is used as a shield. The proposed operational time at 300°C can go up to 12 hours. The development of a heat shield for excessive temperature conditions is not an easy task. The quantity of heat outflow through conductive passageways

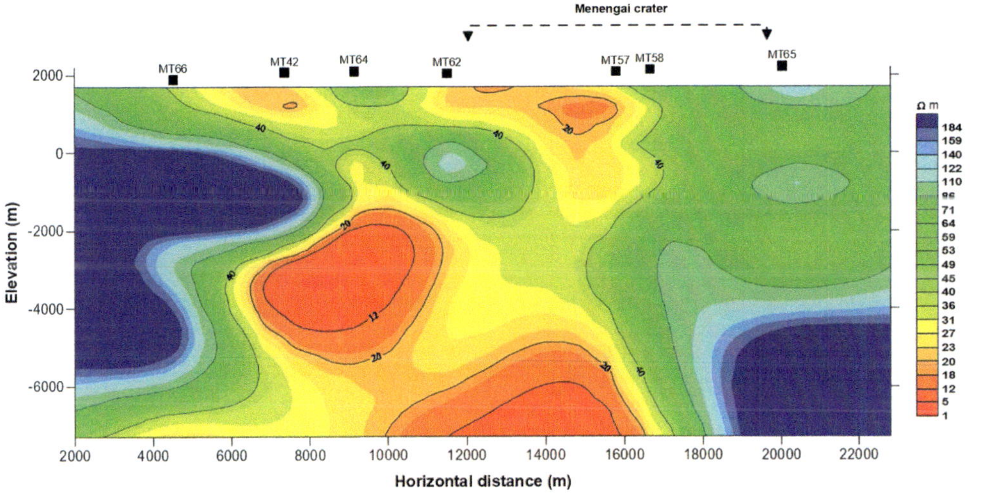

Figure 2.7 Magnetotelluric resistivity cross-section NW-SE [26].

has to be clearly defined. Eutectic materials or thermal absorbing structure inside the tool provide better operational accuracy. These heat absorbing materials generally employ phase change or energy captivation and stocks it into electronics cavity. The instruments with electronics which cannot be shielded are inadequate to be used in geothermal fields.

The high temperature tools are applied for exploration in geothermal wells and in volcanic geothermal systems where they are estimated to touch supercritical fluids (conditions over 374°C temperature and 22 MPa pressure). The requirement for new measurement and logging tools are for higher temperature and pressure readout tolerance and moreover for enhanced temperature tolerance of current formation evaluation tools. Furthermore, casing assessment during cementing and late inspections are necessary at high temperature systems.

Formation characterization is a must to detect the finest potential section in the well for geothermal systems. Characterization is required as various different formations will be penetrated in a well. To identify these formations properties are not only important for drilling integrity but also for stimulation purposes. In several geothermal projects, a hot granite formation was targeted in order to obtain fast rate of investment returns. Besides, the properties of the leading flow fractures such as dimension, positioning, mechanical properties of the surrounding formations have to be identified.

To identify these properties, there are several tools and logging units which are still in use. To define the crossings of the formations in the well, some high and low-frequency acoustic instruments are run along with different resistivity probes including micro-electrical formation imaging and natural gamma ray detectors. There are some MWD Logging While Drilling Tools and some Temperature Logs to provide data

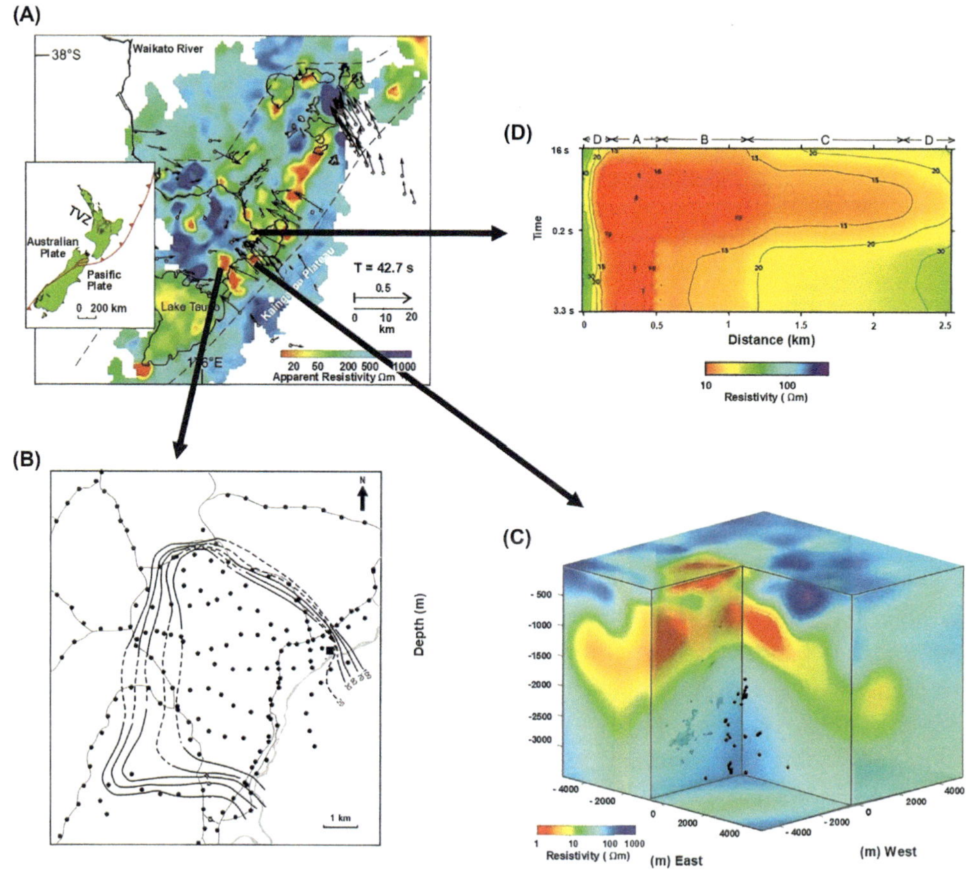

Figure 2.8 Electromagnetic studies in Taupo volcanic zone (TVZ) [28]. (A) Schlumberger DC apparent resistivity map of the TVZ [29]. (B) Contour lines of Schlumberger DC apparent resistivity of the Wairakei geothermal field [30]. (C) 3D resistivity model from 3D inversion of magnetotelluric (MT) data in the Rotokawa geothermal field [31]. (D) Distance-time apparent resistivity pseudo section from time domain electromagnetic method data in the Ngatamariki geothermal field [32].

from formations in real-time. To obtain thermal conductivity, chemical and other geophysical properties, Temperature Gradient Tools, Fluid Samplers and Gamma and Sonic Rock Density Gauges are currently used. This section summarizes the recent improvements on these tools.

For formation characterization, there are too few apparatuses to work over 300°C. Some tele-viewers are proposed to work within spectral gamma ray tool at 300°C. There exists a Vertical Seismic Profiling tool to be able to operate at temperatures up to 240°C [36]. Another advanced seismic tools generated to withstand 210°C with enhanced data transmission rates via Fiber Optic Wireline above 1 Mbit/s [37]. These apparatuses are advantageous in the examination of induced fracture and micro-seismic applications. Furthermore, a Dual Laterolog Resistivity tool has been introduced to work over 300°C.

Another technique to provide the usage of conventional tools in geothermal wells is to cool down. The advanced formation characterization tools used in geothermal environment proposed to be applicable in cooled regions of the well for temperatures up to 300°C. However, for bearance further than the 400°C is needed for well recovery and monitoring. The only possible tools that can be run are memory-tools as stated earlier.

The Managed Pressure Drilling is an application that the drilling operation run under pressure measured at the bit. This is an application of underbalanced drilling. There are some tools to control the drilling operation and measure the pressure in order not to lose the control of the well. There is also some underbalance drilling applications in geothermal systems. Underbalance drilling correspondingly benefits to preserve the geothermal production formations. It may also affect the fracture conductivity via persuading supplementary compressive stresses. Underbalance drilling applications are common in Enhanced Geothermal Systems (EGS). Besides pressure drilling uses, the directional drilling may be more practical in as much as it surges injectivity since the borehole positioning upsets shear fracturing and enlargement.

There are new developments in the real-time angle measurement at high temperature environment. The newly introduced measurement tools are operational at temperatures up to 300°C. This allowed real-time navigating of the Downhole Motor Drilling Assembly [38]. Several inventions have been established. Automatic navigating tools that can keep verticality or ensured in an indicated track today exist. However, none of them are operational at temperatures beyond 175°C.

Formation imaging is very crucial for geothermal applications. There exist some acoustic imaging tools by which natural fracturing and some rock properties can be determined. Besides, Acoustic Imaging Tools, there are some Gamma Ray Tools operational at temperatures up to 300°C. The total count Gamma Ray Detectors can be applied to detect the locations of high temperature formation boundaries. Comparing to oil and gas applications, the range of measurement is very low. Nevertheless, Spectral Gamma Ray Tools are more sensitive information description in geothermal applications since some of them employ carbon dioxide to cool down the detector crystals before utilization in the wells.

Getting in situ samples from the wells is always a challenging task for the service providers. It is even harder in geothermal wells. A novel fluid sampler has been established to withstand temperature up to 300°C. It can get samples from deep borehole fluids at high temperature [39,40].

The use of mechanical spinners provides the flow measurements in geothermal wells. These deliver simply a velocity measurement, not mass flow. Furthermore, reservoir flow models can be achieved with high temperature tolerant tracers. There are some organic naphthalene disulfonate tracers previously validated in a magmatic reservoir. Moreover, neutral density and chemically inert tracers can be used for flow measurement in supercritical reservoirs.

To detect Tracer at high temperature conditions (>150°C), there are some researches carried out to establish a downhole Fluorimeter. The tool is aimed to function under both laminar and turbulent flow characteristics. The tool can precisely work either at enlarged borehole diameters by volumetric flow rate measurement. There exists a single-conductor wireline in the tool and it is heat shielded. An Optical Fiber bundle will assist to carry the excitation light signal of the tool. It also catches the returned emission signal from the Fluorescent Dye. This dye is presented from a pump placed at the end of the tool. Temperature and pressure measurements can also be taken via Deep Resistivity Log (RTD) and Pressure Transducers. The stimulation effect of the fractures at higher temperatures can be monitored by using this tool. It also has benefit on determining the flow paths at high temperature geothermal environment [41].

2.2.3 Pressure/temperature sensors and monitoring materials

2.2.3.1 Fiber Optic Sensors

Fiber Optic Sensors are crucial players of oil and gas as well as geothermal energy industry. They are trustworthy and robust instruments which have been used in many projects that generate electricity from geothermal power. Fiber Optic Sensors have more than 20 years background of industrial usage and they firstly started to be used in offshore oil and gas wells. By using new technology materials such as Fiber Coatings, the operational ranges of Fiber Optic Sensors have been increased and they also started to be used in high-temperature environments such as geothermal sources.

Fiber Optic Sensors such as; Acoustic Tools, Point Pressure Gauges, DTS and Distributed Thermal Perturbation Sensor (DTPS) are developed for the aim of collecting subsurface fluid and rock properties in different phases of a Enhanced Geothermal System (EGS) project. Herein, the accuracy of data and decreasing the data uncertainty with less error percentage is crucial to estimate the potential and characterize the reservoir by using reservoir modeling techniques. This better understanding paves the way of taking the right actions and ensures the effectiveness of stimulation, fracturing, steam optimization and water injection processes with an improved reservoir monitoring [42].

Additionally, Fiber Optic Technology can be used to provide a detailed high-resolution image as used in Fiber Optic Seismic Sensors. Provided images are vital in early times of a geothermal well as the dominated flow comes from fractures. It gives the chance to describe the fracture location and depth as well as properties in details [43] (Fig. 2.9).

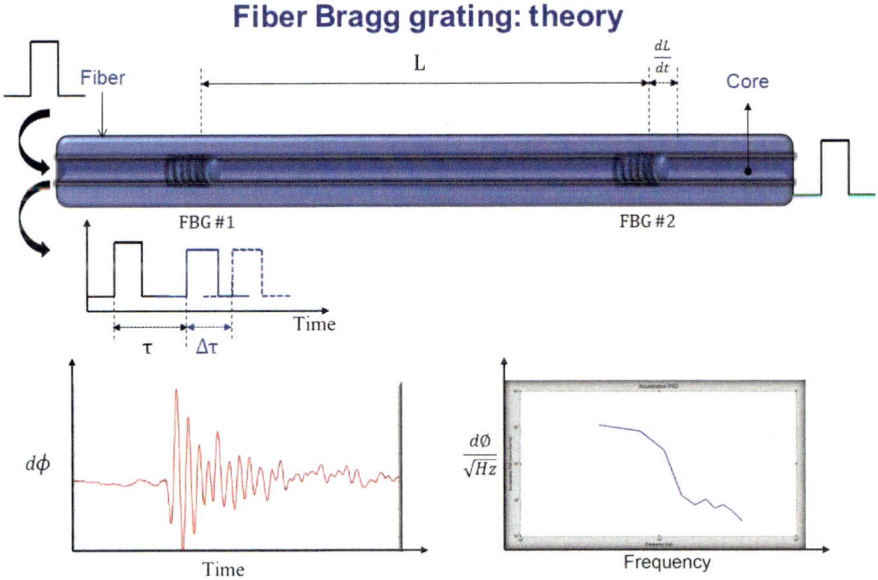

Figure 2.9 Fiber Bragg grating theory [43].

Fiber Optic Seismic Sensors are based upon Fiber Bragg Grating (FBG) theory and use the angle values of two different reflections. Herein, the Fiber Optic sensor is fabricated by thin silica glass fiber. Two or more FBGs which are formed by etching the internal part of the core by using an ultraviolet laser are being used as a reflector. It also helps to apart the core into different sections. A light pulse (I_o) is sent to the core. The theory assumes that the Fiber Bragg Grating applied parts have lower reflectivity. Thus, it allows to measure with large number of sensors. A pulse of light sent at a certain time and the reflections of different FBGs are recorded.

The incident light (I_o) which is sent inside the core is reflected by FBG sections with the wavelengths that are described by Bragg equation;

$$\lambda_B = 2\eta_{eff} \cdot \tau \tag{2.6}$$

Herein, η_{eff} = core material's refractive index, τ = grating period

Light interference function can also be calculated as;

$$I_R = I_o \cdot 2R(1 + \cos \varphi_c) \tag{2.7}$$

where, R = reflection coefficient of glass/air and air/glass interface. φ_c = reflected light waves phase shift

$$\varphi_c = \frac{4\pi nL}{\lambda} \tag{2.8}$$

λ is the optical wavelength of free space, and L is the cavity length of the interferometer.

2.2.3.2 Distributed Temperature Sensing Systems and Distributed Thermal Perturbation Sensor

Measurement of transient wellbore temperatures is crucial to characterize a geothermal reservoir and design a production strategy for an optimized energy production. Temperature data could be achieved by getting static formation temperatures by transient wellbore temperature measurements after the drilling operation [44]. Additionally, temperature changes are crucial parameters to describe with the temperature data during the production or injection process in a geothermal well [45,46].

DTS systems are effective and most popular systems that helps to understand the hydrological cases such as; continuous monitoring of streams, lakes, atmosphere and oceans (Fig. 2.10).

DTS systems were firstly developed to enable monitoring for fires and describe the temperature distribution for pipelines and other industrial materials in 1980s by various researchers [47–51]. In mid of 1990s, these Fiber-Optic Systems started to use in geothermal applications in temperature monitoring of shallow boreholes [52] as well as monitoring during injection tests [53]. Following technological developments, new DTS systems with higher resolution (0.01°C/m) for deeper wells (with Fiber Optic Cable length up to 10,000 m) for higher temperature ranges (up to 400°C) with different coating materials [54] were started to be used [55–59].

A DTS system mainly consists of a DTS unit that contains a Fiber-Optic Probe up to 20 km in length, an Optical Receiver, a Data Processing Module and one or more Lasers. The Fiber-Optic Probe is fused with silica. A laser light pulse launched into the Optical Fiber and the interaction with silica along the full length of Fiber-Optic Probe

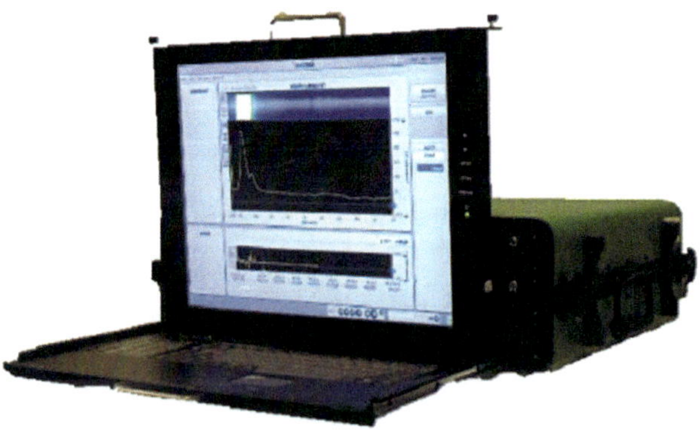

Figure 2.10 A portable Distributed Temperature Sensing system [46].

lets the light to scatter back to the Optical Receiver. Stokes components and anti-stroke ones were obtained by the conversion of small amounts of power that launched as light pulses by Raman Scattering. Measured intensities makes possible to calculate the Stokes/anti-Stokes components ratio. The ratio enables to calculate the temperature by using Eq. 2.9.

$$R(T) = \frac{I_{AS}}{I_s} = \left(\frac{\lambda_s}{\lambda_{as}}\right) exp\left(-\frac{hcu'}{kT}\right) exp(-\Delta az) \tag{2.9}$$

Herein; $R(T)$ = temperature value, I_{AS} = intensity of anti-Stokes light, I_S = intensity of Stokes light, λ_s = Stokes wavelength, λ_{AS} = anti-Stokes wavelength, h = Planck's constant, c = speed of light, u' = wave number separation from the Pump wavelength, k = Boltzmann's constant, T = absolute temperature, Δa = differential optical attenuation between Stokes and anti-Stokes frequencies, z = distance along the optical fiber (Fig. 2.11).

DTS enables real time monitoring of distributed subsurface temperature (temperature at every meter along the full depth of well) as well as pressure values (point pressure or pressure measured at the bottom of the well) which are crucial to estimate the production potential in a new geothermal well or between new wells. Additionally, shutting-in a well for a while (build-up period) and flowing the well in different choke sizes (draw-down period) changes the bottomhole pressure. These flowing/shut-in tests map real time pressure and temperature changes with time as well as production volumes. It helps to describe the zones and fractures which are producing hot fluids and the effect of production on the pressure within the reservoir zone. It also

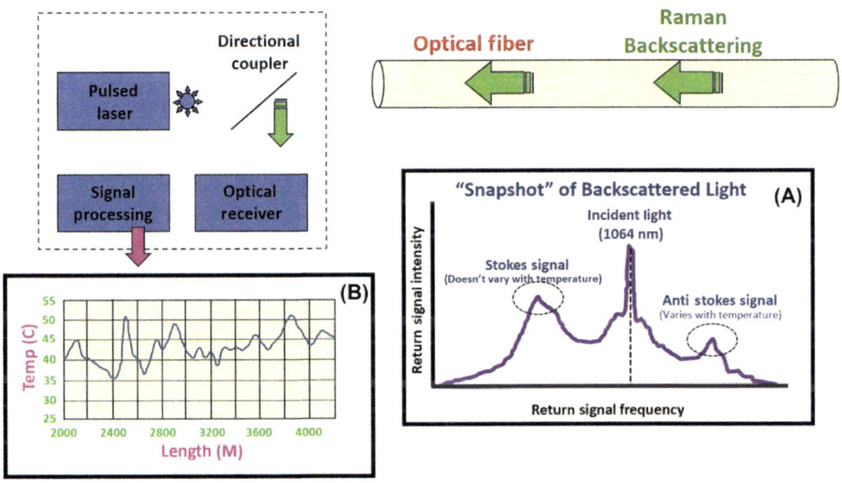

Figure 2.11 Schematic view of a basic Distributed Temperature Sensing System with the snapshot of backscattered light (A) and temperature profile result (B) [46].

paves the way of having a well test interpretation. A successful well test interpretation gives an idea about the reservoir size, wellbore damages caused by drilling, description of flow resistance between wells and their interactions. It also describes the effectiveness of any stimulation operations such as; acidizing, perforation, fracturing operations, and also the success of well completion operation. DTS also can be used in monitoring the development of a fracture during the hydraulic fracturing or a stimulation job which is performed to clean the debris sourced by drilling operation. DTS can identify the changes in temperature when fluids flow to fill the volume of a new fracture. It also helps to monitor chemical injection processes by detecting where chemicals applied during a treatment and which parts of the zones the chemicals penetrated. Furthermore, DTS can provide a permanent monitoring of a production or injection well by allowing the identification of the specific fractures/zones that produce fluids. DTS can perform continuous monitoring for any leaks sourced by casing or tubing parts of the wellbore. It helps to avoid any contamination of groundwater or subsurface aquifer. Apart from downhole applications, a DTS system can also be used in surface piping.

DTPS enables the in situ determination of heat fluxes, thermal conductivities, and imaging flow profiles. Fiber-Optic Distributed Temperature Sensors allow getting the temperature data during the thermal perturbation test and an analysis procedure which consists of forward simulations, Sensitivity Snalysis and Inverse Analysis applied for the inversion of data to develop a template model. Sensitivity Analysis and Inversions are crucial to understand whether the high resolution temperature data are usable for the description of hydrological and thermal properties of interested zone. Thus, DTPS thermal and hydrological profiles are interpreted with acceptable ranges of uncertainty by using inverse modeling of synthetic temperature data [60].

2.2.3.3 Thermal Infrared Remote Sensing

TIR is mainly used to identify the simultaneous thermal circumstance of a certain area. It helps to describe wide range of parameters in various fields such as:

- Description of temperature changes of a seabed/surface or a land,
- Determination of the peat or coal fires as well as fires in forests,
- Determination of water pollution by measuring thermal heat changes,
- Identification of geologic surfaces and facies,
- Analysis of soil moisture,
- Identification of properties for some materials

Geothermal systems consists of three main parameters: a reservoir, a heat source and a fluid which can transfer the heat [61]. These systems are formed in places that have anomalously high crustal heat flow sourced by deeper magmatic materials or hot rock. Hydrothermal systems were formed by the convection of heat from the heat source through groundwater. If hydrothermal water reaches to the surface, it named as

Surface Geothermal Sources. Fumaroles, Geysers, Hotsprings, and Mud Pools can be categorized as Surface Geothermal Sources. Underground hydrothermal systems forms within the fractures and permeable rocks and they can be brought to the surface by drilling a geothermal well.

Thermal Infrared Sensors are effective tools for mapping the geothermal activity by identifying the temperature anomalies in earth. In conventional systems, the reflected data is only able to provide data about the surface characteristics of a particular area, TIR method is also able to provide additional spectral data of subsurface conditions [62]. It is used in many geothermal source exploration projects as support feature for identification of new sources [63–65] as well as in continuous monitoring of hot water sources and magmatic activity [66]. It is also a useful tool to identify heat losses for reservoir models [67]. TIR survey was firstly performed in two geothermal fields of New Zealand and since then, it became one of the essential methods for fundamental research & monitoring of geothermal systems [68].

Compared with Visible-Near-Infrared (VNIR) and Shortwave Infrared (SWIR) tools, TIR mostly requires a customization related to the primary interest (chemical & physical properties or temperature of the interested area). In cases like the exploration of geothermal systems, the aim is to identify the temperature anomalies and emissivity spectra which are required to determine best kinetic from the radiant temperature [69]. However, TIR enables a rapid mapping compared with other techniques.

Thermal optical and infrared images could be taken by airborne surveys [66] or Spaceborne Methods [70–72]. Recent developments in sensor and instrument technology increased the number of advanced products used in both techniques. Sensors such as; MAKO, MAGI, OWL, Airborne Gas and Mineral Identification Sensors, Thermal Airborne Spectrographic Imager Tool, HyspIRI (Hyperspectral Infrared Imager), HyTES (Hyperspectral Thermal Emission Thermometer), ASPER (Advanced Spaceborne Reflection and Emission Radiometer) and the airborne version of that; MASTER (MODIS-ASPER Airborne Simulator) increased the interest to TIRS systems [73].

The most common tool for airborne surveys is the airborne visible/IR image spectrometer (AVIRIS) which is a hyperspectral airborne instrument that is flown in a NASA aircraft [74–76]. There are advanced instruments such as; AVIRIS (used by airborne and includes five electrical and six optical subsystems) and ASTER (used by satellite systems) which could be used for applications like mineral mapping and vegetation survey. It actually includes three different subsystems that could work under different wavelengths (VNIR, SWIR and TIR) with separate telescopes [77]. Wavelength intervals are between 0.52 and 0.86 μm with 15 m^2 pixel size for VNIR, 1.6 and 2.43 μm with 30 m^2 pixel size for SWIR, 8.13 and 11.65 μm with 90 m^2 pixel size for TIR.

Mainly there are three methods of estimating the surface and subsurface temperatures such as; three dimensional geostatistical estimation, physical basis of land surface data, and usage of Land Surface Temperature (LST) algorithm [78] (Fig. 2.12).

In 3D geostatistical estimation, it is possible to form a spatial model for the data of local areas that includes significant spatial trends by using KED. KED is described by;

$$\hat{A}(S_0) = \sum_{l=0}^{L} a_1 f_1(S_0) + \sum_{i=0}^{n} \lambda_i (A(S_i) - \mu(S_0)) \tag{2.10}$$

herein, S = Coordinate in N dimension, A = Target variable under consideration, $\hat{A}$ = Estimated Value of Variable Z at an unsampled point (S_0), n = Number of sampled points used for the estimation, λ_i = Kriging Weight, $\mu(S_0)$ = Mean sample data within the search window.

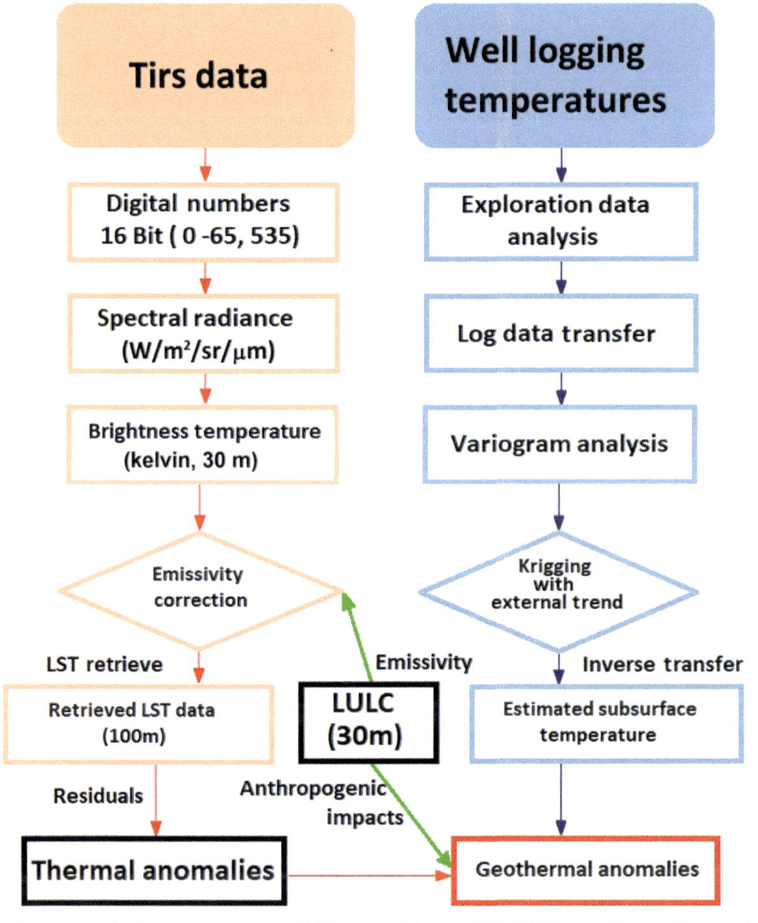

Figure 2.12 An example temperature modeling workflow with Kriging with an External Drift (KED) for TIRS image and well-logging temperature data [78].

Land Surface Data description is based on the Planck's Law and describes a spectral radiance, B(λ,T) that is the function of the Temperature (K) and wavelength of the electromagnetic waves (λ (m)). The equation can be written as;

$$B(\lambda, T) = \frac{2hc^2}{\lambda^5}\frac{1}{exp^{\frac{hc}{\lambda kT}} - 1} \tag{2.11}$$

whereas, $h =$ Planck constant (6.6256×10^{-34}Js), $c =$ Velocity of electromagnetic waves (2.9979×10^8 m/s), $k =$ Boltzmann constant (1.38×10^{-23} J/K).

The equation also can be written in a simplified form as;

$$B(\lambda, T) = \frac{\lambda^{-5} c_1}{\pi(exp(c_2/T) - 1)} \tag{2.12}$$

where, $c_1 = 2\pi hc^2 = 3.7418 \times 10^{-16} W/m^2$, and $c_2 = hc/k = 1.4393 \times 10^{-2} mK$

Temperature value in a certain area can also be described as (Fig. 2.13);

$$T_B = \frac{c_2}{\lambda \ln(c_1/\pi\lambda^5 B + 1)} \tag{2.13}$$

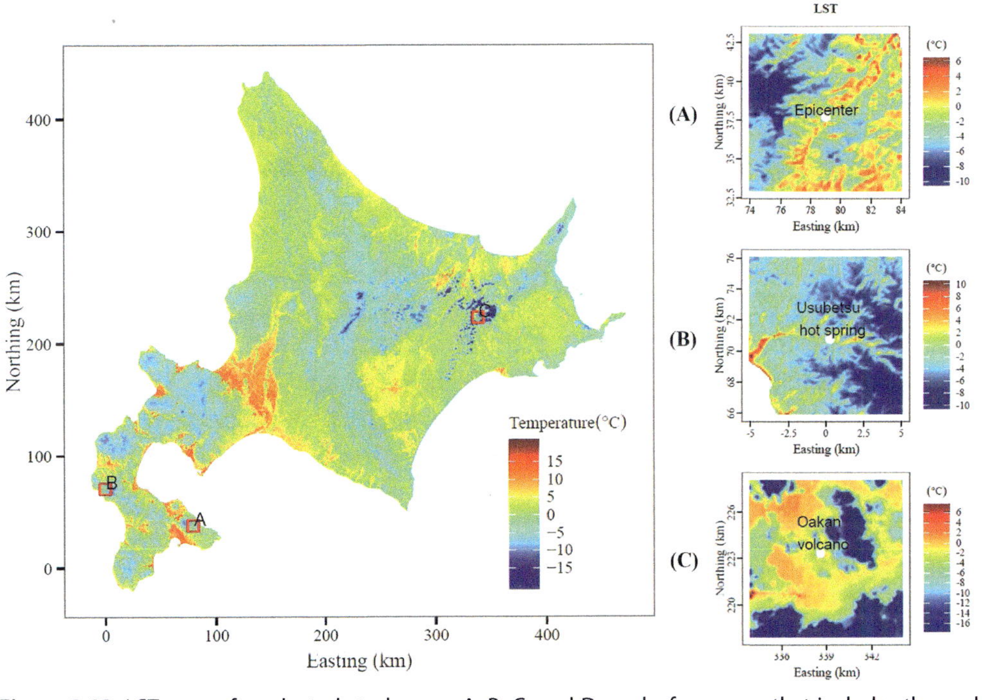

Figure 2.13 LST map of a selected study area. A, B, C, and D marks four areas that includes thermal anomalies [79].

Stefan-Boltzmann Law (Eq. 2.6) and Wien's Law (Eq. 2.7) also describes the electromagnetic radiation that the blackbody emitted, and it can be written as the function of temperature [69,80].

$$T_{Rad} = \sigma \cdot T_{kin}^4 \tag{2.14}$$

and;

$$\lambda_{max} = \frac{A}{T} \tag{2.15}$$

Where; T_{Rad} = radiant flux of the black body (W/m^2), T = absolute kinetic temperature (K), σ = Stefan-Boltzmann constant (5.6697×10^{-8} W/m^2/K^4), λ_{max} = maximum spectral radiant exitance (μm), A = Wien constant (2897.8(μm K)), T = absolute kinetic temperature (K).

In the literature, many algorithms based on various assumptions have been described for the identification of Land Surface Temperature by using TIR images. The most robust algorithm is the "Split Window Algorithm" which could apply for cases that the date recorded by two or more thermal channels [79]. Another common algorithm that was suggested by Artis and Carnahan [81] in 1982 is single-channel algorithm.

2.2.3.4 Airborne imaging with technological devices and vehicles

TIR by using airborne instruments is one of the proven and economical methods that could provide high resolution image data. It is an older method compared with spaceborne methods and includes several applications from different projects in the literature. Hellman and Ramsey [82] used AVIRIS data and compared with spaceborne methods for characterizing the Yellowstone hot spring deposits of Lower, Midway and Upper Geyser Basins in 2003. Haselwimmer [66] used Airborne Thermal Imagery to identify the locations of Pilgrim Hot Springs of Alaska. Vaughan et al. [72] evaluated TIR image data from MASTER and SEBASS sensors for 12km^2 areas with 5 m spatial resolution. Mutua et al. [83] used high resolution airborne TIRS in Silali Geothermal field of Kenya. Tian et al. [78] combined TIRS data with Well-Logging Temperature data for temperature mapping in Hokkaido, Japan. Gibson et al. [84] combined Full Tensor Gradiometry and Z-Axis Tipper Electromagnetic Deep Penetrating System for the usage of Airborne Geophysical Technologies. In 2015, Calvin et al. [85] used commercial and research Airborne Sensors with spatial resolution range from 2 to 90 m to identify mineral and thermal properties of Nevada geothermal resources by using the surface indicators. However, they commonly requested satellite data for regional overview (Fig. 2.14).

By the development of new technologies, modified unmanned aircraft vehicles and systems such as; Quadcopters are now having the ability to carry on Thermal Infrared

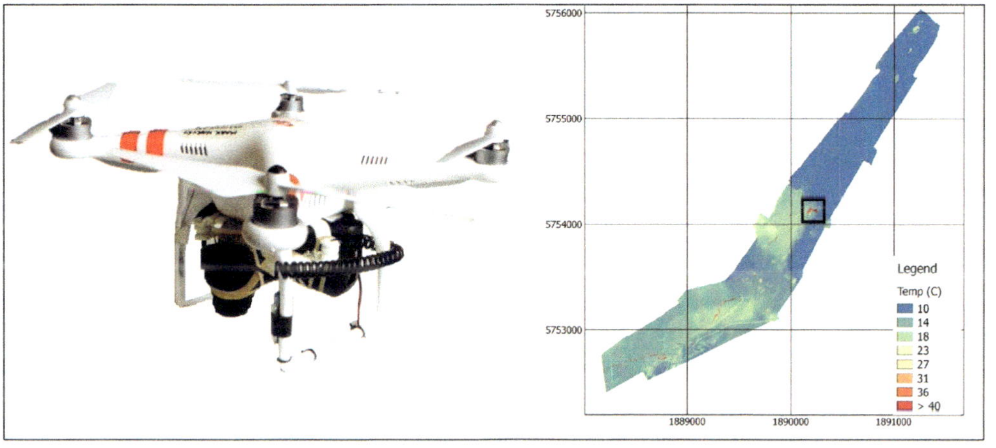

Figure 2.14 Thermal infrared orthophoto taken by DJI Phantom 2 Vision Plus Quadcopter modified with an ICI thermal camera and UAV module, Waikite survey area, New Zealand [86].

Cameras to discover volcanic geothermal activities. Thermal cameras can be installed on Unmanned Airborne Systems (UASs) for different reasons such as; generating real-time geo-rectification imagery. For instance, United States Forest Service flew 14 different UAS sensor missions in corporation with NASA to identify 57 different fires in west part of the United States [87]. It also can be used in cases that need narrowband sensors to identify thermal anomalies in an economical way [88].

2.2.3.5 Spaceborne imaging and remote sensing

Although Thermal Infrared technology has been used for many years on the identification and monitoring of geothermal systems, the success of a project could be limited in airborne imaging due to the day-time heating effects of the sun and it resulted by not detecting the thermal anomaly. Also, in lower resolution devices, impedance sourced by deep water tables could prevent to identify hot spring formations of sub-surface activity (Fig. 2.15).

Especially, in large exploration areas which are arid and have sparse flora, satellite-based TIR is an efficient method for mapping and monitoring due to higher spatial resolution and swath width values compared with airborne imaging for the description of both surface and sub-surface geothermal activity.

The United States has the biggest number of sensors in space. However, Europe and China also have sent several satellites with sensors for spaceborne imaging. For instance, the earth observation satellites called "Sentinels" sent by Europe for the aim of measuring the topography of sea surface, identifying ocean and land surface colors with high accuracy to support ocean forecasting systems, measuring sea and land temperatures, environmental

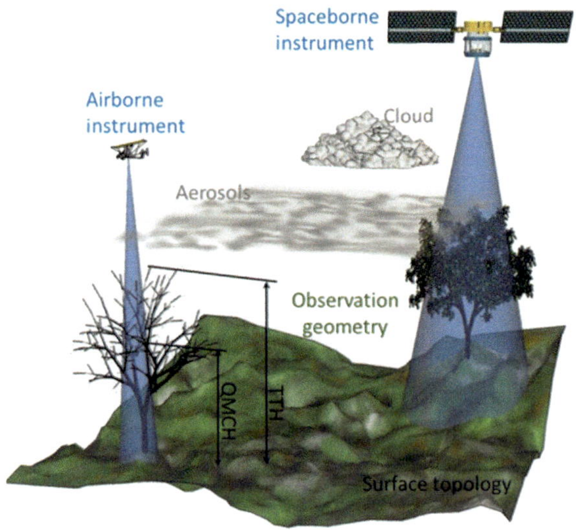

Figure 2.15 Airborne and spaceborne comparison [89].

monitoring and climate monitoring. The latest Sentinel satellite, Sentinel 3B includes multiple sensing instruments such as; Sea and Land Surface Temperature Radiometer, Synthetic Aperture Radar Altimeter (SRAL) (provides accurate topographic measurements), Doppler Orbitography and Radiopositioning Integrated by Satellite (DORIS) (a receiver for orbit positioning), Microwave Radiometer, Laser Retroreflector, Global Navigation Satellite System, and Ocean and Land Color Instruments which includes five cameras for imaging in a wide view and 21 spectral bands that ranges from near-infrared to optical.

Table 2.1 lists the platform and satellites that use Thermal Infrared Remote Sensing Sensors, and the sensor characteristics such as; spatial resolution, revisiting time, Swath width, agencies.

2.2.3.6 Tracers

Tracers are very useful materials for the geothermal exploration to delineate subsurface and re-injection fluids movement. They are mainly used for the determination of hydraulic connectivity between the wells and fluid volume of the reservoir which are crucial parameters for the feasibility studies of the reservoir management for a geothermal system. The main advantages of the tracers are that they are very stable and cost-effective [90].

The main idea in a tracer test is to inject a certain amount of tracer (generally 100 kg or less) into an injection well and to monitor the tracer recovery from the adjacent wells. If the returning tracer decreases, it does not mean every time that the connectivity between the wells is not good enough. It may indicate thermal

Table 2.1 Typical sensors used in spaceborne imaging [69].

Sensor	Spatial resolution	Revisit	Swath width	Platform/satellite	Agency	Launch year
ETM +	60 m	16 days	185 km	Landsat-7	USGS, NASA	1999
TM	120 m	16 days	185 km	Landsat 5	USGS, NASA	1984
TIRS	100 m	16 days	185 km	Landsat 8 (LDCM)	USGS, NASA	2013
ASTER	90 m	4–16 days	60 km	Terra	NASA	1999
IRMSS	160 m	26 days	120 km	CBERS-1, 2, 2b	CRESDA, INPE	1999–2003, 2003, 2007–2010
IRSCAM	80 m	26 days	120 km	CBERS-3 and 4, 4b	CRESDA, INPE	2012, 2014, 2016
MERSI	250 m	1 day	2800 km	FY-3A, FY-3B	NRSCC, CAST, NSM	2008, 2010
InfraredCam	300 m	31 days	720 km	HJ-1B	CRESDA, CAST, NRSCC	2008
NIRST	351 m	<1–2 days	182–1060 km	Aquarius	NASA, CONAE	2011
BIRD	370 m	10 days	190 km	BIRD	DLR	2001–2001
TET-1	356 m	10 days	180 km	TET-1	DLR	2012
VIIRS	1.6 km	<1 day	3000 km	Suomi NPP	NASA/NOAA	2011
CrIS	14 km	<1 day	2200 km	Suomi NPP	NASA/NOAA	2011
CERES	20 km	<1 day	3000 km	Suomi NPP	NASA/NOAA	2011
IIR	1 km	16 days	64 km	CALIPSO	CNES	2006
MODIS	1 km	4 per day	2330 km	Terra, Aqua	NASA	1999, 2002
ATSR-2	1 km	3 days	512 km	ERS-2	ESA, UKSA, CSIRO	1995–2011
AATSR	1 km	35 days	500 km	Envisat	ESA, UKSA	2002–2012
AVHRR/1	1.1 km	<1 day	2600 km	TIROS-N, NOAA 6,8,10	NOAA	1978–1986
AVHRR/2	1.1 km	<1 day	3000 km	NOAA9,10,11,12,13,14	NOAA	1984–2005
AVHRR/3	1.1 km	<1 day	3000 km	NOAA15-19, Metop A,B	NOAA, EUMETSAT	1998, 2000, 2002, 2005, 2006, 2012
MSG-SEVIRI	1–3 km	<1 day	Full earth disk	Meteosat-8/9/19	ESA/EUMETSAT	2002, 2005, 2012
MVIRI	5 km	Every 30 min	Full earth disk	Meteosat-3/4/5/6/7	EUMETSAT, ESA	1988, 1989, 1991, 1993, 1997

(Continued)

Table 2.1 (Continued)

Sensor	Spatial resolution	Revisit	Swath width	Platform/satellite	Agency	Launch year
MSU-MR	1 km	37 days	3000 km	Meteor 3M, and −M N1	ROSHYDROMET a.o.	2001, 2009
MSU-GS	4 km	Geostatically <1 day	Full earth disk	Elektro-L N1	ROSHYDROMET a.o.	2011
IMAGER	4 km	Geostatically <1 day	Full earth disk	MTSAT-1,2,3	JMA	1999, 2006, 2013
MVISR	1.1 km	3−4 days	3200 km	FY-1C, 1D	CMA, NRSCC	1999, 2002
IVISSR	5 km	1 day	Full earth disk	FY-2C, 2D, 2E, 2F	NRSCC,CAST, NSMC	2004, 2006, 2008, 2012
VHRR	8 km	Every 30 min	Full earth disk	Insat-2A,B,E, -3A	ISRO	1992, 1993, 1999, 2003
IASI	25 km	29 days	2052 km	MetOP-A/B	CNES, EUMETSAT	2006, 2012
HIRS/3	20.3 km	<1 day	2240 km	NOAA 15,16,17	NOAA	1998, 2000, 2002
HIRS/4	20.3 km	<1 day	2240 km	NOAA 18/19, MetOP A/B	NOAA, EUMETSAT	2005, 2006, 2009, 2012
ScaRaB	40 km	1 day	2200 km	Meteor-3, Resurs-01, Megha Tropiques	CNES	1994, 1998, 2011

degradation or adsorption process onto the subsurface rocks. Therefore, laboratory simulation studies which describe the tracers need to be chosen before conducting the tests on the reservoir [90]. Dr. Peter Rose from the Energy and Geoscience Institute at the University of Utah claimed that UV-fluorescent polyaromatic sulfonates are to be perfect tracers for the high-temperature geothermal systems due to its environmentally friendly character, having possibility to detect by using fluorescence spectroscopy, inexpensive price, and thermally stable structure. His research group has shown that naphthalene sulfonates are favorable to use in geothermal systems up to 340°C as a conservative tracer [90]. Aliphatic alcohols and certain types of freons are the other tracers that can be counted. Since these types of tracers are associated with the steam phase of geothermal fluid, they are commonly used for the steam-dominated geothermal reservoirs such as the Geyser in California, United States. However, their usage is not that easy compared to the other chemicals mentioned above [90].

Frane et al. improved a methodology regarding the testing of the smart encapsulated tracers that is relatively new technology than the conventional ones. These tracers have been used in this study to map the distribution of the flow field along with the flow path and specific temperature inside the reservoir [91]. Although it is stated that promising results can be obtained from the tests, the method needs further development for wide use in the geothermal exploration.

2.2.4 Advanced drilling fluids and applications in geothermal fields

Drilling fluids have to provide proper rheological and filtration properties in order to deliver well stability at high temperature conditions. Moreover, it must be environmentally friendly [92]. There are recent improvements with conventional water and oil-based drilling fluids. These applications require usage of some chemical additives to improve the properties of drilling fluid in high temperature environment. However, the proposed drilling muds have limited use due to their bentonite clay and chemical ingredients [93]. These additives may increase the density and also have a positive effect against corrosion. They may also develop proper viscosity and prevent the development of bacteria in high temperature and high-pressure environment [94].

Nonetheless, the temperature and pressure conditions for geothermal and deep oil and gas wells are very elevated. The conventional drilling fluids may not provide the desired properties and heat transfer requirements in these extreme conditions [95]. Designing a functional drilling fluid in high temperature conditions necessitate to improve the thermal properties of the drilling mud.

The characteristic high temperature-resistant drilling fluid structures comprise Sulfonated and Polysulfonate Drilling Fluids. The treating agents are used in the Sulfonated Drilling Mud to provide noble thermal stability, rheological and filtration properties in high temperature and high-pressure environment. They also may help to improve the resistance against tough salty environments. The Polysulfonates are Polymer Drilling Fluids associated with Sulfonated Drilling Fluids. They have benefits in increasing penetration rate and improving the well stability [96].

Currently, the Water-Based Drilling Fluids with improved properties by the additives can be used up to 260°C. Besides, the Oil-Based Drilling with additives provides desired needs up to 290°C. The application and conditions of some field operations are tabulated in Table 2.2 [97].

Another approach to deal with the problems originated from geothermal environments is to use high temperature foam drilling fluids. The application of this type of drilling fluids consists of using high temperature agent within the conventional foam drilling fluids.

Foam drilling fluids with enhanced properties with additives can be used at temperatures around 300°C. There are some field applications where it is reported that the HighTemperature Foam Drilling reached to temperatures up to 350°C [98].

An advanced method to battle with high temperature environment in terms of drilling fluids is using nanotechnology. The launch of nanotechnology has transformed the science and engineering approaches in different sectors out of it's wide variety of uses. There are different nanoparticles applications in the oil and gas industry [99]. Nanoparticles are used in Drilling Fluids to enhance the Drilling Fluids properties and stability in harsh environment. Most of the studies are carried out in laboratory conditions these days. However, with the technological development, the application of nano-based drilling fluids may be widespread in the future.

The drilling fluids need to be light-weighted, tough and resistant to corrosion. Using Nano-Based Drilling Fluids may let the engineers handle these properties [100]. The use of Nanoparticles within Drilling Fluids will allow the drilling industry to regulate the mud's rheology by adjusting the composition, type, or size distribution of Nanoparticles in Drilling Fluid to deal with any exceptional conditions [99]. The nanoparticles are too much smaller than the fine particles used in the industry. There are some studies that define Nano-Based Drilling Fluids as Drilling Fluids in which there exist additives having particle sizes between 1 and 100 nm [101].

Table 2.2 Applications of drilling fluids in some deep wells [97].

Well depth, m	Temperature, °C	Location
6981	232	Lake Pontchartrain (United States)
7265	290	Texas (United States-oil-based)
7380	200	Xinjiang (China)
7026	235	Shengli (China)
6005	236	Henan (China)
5300	213	Liaohe (China)

The manufacturing process is very hard. However, the manufactured Nano-Products are more impregnable and reactive than the other fine particles. They also have effective heat conductivity [102]. The effectiveness of these Nanoparticles is originated from the improved surface interface. Once the size of the particles in a fluid system decreased, there are more surface to tolerate the heat generated from the surroundings [93].

2.2.5 Advanced coating and composites in geothermal systems

The interaction with geothermal fluids with the well, the well-head and the production components affects their metallurgical stability as the geothermal fluids are highly corrosive. Carbon Steel items coated with Anti-Corrosion Materials are generally preferred for high temperature environments. The coatings increase the endurance of components like Heat Exchanger, Casings and Liners in the well, Wellheads and Condensers.

Carbon Steels or Low Alloy Steels with less than 0.25% C (Carbon) and 0.4%–1.5% Mn (Manganese) are the prevailing materials used for pipes and equipment in geothermal wells. Generally, this material has presented to be both economical and enduring for the geothermal applications. Similar behavior can be experienced from the corrosion testing experiment where the Carbon Steel Samples are in direct contact to the superheated steam at 360°C. Besides, there are some researches where the Carbon Steels have greatly affected from erosion-corrosion problems [103]. Regardless, Carbon Steels are one of the favorite materials to be used in situations where condensate can be evaded and where erosion-corrosion is not that problematic.

Another candidate material which can resist high temperature environment is Stainless Steel. Stainless Steels are Steels having more than 12% Cr (Chrome). The application of Stainless Steels is generally with Ni (Nikel), Mo (Molybdaenum) and N (Nitrogene) alloys. Stainless Steels have varied practices in different equipment's and piping. Conventionally, this material group can be separated into two main groups in geothermal applications as Austenitic Stainless Steels and Duplex Stainless Steels. The Duplex type is not recommended for temperatures beyond 250°C as the material tensile strength weakens above that temperature [104]. Below 250°C the Duplex types have outstanding resistances to corrosion.

There are some lower alloyed categories of Austenitic types as graded by American Iron and Steel Institute (AISI) as AISI 304 and AISI 316 and higher alloyed as S31254, N08904 and N08028 [105,106].

Grade 304 stainless steel is the most mutual Stainless-Steel configuration. It comprises of 16% and 24% Chromium and up to 35% Nickel, along with small amounts of Carbon and Manganese. Grade 316 is the second-most common form of Stainless

Steel. It has practically physical and mechanical properties as similar as Stainless Steel grade 304. The main alteration is that 316 Stainless Steel integrates about 2%–3% Molybdenum. Both Stainless Steel types are known to be corrosion resistant at surface applications. However, 304/316 types have partial resistance to corrosion at steam-dominated high temperature environment.

The higher alloyed Austenitic Steels have better resistance against geothermal fluids in terms of corrosion, particularly the S31254 type. S31254 type is a Super Austenitic Stainless Steel with some high constituents of Molybdenum and Nitrogen. It delivers high resistance to pitting and crevice corrosion and higher strength than those of conventional Austenitic Stainless Steels such as 316L [104]. Herein, higher Austenitic graded Stainless Steel can generally be used in high temperature geothermal steam with temperatures above 250°C.

Another material considered as a good candidate for geothermal applications is Nickel-Based Alloys. Nickel-Based Alloys with Nickel content below 60% have been used in harsh environments such as geothermal environment for a long period of time. However, there are some inappropriate consequences gathered from some geothermal field applications of the Nickel Alloys. When temperatures exceed 300°C, the Nickel Alloys suffered from corrosion pitting and some small corrosion damage was observed. More testing of the Nickel Alloys is required for improved interpretation of their use in high temperature geothermal steam.

Titanium alloys are another material that is used in geothermal environment. The Titanium alloys show better resistance to corrosion at high temperatures when they are used with Palladium (Pd) alloys. It can be more useful at temperatures over 300°C. However, the concern of strength for Titanium Alloys can limit their usage at temperatures more than 400°C.

There are Polymeric Coatings which are used in geothermal environments. A Semi-Crystalline Polymeric Substance that is called Polyphenyledesulfide (PPS) is a good resistant against high temperature hydrothermal oxidation. Some early studies revealed the occurrence of oxidation on PPS coatings once exposed to acidified brine at 200°C. Even so, PPS coatings achieved shielding the carbon steel heat exchanger materials. [107–112].

The subsequent studies proposed that PPS-Coated Carbon Steel Components could be a good alternate material to be used in lieu of Titanium Alloys, Nickel Alloys, and Stainless Steel. Several filling ingredients can be put on the coating system to improve external hardness, thermal conductivity, and mechanical assets. There are some studies which suggested the use of the Ceramic Calcium Aluminate (CA) within PPS coatings to increase wear resistance and high temperature stability [111]. Just like the use of the ceramic CA within PPS coatings, adding Carbon Fiber on PPS coating systems enhance thermal conductivity and mechanical properties. For current applications, thermal conductivity and wear resistance are vital in heat exchangers. Using a PPS coating system with Carbon Fiber and CA fillers could be beneficial.

Besides, Casings and Heat Exchanger units, Well-Head components also need to battle harsh geothermal environment. The flow velocity can be up to 3 m/sec with a temperature over 250°C. The geothermal Well-Head components were generally made of Carbon Steel and Titanium Alloy based materials with different coatings against high temperature and corrosion. PPS is also used as coating in well-head components. PPS can withstand temperature up to 200°C as stated previously. To increase the melting point of PPS to temperatures over 250°C is very tough development in order to protect Carbon Steel components used in the geothermal fields. So as to do that, Polymer/Clay Nanocomposite Technology was adjusted to use Montmorillonite Clay as the Substitute Nanoscale Filler. The adaptation of the Montmorillonite Nano-Filler occurred in three stages. At the first stage, it increases its melting point up to 290°C. In the second stage, it raises its crystallization energy by designating exceptional adherence of the exteriors of nanofillers to PPS by means of a robust interfacial connection. At the final stage, it decreased the degree of hydrothermal oxidation as a result of Sulfite connected alterations (Fig. 2.16).

The thickness of the Polymer Clay Nanocomposite should be approximately 150 μm at 300°C for an effective protection against hot brine originated corrosion whilst it is used with Carbon Steel. The protection is weaker in applications with clay-free PPS coatings having the same thickness [114].

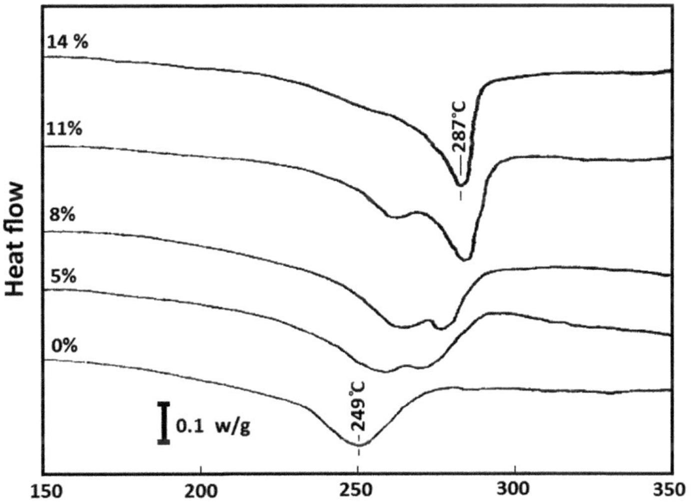

Figure 2.16 Scanning calorimeter curves for Montmorillonite (MMT) nanofillers-filled by polyphenyledesulfide coatings [113].

2.2.6 Advanced cement applications in geothermal fields

The main target of cement usage in oil/gas and geothermal fields is to provide zonal isolation. The application of several different grades of cement for this purpose must also deliver support for the casings and tubing in the wells. Any failure in cement may cause significant environmental and financial problems. Several alternative cement slurry compositions have been proposed for the hydrocarbon exploration. Most of these compositions are not suitable for geothermal applications due to the negative effects of high temperature environments on mechanical and chemical properties of cements. This chapter summarizes the advanced material applications for Cement Slurry design which can be appropriate in geothermal fields.

The zonal isolation of the oil, gas, and geothermal production well is a considerable environmental and operational concern. Setting Cement in the annular section of the well accommodate this vital duty [115]. The set Cement must not allow any fluids migration from the formation, also any fluids invasions from the wellbore through aquifers and other sensitive geological features. High temperature environments deteriorate the properties of cement, as consequently cement may not provide its wellbore sealing task. High temperature (up to 400°C) and chemically corrosive (typically Hypersaline, CO_2, and H_2S rich) geothermal production environments cause the failure of cement [116].

Conventional well cements consist of Silicate Hydrates (CaO-SiO_2-H_2O structure) and Calcium Aluminum Silicate Hydrates (CaO-Al_2O_3-SiO_2-H_2O structure). The achievement of zonal isolation and casing supporting duty of these cements is very low in high temperature environment. They cannot also avoid the corrosion at the pipes due to the corrosive geothermal fluid. These situations may get even worse in deeper wells (>5 km) experiencing high temperature (>250°C) with intense saline and CO_2 rich (>40,000 ppm) environments [113].

The acidic environment deteriorates with carbonation and degrades the strength of conventional Cement by causing an increase in the cement matrix permeability and a decrease in the casing support [117]. Besides these consequences, the failure may result in debonding at formations well as at casing, shear weakening and tensile deficiencies. The conventional Cement has high compressive strength. However, comparing the compressive strength, they have weaker tension and adhesion. In geothermal applications, cement hydration causes segregation between the formation and cement matrix [118–120].

There are several approaches to overcome these problems. These methods generally propose using different cement materials in the slurry compositions. The application of the Phosphate Bonded Cement, Foam Cements [121], CO_2 resistant Cements and Cements with self-healing capabilities are currently in use. In this section, some of these innovative cement applications are described.

2.2.6.1 Foam cements

The purpose of using foam cement is to obtain a low-density cement slurry to be able to place it in problematic formations. Besides cement and water, Foam Cement generally consists of some additives, Foam Stabilizers and a Gas (generally N_2). It is accessible to provide a settlement at weak formations to get over the excessive loss and fallback problems. The usage of cement delivers relatively high compressive strength and low permeability to avoid gas migration through cement matrix [122].

The temperature alteration at the surrounding area causes extra stresses around the wellbore in the geothermal fields. These extra stresses and the existence of high temperature not only deteriorate the cement body but also cause changes in casing diameter.

Foam Cements proposed to be more elastic and are able to bear the thermal expansion and shrinking effect compared to conventional cement applications [123].

2.2.6.2 Phosphate bonded cement

In geothermal applications, cement exposed to extreme thermal resistance. In some applications, it should withstand to thermal shock of up to 600°C. It should have improved toughness as a mechanical property, also must resist to mild to strong acidic environment. In order to achieve these goals, several phosphate bonded cements were proposed [124,125].

One of them is CA hPosphate Cement (CaO-Al_2O_3-P_2O_5-H_20) which has two basic ingredients besides water as CA cement and Sodium Polyphosphate. It was considered to be used in slightly acidic (pH ~ 5.0) and carbon dioxide rich environments. This type of cement proposed to be economical as a result of the employment of low-cost cement-forming by products from several different manufacturing processes [113].

An improved application of Phosphate Cement is using Sodium Silicate activated Calcium where the cement is exposed to high temperature environment (Na_2O-CaO-Al_2O_3-SiO_2-H_2O). This type of cement also called Thermal Shock Resistant Cement. The compressive strength properties of the cement are stable even at temperatures as high as 600°C [126].

2.2.6.3 Self-healing cements

Self-healing materials applications to different conditions are a popular topic in the past decade. There are several different applications in terms of well cements as well [127,128]. In recent Self-Healing Cement development focuses to enhance adhesion in both the casing and the formation boundaries at high temperature conditions like geothermal applications. Usage of Self-Healing Cement in geothermal fields is to provide long-term integrity in the wellbore. To battle the high temperature conditions, some Self-Healable Polymers are added to cement to improve cement bonding properties. There are several proposed polymer-cement mixtures used in oil and gas industry with self-healing capabilities. However, the early proposed compositions were not suitable for geothermal well surroundings [129–138].

In recent studies, a blend of self-healable resins and cement suggested to obtain a thermally stabilized polymer-cement composition. This mixture could be a novel approach in order to engage wellbore failures, production difficulties and reservoir intervention at geothermal fields [139]. Polymer Cements help to deliver necessary assets such as better fracture strength, better zonal isolation and chemical endurance [140,141].

Self-healing process can be based on microcapsules. These microcapsules consist of polymeric monomers. These monomers are sheathed and inserted into the cement structure and repair the damage in case of weakening cement stability. Another self-healing process can be achieved by redeployment of covalent bonds or molecular interfaces leading to heal [142,143]. The sheathed monomers process usually yields in better healing efficiencies for one damage period. However, if multiple damage periods occurred, they have restricted healing capability. The preparation of microcapsules to bear the thermal and mechanical stresses in geothermal environment is also not an easy task. The production of materials being able to provide healing in multiple damage periods is beneficial and more accurate process. Some researchers suggest using as monomer Self-Healing Epoxies with Disulfide Bonds. To deliver self-healing performance, multifunctional thiols cross-linkers are used. The chemical interactions of these materials are given in Fig. 2.17 [144–146]. The major repair structure is a free thiolate group with a disulfide. The reactions ended with forming a new Disulfide group and freeing thiol [146].

The self-healing cement approach needs improvements to be applicable at temperatures over 250°C. The healing process is not accurate over 250°C. Besides, the polymers are affected from the stresses caused by high temperature. There are

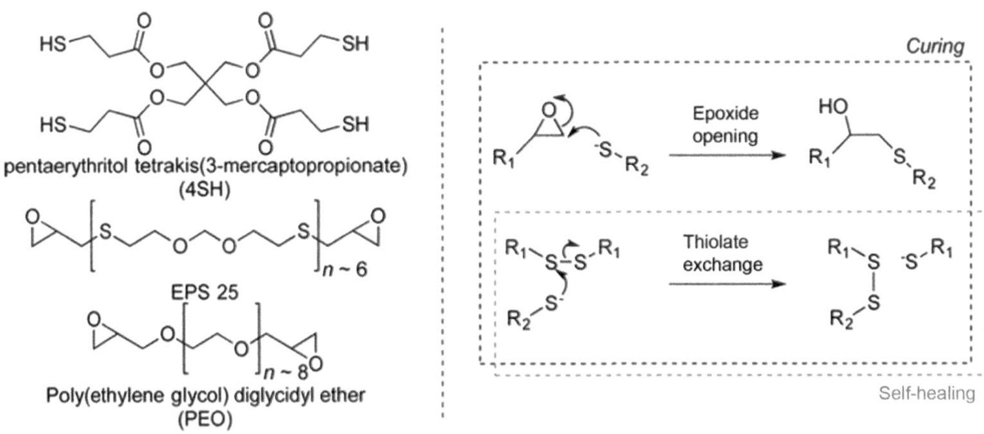

Figure 2.17 Chemical structures of monomers and mechanisms of polymer curing and self-healing [144–147].

Step 1. Hydrolysis of reactant:

$$CaO.Al_2O_3 + 4H_2O \rightarrow Ca^{2+} + 2Al(OH)_4^-$$
$$CaO.Al_2O_3 + 7H_2O + 2OH^- \rightarrow Ca^{2+} + 4Al(OH)_4^-$$
$$\text{-[-NaPO}_3\text{-]}_n\text{-} + nH_2O \rightarrow nNaHPO_4\text{-} + nH^+$$

Step 2. Acid-base reaction:

$$Ca^{2+} + nNaHPO_4\text{-} \rightarrow Ca(HPO_4)$$
$$Al(OH)_4\text{-} + H^+ \rightarrow AlH(OH)_4$$

Step 3. Hydrothermal hydration:

$$Ca(HPO_4) + 4Ca^{2+} + 2NaHPO_4\text{-} + xH_2O \rightarrow$$
$$Ca_5(PO_4)_3(OH) + 2Na^+ + xH^+ + yOH^-$$
$$AlH(OH)_4 \rightarrow \gamma^-AlOOH + 2H_2O$$

Figure 2.18 Calcium aluminate cement synthesizing step reactions [3].

ongoing researches to achieve the production and testing of self-healing materials to be used in higher temperature conditions.

2.2.6.4 CO_2 resistant cement

One of the problems in the geothermal environment is the existence of CO_2. When CO_2 gets in contact with the well cement, causes cement matrix conductivity, and increases in terms of porosity and permeability and also a decrease in bonding and supporting ability of cement. There are different proposed cement slurry designs. The prominent one is CA cement synthesizing.

The CA Phosphate Cement was synthesized from CA cement and Sodium Polyphosphate solution [148]. The synthesis starts with hydrolysis of reactant, following with acid-base reaction and ended with hydrothermal hydration given in Fig. 2.18 [113].

These reactions yield of synthesis of the CA Phosphate Cement. The obtained product has high compressive strength ($>$90 MPa at 150°C–300°C temperature range) and low water permeability ($<1 \times 10^{-4}$ Darcy). Such properties are required to achieve long-term integrity in terms of cement in a geothermal field.

2.3 Advanced materials used in geothermal heat transfer and conversion

2.3.1 Geothermal heat pumps and exchangers

Geothermal Heat Pumps and Heat Exchangers are encased assembly units that perform a heating and a cooling process by using geothermal energy. Geothermal Heat Pumps

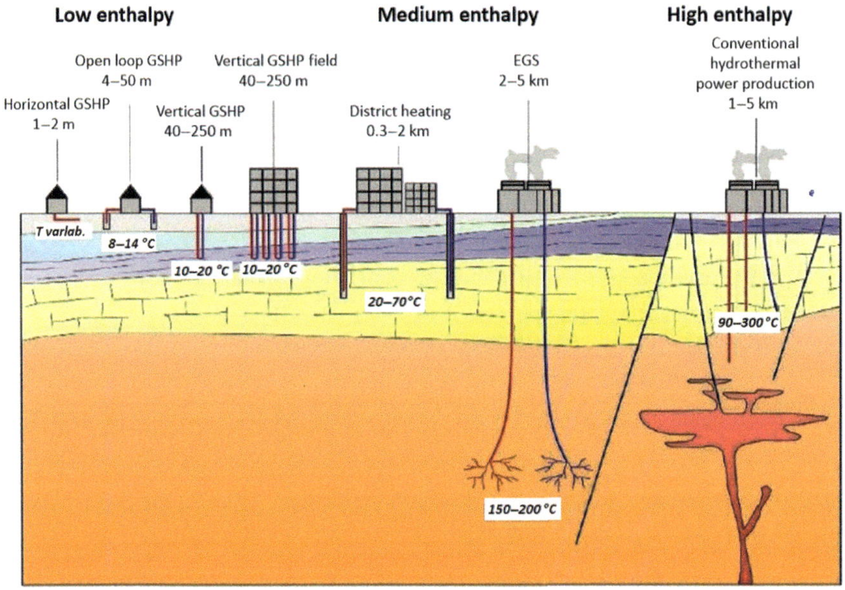

Figure 2.19 Classification of geothermal system as per enthalpy content [149].

are extremely appealing option compared to conventional electric or fossil fuel space conditioning equipment by providing a cost advantage on first cost basis as well as in operation and maintenance. Geothermal fluids can be extracted from the ground in various temperature range. They can be used directly as per heat content quality or to produce electric power. Recently, Ground-Source Heat Pumps that aims to extract or store heat from shallow depths with low temperature becomes widespread application for geothermal energy and many studies are performed to increase the efficiency of the systems. In this chapter, Heat Pumps and Heat Exchangers mainly used in shallow geothermal energy applications having low enthalpy (temperatures less than 100°C) are discussed. Fig. 2.19 gives the classification of geothermal systems and summarizes the overall Heat Pump Systems due to installation type.

2.3.1.1 Ground-source heat pumps

The GSHPs are the Geothermal Heat Pumps and earth or groundwater are used as the source. They [150,151] involve the coupling of low-grade thermal energy from earth sources to a Heat Pump. The groundwater-source type is the most common Geothermal Heat Pump type which uses geothermal water coming from aquifer, springs or drilled wells. The direct exchange heat pump circulates refrigerants and its refrigerant lines are buried directly in the ground in a closed-loop configuration which leads them to compensate only small installations. Geothermal Heat Pump systems are

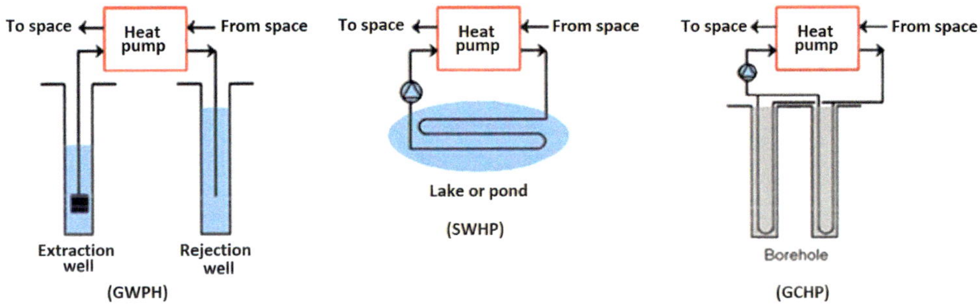

Figure 2.20 Schematic of different Ground-Source Heat Pumps [155].

promising energy source with its benefits such as being environmentally friendly, having low life-safety risk and low maintenance cost [152,153].

Ground-Source Heat Pumps (GSHP) put through the usage of the stable temperature of the shallower ground with Horizontal Heat Exchanger Systems and deeper ground with the depths of up to 150 m using Vertical Heat Exchange Boreholes. They are mainly used in buildings for Space Heating, Cooling, and sometimes domestic hot water maintenance. Their initial capital cost is high due to the extra expense for Buried Heat Exchangers in the earth or providing a well for the energy source but after installation is completed the annual cost is low throughout the system's life which results net savings [154]. The GSHPs working principle is shaped based on Groundwater Heat Pump (GWHP), Surface Water Heat Pump (SWHP) or Earth–Ground Coupled Heat Pump (GCHP) as illustrated in Fig. 2.20 [155,156].

The GWHP system uses groundwater as heat source or sink whereas in SWHP system, heat is transferred by circulating working fluid through pipes located at an optimal depth within a lake, pond, aquifer, reservoir or open channels [155,156]. GCHP system consists of a coil with circulating fluid in a closed loop of horizontal or vertical pipes buried into ground to interchange thermal energy to and from the earth [157].

GSHP system contains three parts [155,158,159] as illustrated in Fig. 2.21;

Heat Pump: transfer heat between building and ground and alters its temperature.

Ground Heat Exchanger: an underground piping system that enables underground heat extraction.

Heat Distribution System: Ductwork or Piping System to deliver heat throughout the building.

2.3.1.1.1 Principles and thermodynamics of the heat pumps

At shallow depths, the temperature in the ground fluctuates with outside air temperature (Fig. 2.22). It becomes more stable with increasing depth and sustains near-constant temperature approximately equivalent to the average annual air temperature at around 5–10 m depths. At those intervals, ground temperature is

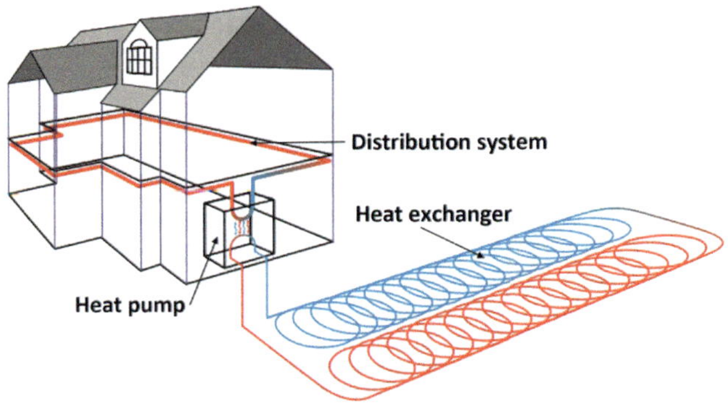

Figure 2.21 A typical GSHP system [160].

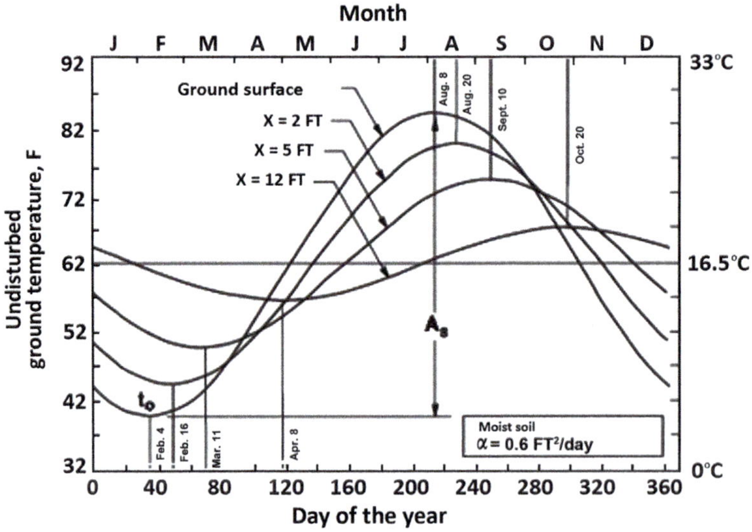

Figure 2.22 Variations in the annual temperature of the soil belong to Stillwater, OK [161].

warmer during winter and cooler in summer compared to ambient air temperature. Earth's temperature starts to increase with depth due to geothermal gradient as goes down deeper [162].

Thermal stability of subsurface is used for heating in the winter by extracting heat from the ground which has a higher temperature than air and cooling in the summer by re-injection of heat into the ground which has a lower temperature than air [162] An efficient heat transfer between ground and buildings is provided in this manner as illustrated in Fig. 2.23.

Heat Pumps are based on a reverse Carnot Thermodynamic Cycle, which consumes drive energy and produces a thermal effect. The Carnot Cycle is a hypothetical

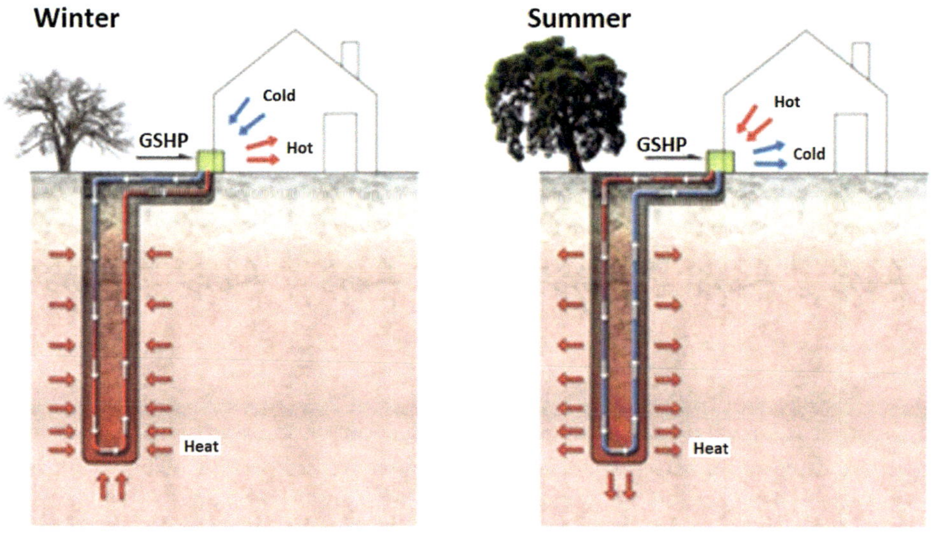

Figure 2.23 A shallow geothermal energy system in heating and cooling modes [163].

thermodynamic cycle for Heat Pumps that specifies the ratio of the heat transfers in a fully reversible cycle, and depends only on the temperatures of the source.

$$\left(\frac{Q_1}{Q_2}\right)_{reversible} = \frac{T_1}{T_2} \tag{2.16}$$

Where Q_1 *-and* T_1, Q_2 *and* T_2 are energy and temperature of the hot and cold sources, respectively. Thermal energy produced in Heat Pumps depends on the temperatures of the working fluid. If there is small temperature divergence between the heat source and the heat sink, compressor needs to do less work to expand the temperature of the working liquid [164]. A commonly used measure for the efficiency of the Heat Pump is the Coefficient of Performance (*COP*) which corresponds to the ratio between the delivered heat energy (Q_{HP}) and the total electrical input to the compressor of the Heat Pump (E_{HP}) in kW [165].

$$\varepsilon = COP = \frac{Delivered\ heat\ energy}{Total\ electrical\ input} = \frac{Q_{HP}}{E_{HP}} \tag{2.17}$$

COP depends on the Heat Pump's design and the operating characteristics of the working fluid. Since Heat Pump Cycle is the reversed Carnot Cycle, the efficiency of the system can also be calculated from the thermodynamic cycle using the temperature difference between the heat source and the heat sink. The maximum *COP* or Carnot efficiency gained from a heat pump in heating and in cooling modes can be calculated by below equations, respectively [165].

$$COP_{max,h} = \frac{T_H}{T_H - T_C} \quad \text{and} \quad COP_{max,c} = \frac{T_C}{T_H - T_C} \tag{2.18}$$

For Ground-Source Heat Pumps the value of *COP* is normally changing between 3 to 6 due to the earth connection setups, system sizes, earth characteristics, installation depths, and regional climate [153,156,166].

2.3.1.1.2 Basic components of heat pump

The Geothermal Heat Pump is mainly comprised of a compressor, an evaporator, a condenser, and an expansion device as illustrated in Fig. 2.24. The compressor increases the working fluids temperature by pressurizing it and then drives it to the condenser. The evaporator and condenser are Coiled Heat Exchangers that contain refrigerants. In the condenser, refrigerant losses heat and becomes high-pressure, mild-temperature liquid. The condensed liquid goes through an expansion valve where its pressure decreases and then moves to the evaporator where it gains heat and becomes a low-temperature vapor. Finally working fluid turns back to the compressor and cycle starts over [167,168].

Heat pumps use working fluid referred to as refrigerant which is responsible from the heat transfer. Refrigerants are volatile evaporating and condensing fluid that has low boiling points which can be easily changed by pressure changes, high critical temperature, and high specific heat capacities. They should have a moderate density in liquid form and a relatively high density in gaseous form. The regular refrigerants can be classified into different groups:

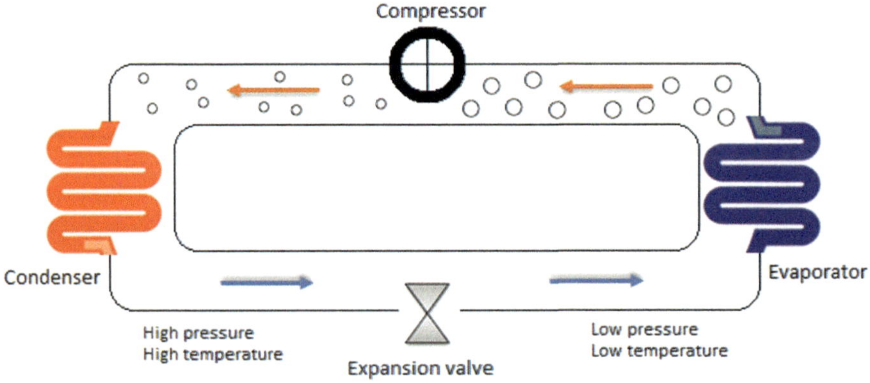

Figure 2.24 Configuration of heat pump system.

Freons (chlorofluorocarbons such as R-11, R-12, R-114, hydrochlorofluorocarbons such as R-22, hydrofluorocarbons (HFCs) such as R-134a, R-32, R-407C, R-410A, R-404A and natural refrigerants such as propane, butane, ammonia, water, CO_2 [165,169].

2.3.1.2 Geothermal heat exchangers

Geothermal heat exchangers (GHEs) is an important part of the GSHP system. They are comprised of a collection of pipes that transfer fluid between the ground and heat pump unit.

2.3.1.2.1 Heat transfer in heat exchanger

A Counter-Flow Heat Exchanger is preferably used in exchanging energy from the geofluid to water in the geothermal system applications. In practice, there is no perfect heat exchange due to heat loss occurred during transfer. The actual heat transfer in the Heat Exchanger can be computed with Eq. 2.19.

$$P_x = E_x M_x (T_{hi} - T_{ci}) \tag{2.19}$$

where P_x is the actual rate of heat transfer (W), E_x is the Heat Exchanger effectiveness, M_x is the smaller of the thermal capacities of the flows (W/°C), T_{hi} is the inlet temperature of the high-temperature fluid (°C) and T_{ci} is the inlet temperature of the low-temperature fluid (°C). The temperature changes which occur in the fluid streams depend upon the relative sizes of the fluid flows and the efficiency of the heat exchanger. The Heat Exchanger effectiveness is the ratio of heat transfer rate to the maximum heat transfer rate formed from the geofluid to the circulation water if the heat transfer area is infinite. It can be calculated from the number of heat transfer units, NTU, and the flow ratio, R which is the ratio of the smaller to the larger heat flow capacities through the heat exchanger [170].

$$E_x = \frac{\dot{Q}}{\dot{Q}_{max}} = \frac{1 - e^{NTU(1-R)}}{R - e^{NTU(1-R)}} \tag{2.20}$$

Heat Exchanger performance is affected by the structural and geometric configuration of the heat exchanger, pipe material, length, the ground temperature distribution, soil moisture content and its thermal properties and working fluid. Generally, water is used as the base working fluid as it is widely available and cheaper. Alternatively, ethylene or propylene glycol having low volatility and relatively low corrosivity can be used [168]. In recent days, nanofluids which are more improved types of heat transfer fluids are being used as working fluid in heat pumps and heat exchangers.

2.3.1.2.2 Types of heat exchanger

Direct exchange A direct-exchange loop is a GHE that is in direct contact with the GSHP by circulating the refrigerant in the ground. The heat is exchanged directly between the refrigerant and the ground through copper tubes. Direct Exchange (DX) system is more efficient than indirect system because of the better thermal interaction with the ground. The installation cost is low due to the shorter ground coil and absence of circulation pump. But they are suitable for smaller domestic applications [151,170].

Open-loop The open system uses groundwater from a conventional well as a heat source. Groundwater from a water-bearing layer is pumped from an aquifer through one well, passed through the Heat Pump or Heat Exchanger where heat is added to or extracted from a heat carrier and then discharged either into the surface stream, pond, lake, sewer or re-inject to an aquifer as given in Fig. 2.25A. The water supply and discharge are not connected, that's the reason the loop is referred as open [151,170–172].

Closed loop A closed-loop system extracts heat from the ground itself using a continual loop of plastic pipe system placed horizontally or vertically in the ground or

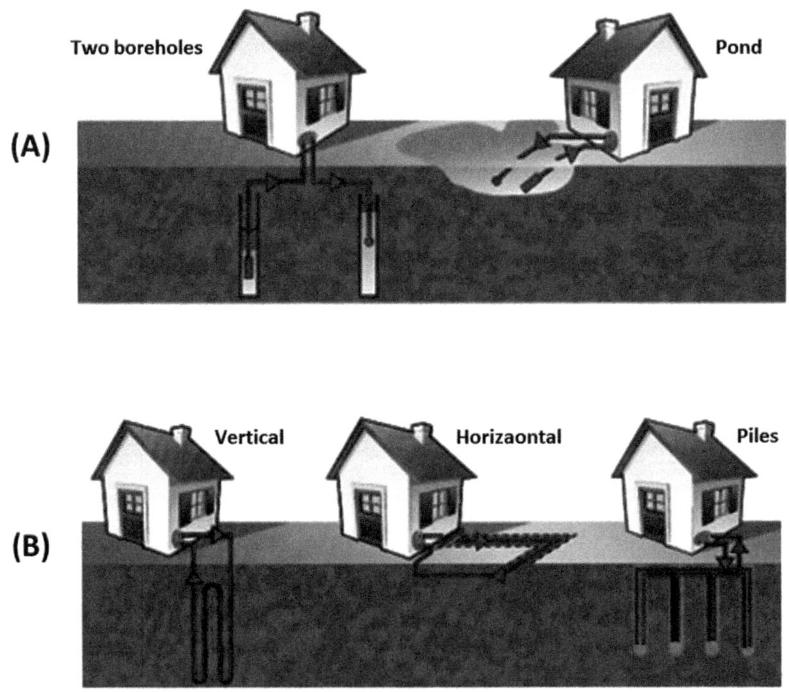

Figure 2.25 Open-loop (A) and close loop (B) systems [26].

submerged in a pond with both ends connected to the heat pump (Fig. 2.25B). In a Horizontal Ground Heat Exchanger, pipes are connected either in series or in parallel in a horizontal trench and buried at a depth of between 0.8–2 m. In some cases, the pipes are extended by modified slinky shape to reduce the space and enhance the performance [151,170–172].

In a vertical ground Heat Exchanger, plastic pipes are inserted in either a U-shape or coaxial form into a borehole which is constructed vertically in the ground (Fig. 2.26). The borehole round the pipes is backfilled with a material of high thermal conductivity called grout which is a mixture of sodium bentonite and silica sand.

The heat carrier fluid is usually water or water mixed with antifreeze and flows down to the bottom usually 20–300 m deep with a diameter of 10–15 cm. Compared with Horizontal Heat Exchangers, vertical loops aren't economical due to its higher installation cost. However, for a given heating and cooling load, they require less piping as the deep ground temperature remains cooler in the summer and warmer in the winter than near surface ground [151]. Vertical ground Heat Exchangers also referred as Borehole Heat Exchangers (BHEs). The BHE is composed of single or double U-tube High-Density Polyethylene pipes that extract geothermal heat from the shallow hot dry rocks without geofluid production. The most widespread technology to utilize the shallow geothermal resources is BHEs equipped with geothermal heat pumps [150,151,171]. They are suitable for a single residence to large commercial buildings where several ground heat exchangers are used.

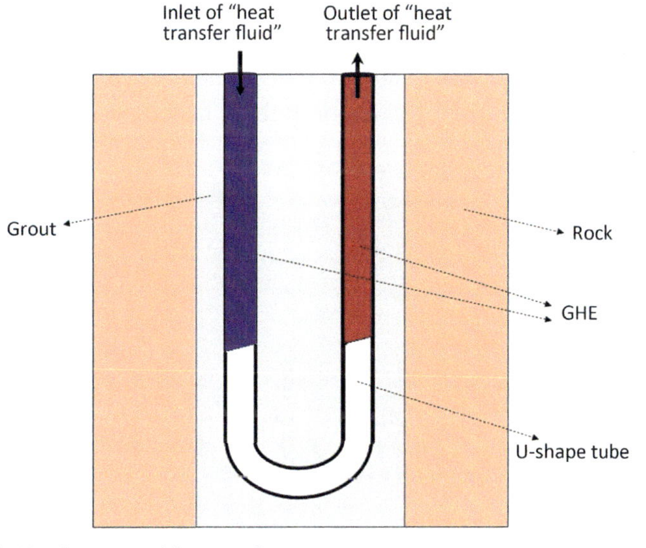

Figure 2.26 Single U-tube ground heat exchanger.

For deeper application when higher output temperature is needed, Deep Borehole Heat Exchangers (DBHEs or DBHs) are utilized. They are transformed from dry inactive boreholes. In this case the depth is the same as geothermal wells, the heat transfer area is larger, the temperature of the surrounding rock is higher, and the heat conductivity of the deep compacted rocks is also higher than the shallow region, but thermal power capacity is lower than thermal water well. The amount of heat from DHE system is related to bore and casing diameter, length of DHE, tube diameter, number of loops in the well, fluid's flow rate and fluid's temperature [150].

2.3.1.3 Hybrid ground-source heat pump systems

Hybrid Ground-Source Heat Pump (HGSHP) systems offers the combination of two or more energy sources or supplemental equipment for heating and cooling processes to increase the system efficiency while decreasing the capital and operating costs. Cooling tower, solar collector, waste heat source and boiler can be used as supplemental heat source. Fig. 2.27A and B illustrates the case where a cooling tower is added to Ground-Source Heat Pump System for heat rejection or a solar collector schema that can be used as an extra heat source by converting solar energy into heat and transfer to the ground-source Heat Pump System. [168,173,174].

Several simulation and experimental studies can be found related to HGSHP system. Hackel and Pertzborn [175] monitored the actual HGSHP systems. Their results show that HGSHP systems are more economically and environmentally improved than conventional GSHP systems.

2.3.1.4 Nanofluids

The Nanofluid is a new dimensional heat transfer fluid that consists of very small quantities of solid nanoparticles uniformly and stably suspended in a base liquid. Nanoparticles mainly made of metals, oxides, carbides or carbon nanotubes and ability to change

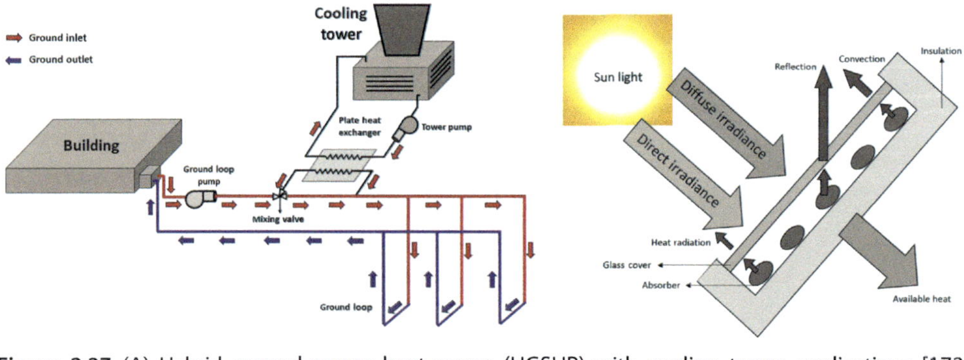

Figure 2.27 (A) Hybrid ground-source heat pump (HGSHP) with cooling tower applications [173] and (B) solar panel in HGSHP system for heating applications [174].

thermophysical properties of the base fluid dramatically. Nanofluids have high thermal conductivity, stability and homogeneity [176]. The nanoparticles have dimensions in the 1-100 nm range, so they act like base fluid molecules, thereby, don't cause blockage in the flow path. The increase in thermal conductivity depends on the type of nanoparticle and particle concentration present in the nanofluid. Nanofluids are produced by the suspension of nanoparticles in heat transfer base fluids such as water, ethylene glycol, propylene glycol or engine oil [177]. It is reported that there can be up to 40% increase in thermal conductivity with the addition of %0.3 (vol.) copper nanoparticles to ethylene glycol [178]. Similarly, there is a 150% thermal conductivity enhancement of poly-olefin oil by adding 1%1.0 (vol.) multiwall carbon nanotubes [179]. High thermal conductivity is a must in the development of energy-efficient heat transfer equipment and due to their higher heat transfer feature than regular heat transfer fluids, nanofluids are started to be used in the GSHP systems. Nanofluids increase heat transfer and stability, minimize clogging, miniaturize heat exchangers with microchannels, extract more heat from the ground in shorter tube lengths so as to reduce the pumping power. They are more efficient than conventional fluids and lead to cost and energy savings in heat transfer.

2.3.1.4.1 Properties of nanofluids

Nanofluids can consist of nanoparticles, such as metals (Al, Cu, Ag, Au), metal oxides (Al_2O_3, ZnO_2, CuO_2, TiO_2) and carbon-based materials (nanotubes, graphite, nanodiamonds). Nanofluids influence several thermophysical properties of the base liquids. Alteration depends on the particle shape and size, material type and physical properties, and amount of nanoparticle dispersed in the base liquid. Thermophysical properties of several nanoparticles and water are tabulated in Table 2.3. The thermophysical properties of nanofluids can be calculated using the correlations available in the literature [179,180].

Density of nanofluids Nanoparticles increase the density of the base fluid. Density of nanofluid can be calculated from Pak and Cho's [181] correlation (Eq. 2.21) for variable volumetric concentrations.

$$\rho_{nf} = \phi\rho_{np} + (1 - \phi)\rho_f \tag{2.21}$$

Table 2.3 Thermophysical properties of nanoparticles and water [179,180].

	ρ, kg/m^3	*K, W/mK*	*Cp, J/kg.K*
Water	1000	0.63	4190
Aluminium (Al)	2700	237	877
Copper (Cu)	8933	401	385
Silver (Ag)	10490	450	235
Alumina (Al_2O_3)	3970	40	775
Copper Oxide (CuO)	6500	32.9	525

where, ρ_{nf}, ρ_{np} and ρ_f are the densities of nanofluid, nanoparticle and base fluid respectively and ϕ is the volumetric concentration of nanoparticle.

Thermal conductivity of nanofluids Since nanofluids have high impact especially on thermal conductivity and heat transfer capability, several studies on nanofluids have focused on thermal conductivity. Initially, Maxwell derived Eq. 2.22 for effective thermal conductivity for a solid-liquid mixture;

$$k_{eff} = \frac{k_{np} + 2k_f + 2(k_{np} - k_f)\phi}{k_{np} + 2k_f - (k_{np} - k_f)\phi} k_f \tag{2.22}$$

where, k_{eff} is Nanofluid thermal conductivity; k_{np} is Nanoparticle thermal conductivity; k_f is based fluid thermal conductivity and ϕ is volume fraction of particles. Later, Hamilton and Crosser rearranged the equation by including a particle shape factor (n) given in Eq. 2.23.

$$k_{eff} = \frac{k_{np} + (n-1)k_f - (n-1)(k_{np} - k_f)\phi}{k_{np} + (n-1)k_f + (k_{np} - k_f)\phi} k_f \tag{2.23}$$

$n = 3/\psi$ and ψ is the ratio of surface areas between a sphere and a nanoparticle of equal volume.

Fig. 2.28 shows the comparison of thermal conductivity of Heat Transfer Fluids and Nanofluids [182]. Basically, adding nanoparticle to base fluid increases the thermal conductivity of the fluids. Fig. 2.29 presents the percentage of thermal conductivity enhancement with temperature and volume fraction for CNTs$-Al_2O_3$/water nanofluids [183].

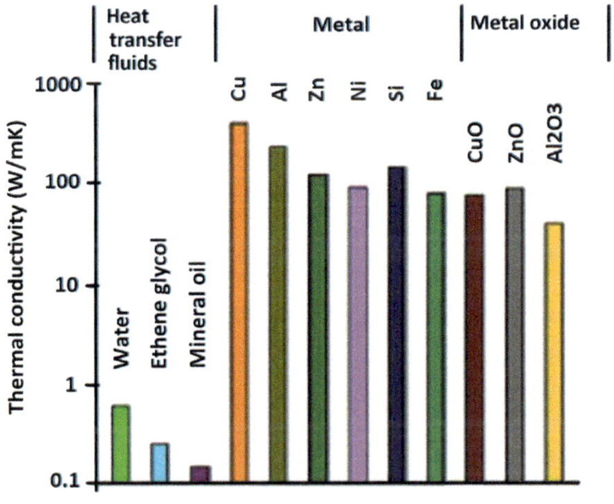

Figure 2.28 Comparison of thermal conductivity of heat transfer fluids and nanofluids [182].

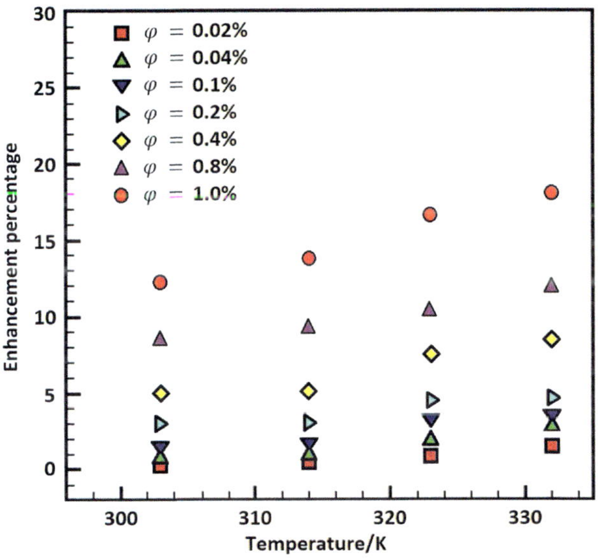

Figure 2.29 Enhancement percentage of thermal conductivity of nanofluid at different concentrations [183].

Specific heat capacity of nanofluids Nanoparticles have lower specific heat capacity than fluid so the amount of heat required for a nanofluid to reach a given temperature is decreased. Eq. 2.24 is used for evaluating the specific heat capacity.

$$Cp_{nf} = \frac{\varnothing \rho_{np} Cp_{np} + (1 - \varnothing)\rho_f Cp_f}{\rho_{nf}} \tag{2.24}$$

herein, Cp_{nf}, Cp_{np}, Cp_f are specific heats of nanofluid, solid particles and base fluid, respectively.

Viscosity of nanofluids Nanoparticles alter the viscosity of the fluid and it affects the convective heat transfer and pressure drop. The viscosity increase in nanofluids depends on the types of the base fluid, types of nanoparticle and nanoparticle concentration. Brinkman's equation given in Eq. 2.25 can be used to calculate the viscosity of nanofluids for different concentrations.

$$\mu_{nf} = \frac{1}{(1-\varnothing)^{2.5}} \mu_f \tag{2.25}$$

where, μ_{nf} is nanofluid viscosity and μ_f is base fluid viscosity.

2.3.1.4.2 Nanofluid applications

In recent years, nanofluids are being used as the working fluid in the GHE in order to improve the performance of GSHP. Countries that predominantly use heat pumps like the United States, Japan, South Korea, China, and Australia have some researches on employing nanofluids in heat exchangers. As per their study, the efficiency of Heat Exchangers is proved to increase significantly than systems with conventional cooling agents [184]. Nanofluids attracted attention priority because of its high thermal conductivity than the base fluid. After further analysis, it is found that it has many advantages over traditional working fluids except one specific deficiency that nanoparticles increase viscosity which leads to higher pumping power and pressure loss. As a result, the optimum nanofluids with lowest viscosity increases and highest fluid convective heat transfer increases should be used in the system.

Initially, Olson [185] proposed the use of nanofluids in the circulation loop of the ground source heat exchanger. Using nanofluids improves the heat transfer thereby reduce the installation and operating cost because the circulation loop and the size of the heat transfer system will be reducing the accompanying reduction in pressure loss and pumping power. Since the system becomes smaller, material volume necessary for a heat exchanger, pump, piping, the amount of working fluid and energy consumption will be also reduced. In addition to this, the system becomes environmental-friendly because of minimized pollution [186,187]. Thermal conductivity is the primary indicator of heat transfer performance of a nanofluid. However, it is recommended to consider the Mouromtseff number (Mo) as figure of merit. It is a function of the density, viscosity, thermal conductivity and specific heat of nanofluid as given by Eq. 2.26. The nanofluids convenient for ground-source Heat Pump applications is characterized by high Mouromtseff number [184–186]. Silver and Copper based nanofluids mainly used in heat transfer applications have high *Mo* number.

$$Mo = \frac{\rho_{nf}^{0.8} K_{nf}^{0.67} Cp_{nf}^{0.33}}{\mu_{nf}^{0.46}} \tag{2.26}$$

The flow type (laminar or turbulent) inside the heat exchanging equipment plays an important role in the effectiveness of a nanofluid. If the heat exchanger operates under laminar conditions the pressure drop is lower and heat flow rate for nanofluid is relatively high as compared with water as given in Fig. 2.30 [187]. As a result, in small-scale application, using nanofluid is a better option. However, higher production cost of nanofluids ought to be additionally considered. As an example, when conventional ethylene glycol water/mixture is replaced with nanofluids with a copper volumetric concentration of 1%, working fluid's cost in overall GSHP system's expense increases from approximately 3.5%–12% [184].

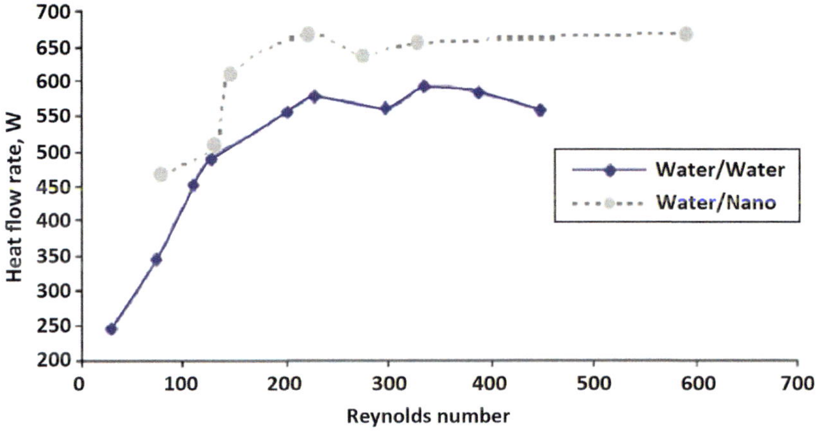

Figure 2.30 Reynolds number vs. heat transfer rate [187].

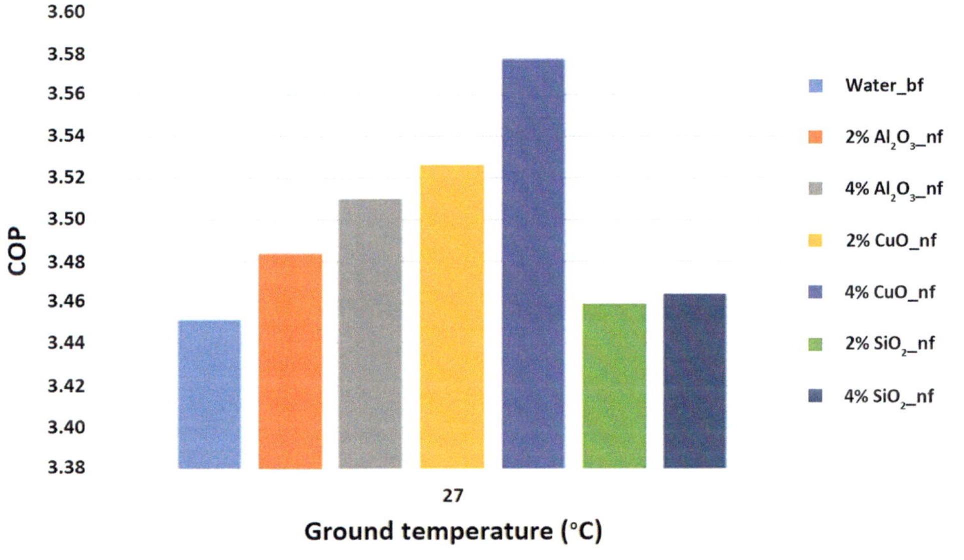

Figure 2.31 COP with base fluid and various nanofluids in the GHE [188].

COP of the system play vital role for the evaluation of the GSHP system's feasibility. Mishra et al. worked on theoretical analysis of employing nanofluids in GSHP [188].

Fig. 2.31 exhibits the COP index of the GSHP system for various nanofluids that have different concentrations at a constant ground temperature. As can be seen from the figure, nanofluids established higher COP than water. According to this study, 4% (CuO) nanofluid having the highest COP seems the best working fluid option for this GSHP system. This study supports the idea of using nanofluids as a heat

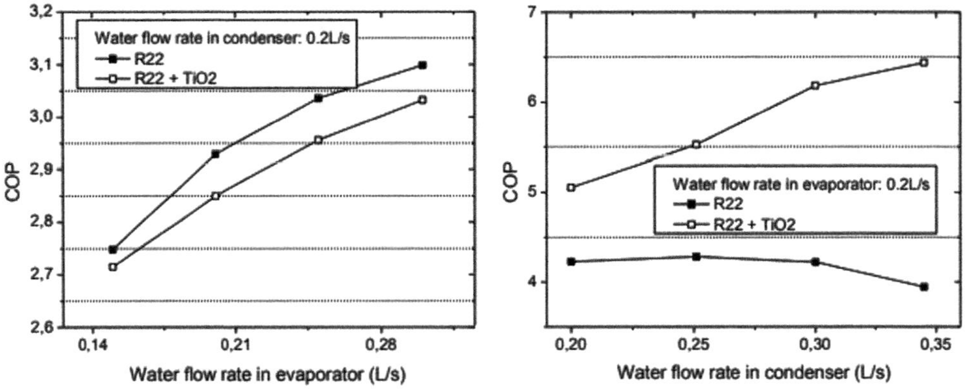

Figure 2.32 COP of the cooling and heating cycle [190].

transfer fluid in GSHP application to improve the performance, but it still requires detailed investigation. Nanorefrigerants are the nanofluids which their host fluid is refrigerant. The heat transfer coefficient of the refrigerant is increased by adding nanoparticles. Therefore, using nanorefrigerants as working fluids can improve the performance of heat pump systems. After characteristics of (R141b + Al_2O_3), nano refrigerant was studied by Mahbubul et al. [189]. The significant effect of volume fraction over the heat transfer and pressure drop characteristics of Nanorefrigerants was stated.

Li et al. [41] compared the heat pump system performance using traditional base fluid pure R22 and R22 + TiO_2 mixtures as refrigerants. According to their results, adding nanoparticles into the working fluid hasn't changed the heat absorbed in the evaporator clearly but has increased the heat released from the condenser. Consequently, as can be seen from Fig. 2.32, there is a slight reduction in COP of the cooling cycle, while the COP of the heating cycle was increased significantly. In addition to this, there is an increase in power consumption of the compressor when nanoparticles dispersed in the working fluid.

As a conclusion, in order to improve the system performance of the heat pump using nanoparticles, characteristics of nanorefrigerants should be evaluated in detail.

2.3.2 Geothermal energy conversion

Deep geothermal reservoirs having high temperature and enthalpy are used for electricity production. Power plants generate electricity by utilizing steam produced from geothermal fluid. Power plant basically consists of turbine, condenser, pump and steam generator (Fig. 2.33). With this system, geothermal heat with high temperature is first converted to mechanical energy and then to electrical energy. There are three geothermal power plant technologies; flash steam, dry steam, and binary cycle as simply

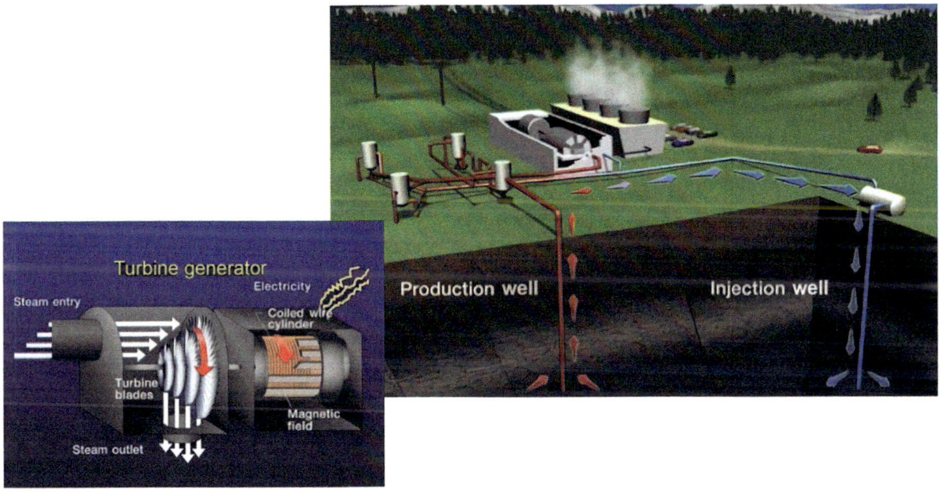

Figure 2.33 Geothermal power plant and its main components [149].

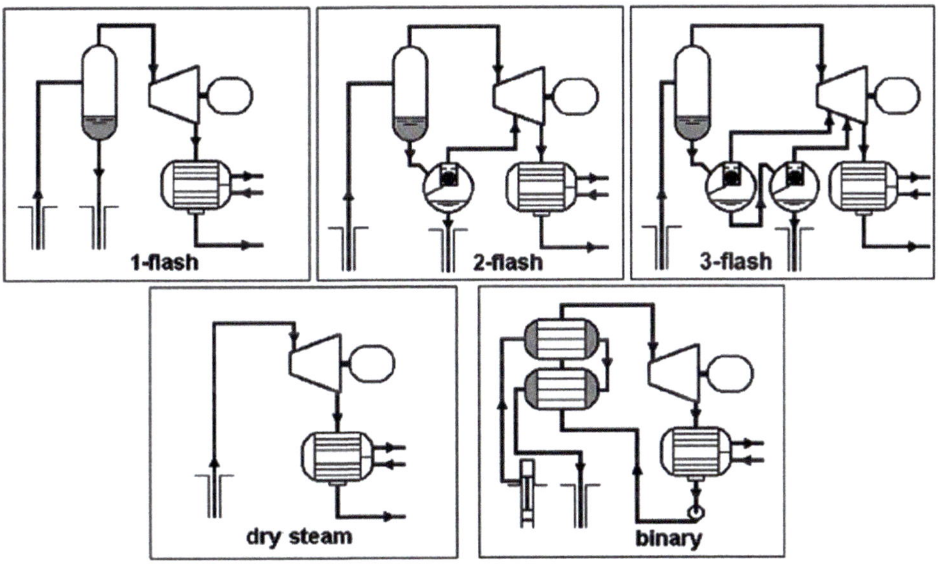

Figure 2.34 Basic geothermal energy conversion systems: simplified plant layouts [191].

illustrated in Fig. 2.34. In flash steam power plants, hot water is flashed into steam and moves to turbine and then condenser where the steam is liquified. Single-flash system refers only one flash process is applied to water where in double or triple flash plants, geothermal fluid undergoes two or three stages of flash, respectively.

Dry steam power plants are the simplest and less expensive one because they directly take the steam through turbine and then forward to the condenser where steam liquifies. In binary system, geothermal fluid transfers its heat to working fluid in the closed loop. The fluid boils and its steam drive the turbine to generate electricity [191–193].

In recent years, advance technology hybrid power plants that combine two different energy system or sources in a single plant are being constructed for higher resource utilization efficiency and obtain a synergistic outcome. Two systems can be combined to form hybrid single-flash, hybrid double-flash and hybrid flash-binary systems; two different sources of energy can be combined such as hybrid fossil-geothermal systems and solar-geothermal. [191].

2.3.2.1 Organic Rankine Cycle (ORC)

In steam-based power plants the Rankine cycle is an idealized thermodynamic cycle of a heat engine that converts heat into mechanical work. Conventional power plants require high temperature ranges and they are not convenient in moderate power (typically less than a few MWe) ranges and for low-temperature application. For lower temperature or low-to-medium grade thermal sources where Rankine cycle doesn't work, ORC that uses organic fluids that have a much lower boiling point can be used (Fig. 2.35).

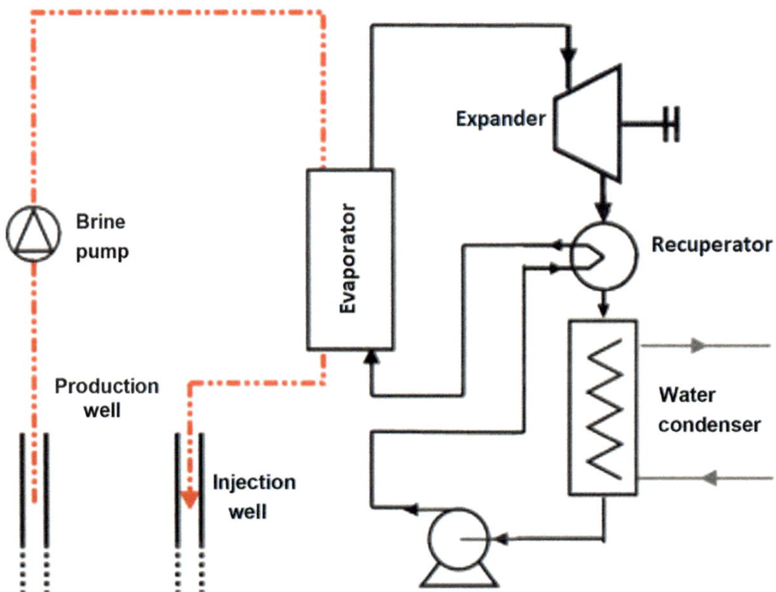

Figure 2.35 Basic working principle of a geothermal ORC system [194].

With the ORC, geothermal (also biomass and solar sources) heat sources having lower enthalpy and temperature can be converted into electricity [191,192,194]. ORC can convert waste heat into electricity where energy is lost to the environment which is called as bottoming cycles enables a better use of the primary energy [194–196]. The advantages of the ORC power plant can be listed as; superheating is not required, lower turbine inlet temperature is suitable, evaporating pressure is lowered while condensing pressure is increased, water-treatment system is removed, and turbine design is simplified [194].

ORC is similar to traditional steam cycle, so the components are same with conventional steam power plant with little modifications. The main difference is the working fluid which is an organic compound characterized by a lower boiling temperature than water and allowing power generation from low heat source temperatures.

In ORC, HFC-134a, HFC-245fa, n-pentane, Solkatherm, OMTS Biomass, and Toluene are the common working fluid instead of water [194]. Fig. 2.36 represents T–s diagram of water and of a few typical organic fluids used in ORC applications. As per the diagram, the slope of the saturated water vapor curve is negative while the curve for organic fluids is much closer to vertical. As a result, the limitation of the vapor quality at the end of the expansion process vanishes and superheating the vapor before the turbine inlet is unnecessary which simplifies the process.

One of the disadvantage of using these kind of fluids is they increase the working fluid pump consumption since the ratio between pump consumption and turbine output power rises with declined evaporation temperature.

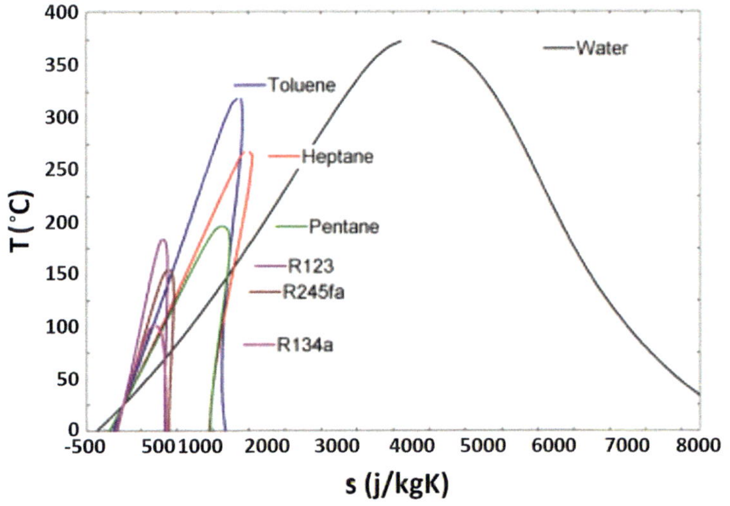

Figure 2.36 T–S diagram the saturation curves of water and of a few typical organic fluids used in ORC applications [194].

2.3.2.2 Thermoelectric applications in conversion of geothermal energy

2.3.2.2.1 Thermoelectric materials

Generally, thermal power generation system needs to transfer thermal energy to mechanical work before electric production. On the contrary, thermal energy can be directly converted to electricity by thermoelectric devices. Materials of the thermodynamic devices facilitate the conversion of heat to electricity based on heat and charge carriers. Electrons in N-Type Materials and holes in P-Type Materials can easily diffuse along metals and semiconductors [197,198]. When there is an applied temperature gradient, electrons and neutrons move in opposite directions and generate electric current [198]. Three thermoelectric effects promote to the direct production electric from heat. Seebeck effect: charge carriers move from hot side to cold side across the terminals of an open metals/semiconductor circuit and produce electric current for power generation because of the temperature difference. Peltier effect: it is the inverse of the Seebeck effect and produces a thermal gradient while electrons absorb heat from the cold side and reject heat at the hot side as electrical power is applied. It is used in solid-state thermoelectric cooling or heating. Thompson effect: expose the relationship between Peltier and Seebeck effects and defines the heat generation or absorption occurred while electric current passes through metals, semiconductors or conductor that has different temperature at its ends.

The device efficiency is based on the material properties of lattice. Thermoelectric material should have ability to minimize the thermal conductivity for the increase in temperature gradient and maximize the electrical conductivity and Seebeck coefficient. The efficiency of thermoelectric materials is described with dimensionless quantity named as ZT (figure of merit) and calculated with equation 4.6 [199,200]

$$ZT = \frac{S^2\sigma T}{\kappa} = \frac{S^2 T}{\kappa\rho} = \frac{S^2 T}{(\kappa_e + \kappa_l)\rho} \tag{2.27}$$

Where, S is the Seebeck coefficient, T is the absolute temperature, σ is the electrical conductivity, ρ is the electrical resistivity and κ is the total thermal conductivity which equals to sum of thermal conductivities of the lattice and electrons. Materials having ZT value higher than 1 are chosen for thermoelectric applications but this value must be much higher for large application. The most well-known thermoelectric materials are chalcogenides alloys specifically based on bismuth telluride (Bi2Te3) and lead telluride (PbTe). Also SiGe based materials have ZT value of around 1 [200–202]. The thermoelectric efficiency of the device is the ratio of electrical power produced to heat production. For power generation efficiency (Eq. 2.28) and for thermoelectric cooling efficiency a separate equation is used [200,203].

$$\eta_p = \left(\frac{T_h - T_c}{T_h}\right)\left[\frac{\sqrt{1 + ZT_{avg}} - 1}{\sqrt{1 + ZT_{avg}} + \frac{T_c}{T_h}}\right] \tag{2.28}$$

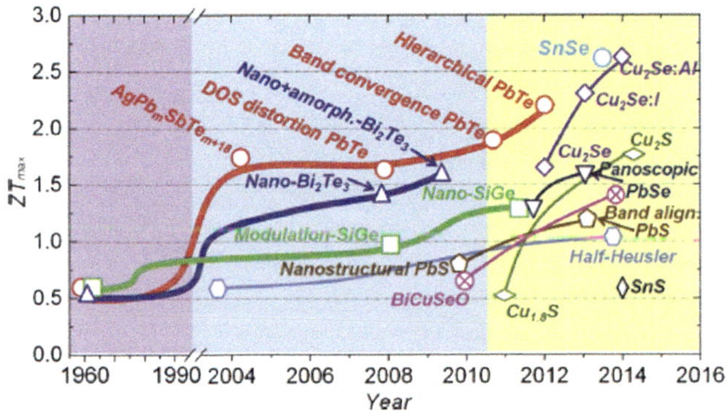

Figure 2.37 *ZT* values of some thermoelectric materials per publishing years [204].

Here ZT_{avg} is the average value of n-type and p-type lattice, T_h and T_c are the temperatures of hot and cold sides. It is a function of the Carnot efficiency (ΔT/TH) and with the current technology it is approximately 1/6 of the maximum Carnot efficiency. To increase the thermoelectric efficiency it is aimed to produce materials that have minimal thermal conductivity and maximal electrical conductivity. Substantially, nanotechnology is one of the options for high efficiency thermoelectric materials generation. Fig. 2.37 gives the *ZT* values of some thermoelectric material found in the literature per publishing years.

2.3.2.2.2 Thermoelectric applications

Thermoelectric tools can operate in two modes. First one is using in Thermoelectric Heat Pump (THP) mode where they produce a thermal gradient according to the Peltier effect by using electric current. Second one is Thermoelectric Generator (TEG) mode where they produce electrical current from temperature gradient with the Seebeck effect. The thermoelectric devices don't have moving parts and they don't require working fluid. They are compact, simplified and have longer lifetime than electric compressors. They are highly reliable, quiet and environmentally friendly. Furthermore, the most attractive point is they require less primary energy so they can operate with different thermal sources no matter they have high or low temperature. There are many theoretical and experimental studies on the application of heat pump for heating and cooling devices for buildings [200,204]. THP is composed of several thermoelectric module located between two heat exchangers as illustrated in Fig. 2.38.

Today, there is an increasing trend in applying thermoelectric technology to improve the efficiency of waste heat recovery from geothermal energy and power plants. TEGs have ability to convert thermal energy to electrical energy directly without need of mechanical work conversion exploiting Seebeck effect. Valid thermodynamic cycles have very high conversion efficiency in comparison with

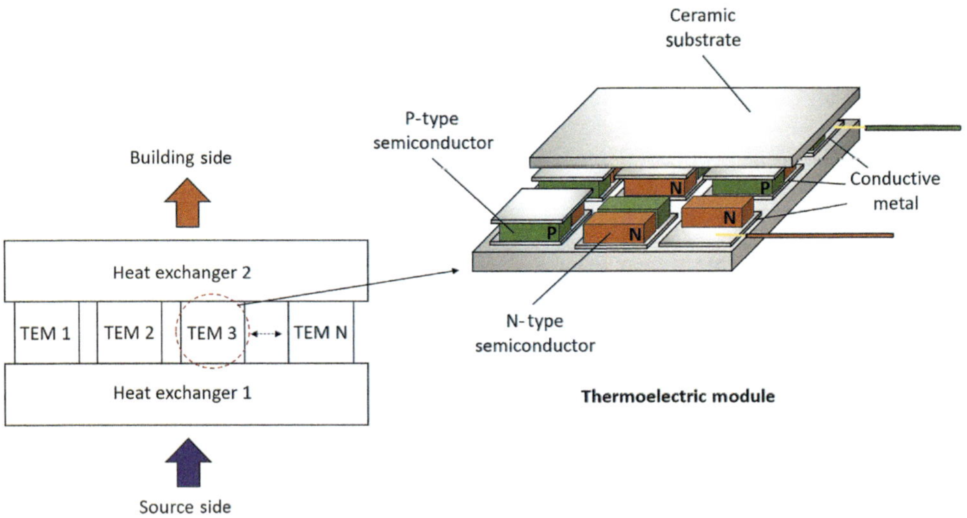

Figure 2.38 Configuration of the thermoelectric heat pump [197].

thermoelectric power generation. However, TEGs have many benefits like being simple, compact and able to work in low operating temperature range. Producing heat at low temperature has specific significance because generally conventional technologies like Rankine Cycle are not able to generate electrical power efficiently in low temperatures. Hence, the vital point in thermoelectric applications is low temperature heat production. In addition to this, Li [205] proposed large scale utilization of TEGs in geothermal application to increase production of geothermal energy.

The Rankine cycle is the most dominant technology used in thermal plant electricity generation in the world. It comprises water's phase change from a liquid state into superheated steam. The expanded superheated steam performs work to surroundings and reverts to its initial liquid phase. In low temperature plant, ORCs takes place where organic fluids with lower boiling point are used instead of water to improve the efficiency. However, the financial cost should be taken into consideration. The Rankine cycle has high COP but since their thermal energy comes from the combustion of fossil fuels which emits large amounts of CO_2, they are not eco-friendly. Its efficiency can be improved, CO_2 emissions reduction can be provided, and also cost-effective enhancements can be ensured by using THPs in the condensation process of the Rankine cycle by reducing the waste heat rejected during the condensation phase. Initial studies, like Li [205] and Kyono et al. [206] focus on the application of TEGs to the condenser of the Rankine cycle. This method increases the electricity production, but it requires impractical number of TGEs which increase the cost of the project. Knox et al. [207] introduced the idea to use THPs inside the condenser in Rankine cycle. As per their study for 600 MWe power plant, the addition of THPs having COP_h of 3 into the cycle there is a 0.4 % increment in cycle efficiency and a 3% reduction in fuel utilization. Siviter et al. [198] reported in their research that the plant fuel requisition is reduced by 1.52 % and overall cycle efficiency increase 0.15 %. In addition to this, since there is a saving in fuel consumption in the modified cycle, the cost of THPs are compensated approximately in 4 years.

2.4 Conclusions

Geothermal energy is one of the largest renewable energy sources. Conventional techniques are started to evolve to new technology versions by the development of advanced materials and tools such as; New Technology Geophysical Tools, Fiber Optic Technology, Pressure and Temperature Sensors with Remote Sensing, Advanced Carriers such as; Airborne, Spaceborne and Satellites, also New Technology Dressing Materials and Nanofluids. It paves the way of reaching data in a robust and efficient way in all steps of geothermal energy applications.

Although, conventional geophysical methods are useful for the geothermal applications most of the time, utilizing only these tools may not be sufficient in many cases. New Generation Tools are making it easier to reach the data in robust and cost-effective way.

Additionally, the harsh environment related to high temperature, acidic and CO_2 rich fluids are very challenging in terms of materials to be used. To provide the proper zonal isolation, the conventional cement slurry designs are not applicable. Some innovative materials were proposed to enhance the cement properties such as; Foam Cement, Phosphate Bonded Cement, Self-Healing Cement, Acid and CO_2 Resistant Cement. These new technology materials and applications made possible to maintain the duties provided by well cement at high temperature conditions as high as 350°C. In some extreme temperature conditions (>500°C), there are efficacious applications of Phosphate Bonded Cement.

Besides, the conventional drilling fluid solutions are enhanced by using Sulfonated and Polysulfonate Drilling Fluids in order to have a harmless and suitable drilling operation in geothermal systems. The Sulfonated and Polysulfonate Drilling Fluids proposed to withstand temperatures up to 290°C. The development of Novel Drilling Fluids are still required for extreme temperature conditions. There are some offered Nano-Based Drilling Fluids which need to be tested in real field conditions as well.

The interaction between Corrosive-Hot Geothermal Fluids and the well and production equipment is also significant issue to be considered in order to conserve wellbore integrity not only during drilling but also during production period. The Coatings are generally used in the equipment which gets in contact with the corrosive geothermal fluid. Carbon Steel Materials with Anticorrosion Coatings are generally preferred. Furthermore, there are also other coatings such as; Nickel-Based Alloys, Titanium-Based Alloys and Polymeric Coatings which is able to protect the Heat Exchanger, Casings and Liners in the well, Wellheads and Condensers up to 400°C.

Logging and Measurement Tools also need protection against hot geothermal fluids. Although the time that tools get in contact with the fluid is shorter compared to well equipment, they need to be sheltered against hot temperature to obtain accurate data. The common method is to practice some heat shield around the electronic parts and sensors of the tools to obtain proper data at temperature conditions up to 300°C. The development of Innovative Logging Tools and Measurement is still vital for the applications in extreme temperature conditions.

Among the numerous geothermal resources, low-temperature resources take great proportion. They are more difficult to extract power due to low temperature and enthalpy that effect on the overall efficiency. However, power generation and electricity unit installations are increasing each year. Accordingly, this section focuses on the heat pump, heat exchanger system and power plants that are used for heat transfer, thermal energy conversion and power generation. The basic principles, components, types and application areas are briefly described. Geothermal energy utilized by ground-source heat pumps and heat exchangers becomes a promising energy source with its benefits such as being environmentally friendly, having low life-safety risk and low maintenance cost. Its installation is quite easy and requires minimal regular maintenance, they are environmentally friendly because they reduce fossil fuel consumption so helps the reduction of greenhouse emission by reducing CO_2 emissions and also they are clean, quiet and odorless. While GSHPs presents very attractive alternative as heating and cooling system, they are still limited by their high initial costs especially negatively affected by the drilling costs of vertical BHE which is the most applied heat exchanger configuration. Today, new technologies are needed for cost reduction and efficiency improvement. Using hybrid system, Nanomaterials and Thermoelectric Material Applications can play an important role.

The hybrid system contains two or more energy sources or supplemental equipment and offers an increase in the system efficiency while decreasing the capital and operating costs. Nanofluids that contain metals, oxides, carbides or carbon nanotubes have the ability to change thermophysical properties of the base fluid dramatically and they have become new trending heat transfer fluid. They are more efficient than conventional fluids and lead to cost and energy savings in heat transfer by extracting more heat from the ground, reducing pumping power and miniaturizing heat exchangers. Since conventional power plants are not convenient for moderate power ranges and low-temperature application, ORC power plants can fulfill their place. Thermoelectric Devices that can convert thermal energy directly to electricity without mechanical work are compact, simplified, have a longer lifetime and don't require working fluid. There are many ongoing studies for the application of thermoelectric materials in heat pumps for direct heating/ cooling or using in power plants. Especially there is an increasing trend in applying thermoelectric technology to improve the efficiency of waste heat recovery from geothermal energy and power plants.

References

[1] W.E. Glassley, Geothermal Energy: Renewable Energy and the Environment, second ed., CRC Press, 2015.

[2] H.K. Gupta, S. Roy, Geothermal Energy: An Alternative Resource for the 21st Century, first ed., Elsevier Scientific Publishing Company, 2006.

[3] J.E. Faulds, M.F. Coolbaugh, D. Benoit, G. Oppliger, M. Perkins, I. Moeck, et al., Structural controls of geothermal activity in the northern hot springs mountains, Western Nevada: the tale of three geothermal systems (Brady's, Desert Peak, and Desert Queen), GRC Trans. 4 (2010) 675–683.

[4] M. Fehler, A. Jupe, H. Asanuma, More than cloud: new techniques for characterizing reservoir structure using induced seismicity, Lead. Edge 20 (3) (2001) 324–328.

[5] N. Xu, C. Tang, H. Li, F. Dai, K. Ma, J. Shao, et al., Excavation-induced microseismicity: microseismic monitoring and numerical simulation, J. Zhejiang Univ. Sci. 13 (6) (2012) 445–460.

[6] N.P. Agostinetti, A. Licciardi, D. Piccinini, F. Mazzarini, G. Musumeci, G. Saccorotti, et al., Discovering geothermal supercritical fluids: a new frontier for seismic exploration, 2017. <www.nature.com/scientificreports>.

[7] E. Gasperikova, G. Newman, D. Feucht, K. Arnason, 3D MT characterization of two geothermal fields in Iceland, GRC Trans. 35 (2011) 1667–1671.

[8] K. Fabian, V.P. Shcherbakov, S.A. McEnroe, Measuring the Curie temperature, Geochem. Geophys. Geosyst. 14 (4) (2012) 947–961.

[9] P.P.G. Bruno, V. Paoletti, M. Grimaldi, A. Rapolla, Geophysical exploration for geothermal low enthalpy resources in Lipari Island, Italy, J. Volcanol Geotherm. Res. 98 (1-4) (2000) 173–188.

[10] D. Castañeda, H. Brunkal, Geophysical techniques for geothermal exploration of Rico, CO. Geothermal Energy Course, Colorado School of Mines, 2009.

[11] J.D. Kana, N. Djongyang, D. Raïdandi, P.N. Nouck, A. Dadjé, A review of geophysical methods for geothermal exploration, Renew. Sustain. Energy Rev. 44 (2015) (2015) 87–95.

[12] E. Ganbat, Geothermal investigations at the Ásgardurfarm, Reykholtsdalur, W-Iceland. Report 6 in Geothermal Training in Iceland, UNU-GTP, Iceland, 83-98, 2004.

[13] K. Árnason, Geothermal exploration and development of the Hengill high-temperature field. Short Course II on Surface Exploration for Geothermal Resources, UNU-GTP, KenGen, Naivasha, Kenya, 2007.

[14] S. Crampin, J.H. Lovell, A decade of shear-wave splitting in the earth's crust: what does it mean? What use can we make of it? And what should we do next? Geophys. J. Int. 107 (3) (1991) 387–407.

[15] W.F. Hanna, Negative aeromagnetic anomalies over mineralized areas of the Boulder batholith, Montana. USGS Professional Paper, pp. 159–167, 1969.

[16] R.E. Criss, D.E. Champion, Magnetic properties of granitic rocks from the southern half of the Idaho batholith: influences of hydrothermal alteration and implications for aeromagnetic interpretation, J. Geophys. Res. 89 (8) (1984) 7061–7076.

[17] R.L. Reynolds, J.G. Rosenbaum, M.R. Hudson, N.S. Fishman, Rock magnetism, the distribution of magnetic minerals in the earth's crust, and aeromagnetic anomalies, Geol Appl. Mod. Aeromagn. Surv. (1990) 24–45. USGS Bulletin 1924.
[18] L.S. Georgsson, Geophysical methods used in geothermal exploration. Presentation in Short Course IV on Exploration for Geothermal Resources, UNU-GTP, KenGen, Naivasha, Kenya, 2009.
[19] H. Eysteinsson, Elevation and gravity changes at geothermal fields on the Reykjanes Peninsula, SW Iceland, in: Proceedings of the World Geothermal Congress, Kyushu-Tohoku, Japan, 2000.
[20] D.W. Strangway, C.M. Swift Jr., R.C. Holmer, The application of audio-frequency magnetotellurics (AMT) to mineral exploration, Geophysics 38 (6) (1973) 1159–1175.
[21] D.B. Hoover, C.L. Long, Audio-magnetotelluric methods in reconnaissance geothermal exploration. USGS Paper, U.S. Energy Research and Development Administration Publication, 2, pp. 1059–1064, 1976.
[22] C. Thanassoulas, Geothermal exploration using electrical methods, Geoexploration 27 (3–4) (1991) 321–350.
[23] K. Vozoff, The magnetotelluric method in electromagnetic methods in applied geophysics, SEG 2 (1991) 641–712.
[24] L.S. Georgsson, K. Saemundsson, H. Hjartarson, Exploration and development of the Hveravellir geothermal field, N-Iceland. Proceedings of the World Geothermal Congress 2005, Antalya, Turkey, 2005.
[25] L. Pellerin, J.M. Johnston, G.W. Hohmann, A numerical evaluation of electromagnetic methods in geothermal exploration, Geophysics 61 (1) (1996) 121–130.
[26] P. Wameyo, Magnetotelluric and transient electromagnetic methods in geothermal prospecting, with examples from Menengai, Kenya. Report 21 in Geothermal Training in Iceland 2005. UNU-GTP, pp. 409-439, 2005.
[27] H.M. Bibby, G.F. Risk, T.G. Caldwell, S.L. Bennie, Misinterpretation of electrical resistivity data in geothermal prospecting: a case study from the Taupo Volcanic Zone. Proceedings of the World Geothermal Congress 2005, Antalya, Turkey, 2005.
[28] G. Munoz, Exploring for geothermal resources with electromagnetic methods, Surv. Geophys. 35 (2014) 101–122.
[29] H.M. Bibby, Electrical resistivity mapping in the Central Volcanic Region of New Zealand, NZ J. Geol. Geophys. 31 (1988) 259–274.
[30] G.F. Risk, Electrical resistivity survey of the Wairakei Geothermal Field. Proceedings of the 6th New Zealand Geothermal Workshop, pp. 123-128, 1984.
[31] W. Heise, T.G. Caldwell, H.M. Bibby, S.C. Bannister, Three-dimensional modelling of magnetotelluric data from the Rotokawa geothermal field, Taupo Volcanic Zone, New Zealand, Geophys. J. Int. 173 (2008) 740–750.
[32] G.F. Risk, T.G. Caldwell, H.M. Bibby, Tensor time-domain electromagnetic resistivity measurements at Ngatamariki geothermal field, New Zealand, J. Volcanol. Geoth. Res. 127 (2003) 33–54.
[33] K. Yasukawa, T. Ishido, T. Kajiwara, Geothermal reservoir characterization by SP monitoring, Proceedings of the World Geothermal Congress 2005, Antalya, Turkey, 2005.
[34] K. Watson, F.A. Kruse, S. Hummer-Miller, Thermal infrared exploration in the Carlin trend, northern Nevada, Geophysics 55 (1) (1990) 70–79.
[35] S.T. Ovnatanov, G.P. Tamrazyan, Thermal studies in subsurface structural investigations, Apsheron Peninsula, Azerbaijan, USSR, AAPG Bull. 54 (1970) 1677–1685.
[36] J.L. Arroyo, P. Breton, H. Dijkerman, S. Dingwall, R. Guerra, R. Hope, et al., Superior seismic data from the Borehole, Oilfield Rev. 15 (1) (2003) 2–23.
[37] J. Wang, H. Wen, The Research of Data Transmission Technology in Measurement While Drilling. Scientific Research Publishing, 2018.
[38] D.W. Brown, D.V. Duchane, G. Heiken, V.T. Hriscu, Mining the Earth's Heat: Hot Dry Rock Geothermal Energy, Springer, 2012.
[39] T. Reinsch, P. Dobson, H. Asanuma, E. Huenges, F. Poletto, B. Sanjuan, Utilizing supercritical geothermal systems: a review of past ventures and ongoing research activities, Geotherm. Energy Sci.Soc.Technol. 5 (2017) 16.

[40] F.S. Simmons, K. Brown, Insights to high temperature geothermal systems (New Zealand) from trace metal aqueous chemistry, Proceedings World Geothermal Congress, Melbourne, Australia, 2015.
[41] M. Mella, P. Rose, S. Olsen, D. Bour, Development of a downhole fluorimeter for measuring flow process in geothermal wellbores. AICHE Annual Meeting, Deer Valley, 2010.
[42] P.E. Sanders, Fiber-optic sensors: playing both sides of the energy equation, Opt. Photonics N. 22 (1) (2011) 36−42.
[43] B.N. Paulsson, J. Thornburg, R. He, A fiber optic borehole seismic vector sensor system for high resolution CCUS site characterization and monitoring, Energy Proc 63 (2014) 4323−4338.
[44] W.L. Dowdle, W.M. Cobb, Static formation temperature from well logs-an empirical method, J. Pet. Technol. 27 (11) (1975) 1−326.
[45] A. Saadat, S. Frick, S. Kranz, S. Regenspurg, Energetic use of EGS reservoirs, Geotherm. Energy Syst Explor Dev Util (2010) 303−372.
[46] M. Jaskelainen, Distributed temperature sensing (DTS) in geothermal energy applications. Spectroscopy vol. 9, 2009.
[47] M.C. Farries, A.J. Rogers, Distributed sensing using stimulated Raman interaction in a monomode optical fibre. In 2nd International Conference on Optical Fiber Sensors: OFS'84, vol. 514, International Society for Optics and Photonics, November, 1984, pp. 121−133.
[48] J.P. Dakin, D.J. Pratt, G.W. Bibby, J.N. Ross, Distributed optical fibre Raman temperature sensor using a semiconductor light source and detector, Electron. Lett. 21 (13) (1985) 569−570.
[49] A.J. Rogers, Distributed optical-fibre sensors for the measurement of pressure, strain and temperature, J. Inst. Electron. Radio. Eng. 58 (5S) (1988) S113−S122.
[50] A. Hartog, G. Gamble, Photonic distributed sensing, Phys. World 4 (3) (1991) 45.
[51] A.A. Boiarski, Distributed fiber optic temperature sensing. In Applications of fiber optic sensors in engineering mechanics, ASCE, 1993, pp. 210−224.
[52] E. Hurtig, S. Großwig, M. Jobmann, K. Kühn, P. Marschall, Fibre-optic temperature measurements in shallow boreholes: experimental application for fluid logging, Geothermics 23 (4) (1994) 355−364.
[53] K. Sakaguchi, N. Matsushima, Temperature logging by the distributed temperature sensing technique during injection tests, in: Proceedings of World Geothermal Congress, Kyushu-Tohoku, Japan, May 2000, pp. 1657−1661.
[54] T. Reinsch, J. Henninges, Temperature-dependent characterization of optical fibres for distributed temperature sensing in hot geothermal wells, Meas. Sci. Technol. 21 (9) (2010) 094022.
[55] J. Selker, N. van de Giesen, M. Westhoff, W. Luxemburg, M.B. Parlange, Fiber optics opens window on stream dynamics, Geophys. Res. Lett. 33 (24) (2006).
[56] M.C. Westhoff, H.H.G. Savenije, W.J. Luxemburg, G.S. Stelling, N.C. Van de Giesen, J.S. Selker, et al., A distributed stream temperature model using high resolution temperature observations, Hydrol. Earth Syst. Sci. Discuss. 11 (4) (2007) 1469−1480.
[57] C.S. Lowry, J.F. Walker, R.J. Hunt, M.P. Anderson, Identifying spatial variability of groundwater discharge in a wetland stream using a distributed temperature sensor, Water Resour. Res. 43 (10) (2007).
[58] R.D. Henderson, F.D. Day-Lewis, J.W. Lane, C.F. Harvey, L. Liu, Characterizing submarine ground-water discharge using fiber-optic distributed temperature sensing and marine electrical resistivity, in: 21st EEGS Symposium on the Application of Geophysics to Engineering and Environmental Problems, April 2008.
[59] K.B. Moffett, S.W. Tyler, T. Torgersen, M. Menon, J.S. Selker, S.M. Gorelick, Processes controlling the thermal regime of saltmarsh channel beds, Environ. Sci. Technol. 42 (3) (2008) 671−676.
[60] B. Freifeld, S. Finsterle, Imaging Fluid Flow in Geothermal Wells Using Distributed Thermal Perturbation Sensing (No. LBNL-4588E). Lawrence Berkeley National Lab.(LBNL), Berkeley, CA, 2010.
[61] M.H. Dickson, M. Fanelli, Small geothermal resources: a review, Energy Sources 16 (3) (1994) 349−376.
[62] A Nishar, S. Richards, D. Breen, J. Robertson, B. Breen, Thermal infrared imaging of geothermal environments and by an unmanned aerial vehicle (UAV): a case study of the Wairakei−Tauhara geothermal field, Taupo, New Zealand, Renew. Energy 86 (2016) 1256−1264.
[63] D.T. Hodder, Application of remote sensing to geothermal prospecting, Geothermics 2 (1970) 368−380.

[64] R.G. Allis, G.D. Nash, S.D. Johnson, Conversion of thermal infrared surveys to heat flow: comparisons from Dixie Valley, Nevada, and Wairakei, New Zealand. Global Geothermal Resources: Sustainable Energy for the Future, Geothermal Resources Council, 1999.

[65] C. Kratt, W. Calvin, M. Coolbaugh, Geothermal exploration with Hymap hyperspectral data at Brady—Desert Peak, Nevada, Remote. Sens. Environ. 104 (3) (2006) 313—324.

[66] C.E. Haselwimmer, A. Prakash, G. Holdmann, Geothermal exploration at pilgrim hot springs, alaska using airborne thermal infrared remote sensing, Geothermal Resource Council Annual Meeting, San Diego, 2011.

[67] C.J. Bromley, S.M. van Manen, W. Mannington, Heat flux from steaming ground: reducing uncertainties, in: Proceedings, 36th Workshop on Geothermal Reservoir Engineering, Stanford University, California, SGP-TR-191, January (2011).

[68] D.J. Dickinson, Aerial infra-red survey of Kawerau, Rotorua and Taupo Urban Areas-1972. Geophysics Division, 1973.

[69] C. Kuenzer, S. Dech, Thermal infrared remote sensing, Sensors, Methods, Applications, Remote Sensing and Digital Image Processing, Springer, 2013, p. 17.

[70] S.J. Hook, R.G. Vaughan, H. Tonooka, S.G. Schladow, Absolute radiometric in-flight validation of mid infrared and thermal infrared data from ASTER and MODIS on the Terra spacecraft using the Lake Tahoe, CA/NV, USA, automated validation site, IEEE Trans. Geosci. Remote. Sens. 45 (6) (2007) 1798—1807.

[71] M.F. Coolbaugh, C. Kratt, A. Fallacaro, W.M. Calvin, J.V. Taranik, Detection of geothermal anomalies using advanced spaceborne thermal emission and reflection radiometer (ASTER) thermal infrared images at Bradys Hot Springs, Nevada, Remote. Sens. Environ. 106 (3) (2007) 350—359.

[72] R.G. Vaughan, L.P. Keszthelyi, J.B. Lowenstern, C. Jaworowski, H. Heasler, Use of ASTER and MODIS thermal infrared data to quantify heat flow and hydrothermal change at Yellowstone National Park, J. Volcanol Geotherm. Res. 233 (2012) 72—89.

[73] C.A. Hecker, T.E. Smith, B.R. da Luz, M.J. Wooster, Thermal infrared spectroscopy in the laboratory and field in support of land surface remote sensing, Thermal Infrared Remote Sensing, Springer, Dordrecht, 2013, pp. 43—67.

[74] G. Vane, W.M. Porter, J.H. Reimer, T.G. Chrien, R.O. Green, AVIRIS performance during the 1987 flight season: an AVIRIS project assessment and summary of the NASA-sponsored performance evaluation, 1988.

[75] R.O. Green, M.L. Eastwood, C.M. Sarture, T.G. Chrien, M. Aronsson, B.J. Chippendale, et al., Imaging spectroscopy and the airborne visible/infrared imaging spectrometer (AVIRIS), Remote. Sens. Environ. 65 (3) (1998) 227—248.

[76] L.C. Rowan, J.C. Mars, C.J. Simpson, Lithologic mapping of the Mordor, NT, Australia ultramafic complex by using the advanced spaceborne thermal emission and reflection radiometer (ASTER), Remote. Sens. Environ. 99 (1—2) (2005) 105—126.

[77] Y. Yamaguchi, A.B. Kahle, H. Tsu, T. Kawakami, M. Pniel, Overview of advanced spaceborne thermal emission and reflection radiometer (ASTER), IEEE Trans. Geosci. Remote. Sens. 36 (4) (1998) 1062—1071.

[78] B. Tian, L. Wang, K. Kashiwaya, K. Koike, Combination of well-logging temperature and thermal remote sensing for characterization of geothermal resources in Hokkaido, northern Japan, Remote. Sens. 7 (3) (2015) 2647—2667.

[79] Z. Qin, A. Karnieli, P. Berliner, A mono-window algorithm for retrieving land surface temperature from Landsat TM data and its application to the Israel-Egypt border region, Int. J. Remote. Sens. 22 (18) (2001) 3719—3746.

[80] J. Walker, Fundamentals of Physics, eighth edn., Wiley, New York, 2008, p. 891. ISBN9780471758013.

[81] D.A. Artis, W.H. Carnahan, Survey of emissivity variability in thermography of urban areas, Remote. Sens. Environ. 12 (4) (1982) 313—329.

[82] M.J. Hellman, M.S. Ramsey, Analysis of hot springs and associated deposits in Yellowstone National Park using ASTER and AVIRIS remote sensing, J. Volcanol Geotherm. Res. 135 (1—2) (2004) 195—219.

[83] J. Mutua, A. Friese, F. Kuehn, T. Lopeyok, M. Mutonga, N. Ochmann, High resolution airborne thermal infrared remote sensing study, Silali geothermal prospect, Kenya, in: Presented at Short Course VIII on Exploration for Geothermal Resources, Organized by UNU-GTP, GDC and KenGen, at Lake Bogoria and Lake Naivasha, Kenya, October 31–November 22, 2013.
[84] H. Gibson, B. Delwiche, D. FitzGerald, S. Wieberg, J. Mims, P. Berardelli, Demonstration of new geophysical methods for geothermal exploration-the technology explained, Proceedings of the World Geothermal Congress, Melbourne, Australia, 2015.
[85] W.M. Calvin, E.F. Littlefield, C. Kratt, Remote sensing of geothermal-related minerals for resource exploration in Nevada, Geothermics 53 (2015) 517–526.
[86] M.C. Harvey, J.V. Rowland, K.M. Luketina, Drone with thermal infrared camera provides high resolution georeferenced imagery of the Waikite geothermal area, New Zealand, J. Volcanol Geotherm. Res. 325 (2016) 61–69.
[87] V.G. Ambrosia, S. Wegener, T. Zajkowski, D.V. Sullivan, S. Buechel, F. Enomoto, ... E. Hinkley, The Ikhana unmanned airborne system (UAS) western states fire imaging missions: from concept to reality (2006–2010), Geocarto Int. 26 (2) (2011) 85–101.
[88] J.A.J. Berni, P.J. Zarco-Tejada, L. Suárez, V. González-Dugo, E. Fereres, Remote sensing of vegetation from UAV platforms using lightweight multispectral and thermal imaging sensors, Int. Arch. Photogramm. Remote. Sens. Spat. Inform. Sci. 38 (6) (2009).
[89] X. Shang, P. Chazette, End-to-end simulation for a forest-dedicated full-waveform LiDAR onboard a satellite initialized from airborne ultraviolet LiDAR experiments, Remote. Sens. 7 (5) (2015) 5222–5255.
[90] M.A. Taylor, The state of geothermal technology part I: subsurface technology. A Publication by the Geothermal Energy Association for the U.S. DOE, 2007, pp. 45–46.
[91] W.D. Frane, J. Vericella, E. Duoss, M. Smith, R. Aines, J. Roberts, Smart tracers for geothermal reservoir assessment, GRC Trans. 38 (2014) 951–958.
[92] S. Agarwal, P. Tran, Y. Soong, D. Martello, K. Gupta, Flow Behavior of Nanoparticle Stabilized Drilling Fluids and Effect of High Temperature Aging. AADE-11-NTCE-3, AADE National Technical Conference and Exhibition, Houston, 12–14 April 2011.
[93] S.N. Shah, N.H. Shanker, C.C. Ogugbue, Future challenges of drilling fluids and their rheological measurements. AADE fluids conference and exhibition, Houston, TX, 2010.
[94] C. Okoro, Aerobic degradation of synthetic based drilling mud base fluids by Gulf of Guinea sediments under natural environmental conditions, Life Sci. J. 8 (2) (2011) 569–576.
[95] D.J. Oakley, K. Morton, A. Eunson, A. Gilmour, D. Pritchard, A. Valentine, Innovative drilling fluid design and rigorous pre-well planning enable success in an extreme HTHP well, IADC/SPE Asia Pacific Drilling Technology, Society of Petroleum Engineers, 2000.
[96] R.G. Yao, G.C. Jiang, W. Li, et al., Research and evaluation on a new high temperature resistance and high density nano- based drilling fluid, Drill. Fluid Completion Fluid 30 (2) (2013) 25–28.
[97] L. Weili, G. Kai, Drilling technical difficulties and solutions in development of hot dry rock geothermal energy, Adv. Pet. Explor Dev. J. 13 (2017).
[98] X.Q. Lai, et al., Foam drilling fluid technology in ultra -high temperature geothermal wells, Drill. Fluid Completion Fluid (2)(2009) 37–38.
[99] J. Abdo, M. Haneef, Nano-enhanced drilling fluids: pioneering approach to overcome uncompromising drilling problems, J. Energy Resour. Technol. 134 (1) (2012) 014501.
[100] R.A.M. Salem, A. Noah, Reduction of formation damage and fluid loss using nano-sized silica drilling fluids, Pet. Technol. Dev. J. 2 (2014) 75–88.
[101] M. Amanullah, A.M.Al-Tahini, Nanotechnology- its significance in smart fluid development for oil and gas field application, in: SPE Saudia Arabia Section Technical Symposium, Society of Petroleum Engineers, 2009.
[102] S. Singh, R. Ahmed, F. Growcock, Vital role of nanopolymers in drilling and stimulations fluid applications, in: Paper SPE 130413 Presented at the SPE Annual Technical Conference and Exhibition, Florence, Italy, 2010, pp. 19–22.
[103] S.N. Karlsdottir, K.R. Ragnarsdottir, A. Moller, I.O. Thorbjornsson, A. Einarsson, On-Site erosion-corrosion testing in superheated geothermal steam, Geothermics 51 (2014) 170–181.

[104] S.N. Karlsdottir, I.O. Thorbjornsson, K.R. Ragnarsdottir, A. Einarsson, Corrosion testing of heat exchanger tubes in steam from the IDDP-1 exploratory geothermal well in Krafla, Iceland. NACE International Conference, Corrosion, 2014.

[105] A.K. Lahiri, Applied Metalurgy and Corrosion Control: A Handbook for the Petrochemical Industry, Springer, Singapore, 2017.

[106] H. Sieurin, R. Sandstorm, Austenite reformation in the heat affected zone of duplex stainless steel 2205, Mater. Sci. Eng. 418 (1−2) (2006) 250−256.

[107] K. Gawlik, T. Sugama, R. Webster, W. Reams, Field testing of heat exchanger tube coatings, Geotherm. Resour. Counc. Trans. 22 (1998) 385−391.

[108] K. Gawlik, S. Kelley, T. Sugama, R. Webster, W. Reams, Field testing of heat exchanger tube coatings, Geotherm. Resour. Counc. Trans. 23 (1999) 65−69.

[109] K. Gawlik, T. Sugama, R. Webster, W. Reams, Development and field testing of polymer heat exchanger tube coatings, Geotherm. Resour. Counc. Trans. 24 (2000) 659−664.

[110] K. Gawlik, T. Sugama, Long-term field testing of polyphenylenesulphide composite coatings, Geotherm. Resour. Counc. Trans. 1 (2003) 577−582.

[111] T. Sugama, K. Gawlik, Filler materials for polyphenylenesulphide composite coatings, Geotherm. Resour. Counc. Trans. 25 (2001) 41−46.

[112] T. Sugama, K. Gawlik, Self-repairing poly(phenylenesulfide) coatings in hydrothermal environments at 200°C, Mater. Lett. 57 (2003) 4282−4290.

[113] T. Sugama, T. Butcher, L. Ecker, Experience with the development of advanced materials g-for geothermal systems, in: Proceedings of Materials Challenges in Alternative and Renewable Energy Conference, Florida, 2011.

[114] T. Sugama, Polyphenylenesulfied/montmorillonite clay nanocomposite coatings: their efficacy in protecting steel against corrosion, Mater. Lett. 34 (2006) 332−339.

[115] E.B. Nelson, D. Guillot (Eds.), Well Cementing, second ed., Schlumberger, Sugar Land, TX, 2006.

[116] E.B. Nelson, V. Barlet-Gouédard, Thermal cements, in: E.B. Nelson, D. Guillot (Eds.), Well Cementing, second ed., Schlumberger, Sugar Land, TX, 2006.

[117] M.H. Ozyurtkan, M. Radonjic, An experimental study of the effect of CO_2 rich brine on artificially fractured well-cement, Cem. Concr. Compos. J. 45 (2014) 201−208.

[118] W. Um, H. Jung, S. Kabilan, C.A. Fernandez, C.F. Brown, Geochemical and geomechanical effects on wellbore cement fractures, Energy Proc 63 (2014) 5808−5812.

[119] H.B. Jung, S. Kabilan, J.P. Carson, A.P. Kuprat, W. Um, P. Martin, et al., Wellbore cement fracture evolution at the cement − basalt caprock interface during geologic carbon sequestration, Appl. Geochem. 47 (2014) 1−16.

[120] W. Um, H.B. Jung, S. Kabilan, D.M. Suh, C.A. Fernandez, Geochemical and Geomechanical Effects on Wellbore Cement Fractures: Data Information for Wellbore Reduced Order Model, Pacific Northwest National Laboratory/U.S. Department of Energy, Washington, DC, 2014.

[121] D.L. Bour, R. Hernandez, CO_2 Resistance, improved mechanical durability, and successful placement in a problematic lost-circulation interval achieved: reverse circulation of foamed calcium aluminate cement in a geothermal well, GRC Trans. 27 (2003) 163−167.

[122] B. Berard, R. Hernandez, H. Nguyen, Foamed CaP cement enables drilling and cementing of geothermal wells: case history, World Geothermal Congress, Bali, Indonesia, 2010.

[123] P. Spielman, R. Hernández, H. Nguyen, Reverse Circulation of Foamed Cement in Geothermal wells; Geothermal Resources Council; San Diego, CA, 12 September 2006.

[124] T. Pyatina, T. Sugama, Acid resistance of calcium-aluminate cement- fly ash F blends, Adv. Cem. Res. 28 (7) (2016) 433−457.

[125] T. Pyatina, T. Sugama, J. Moon, S. James, Effect of tartaric acid on hydration of a sodium-metasilicate-activated blend of calcium aluminate cement and fly ash, Materials 9 (6) (2016) 422.

[126] T. Pyatina, T. Sugama, Cements for geothermal wells, California Geothermal Forum, 2016.

[127] K. Van Tittelboom, N. De Belie, Self-Healing in cementitious materials a review, Materials 6 (2013) 2182−2217.

[128] M. Wu, B. Johannesson, M. Geiker, A review: self-healing in cementitious materials and engineered cementitious composite as a self-healing material, Constr. Build. Mater. 28 (2012) 571−583.

[129] LockCem Cement, Report HO11594, Halliburton, Houston, TX, 2015.
[130] P.H. Cavanagh, C.R. Johnson, S.L. Roy-Delage, G.G. DeBruijn, I. Cooper, D.J. Guillot, et al., Self-healing cement—novel technology to achieve leak-free wells, in: SPE/ IADC Drilling Conference, Amsterdam, The Netherlands, 2007, SPE/IADC: Richardson, TX, 2007.
[131] B.R. Reddy, F. Liang, R.M. Fitzgerald, D. Meadows, Self repairing cement compositions and methods of using same, U.S. Patent 7,530,396B1, 2009.
[132] B.R. Reddy, F. Liang, R. Fitzgerald, Self-healing cements that heal without dependence on fluid contact: a laboratory study, SPE Drill. Completion 25 (2010) 309—313.
[133] S.L. Roy-Delage, M. Martin-Beurel, K. Dismuke, E. Nelson, Self adaptive cement systems, U.S. Patent 8,469,095B2, 2013.
[134] H. Zhao, H. Yu, Y. Yuan, H. Zhu, Blast mitigation effect of the foamed cement-base sacrificial cladding for tunnel structures, Constr. Build. Mater. 94 (2015) 710—718.
[135] D. Snoeck, J. Dewanckele, V. Cnudde, N. De Belie, X-ray computed microtomography to study autogenous healing of cementitious materials promoted by superabsorbent polymers, Cem. Concr. Compos. 65 (2016) 83—93.
[136] H.X.D. Lee, H.S. Wong, N.R. Buenfeld, Self-sealing of cracks in concrete using superabsorbent polymers, Cem. Concr. Res. 79 (2016) 194—208.
[137] J. Todorovic, M. Raphaug, E. Lindeberg, T. Vrålstad, M.-L. Buddensiek, Remediation of leakage through annular cement using a polymer resin: a laboratory study, Energy Proc 86 (2016) 442—449.
[138] W. Li, Z. Jiang, Z. Yang, H. Yu, Effective mechanical properties of self-healing cement matrices with microcapsules, Mater. Des. 95 (2016) 422—430.
[139] M. Ian Childers, M.-T. Nguyen, K.A. Rod, P.K. Koech, W. Um, J. Chun, et al., Polymer-cement composites with self-healing ability for geothermal and fossil energy applications chemistry of materials 29(11) 2017 4708—4718.
[140] R. Morlat, G. Orange, Y. Bomal, P. Godard, Reinforcement of hydrated portland cement with high molecular mass water-soluble polymers, J. Mater. Sci. 42 (2007) 4858—4869.
[141] G.P. Funkhouser, L.S. Eoff, L.R. Norman, Methods and compositions for cementing in well bores, U.S. Patent 7,238,229B2, 2007.
[142] Y. Yang, M.W. Urban, Self-healing polymeric materials, Chem. Soc. Rev. 42 (2013) 7446—7467.
[143] N. Roy, B. Bruchmann, J.-M. Lehn, Dynamers: dynamic polymers as self-healing materials, Chem. Soc. Rev. 44 (2015) 3786—3807.
[144] J. Canadell, H. Goossens, B. Klumperman, Self-healing materials based on disulfide links, Macromolecules 44 (2011) 2536—2541.
[145] U. Lafont, H. van Zeijl, S. van der Zwaag, Influence of cross- linkers on the cohesive and adhesive self-healing ability of polysulfide-based thermosets, ACS Appl. Mater. Interfaces 4 (2012) 6280—6288.
[146] M. Pepels, I. Filot, B. Klumperman, H. Goossens, Self-healing systems based on disulfide-thiol exchange reactions, Polym. Chem. 4 (2013) 4955—4965.
[147] F. García, M.M.J. Smulders, Dynamic covalent polymers, J. Polym. Sci., Part. A: Polym. Chem. 54 (2016) 3551—3577.
[148] T. Sugama, L.E. Brothers, L. Weber, Calcium aluminate cements in fly ash/calcium aluminate blend phosphate cement systems: their role in inhibiting carbonation and acid corrosion at a low hydrothermal temperature of 90°C, J. Mater. Sci. 37 (2002) 3163.
[149] A.M. Manzella, CNR—Institute of Geosciences and Earth Resources Via Moruzzi, PISA, Italy, Geothermal Energy [Power Point slide], 2000. Retrieved from <https://www.sif.it/media/155a6df7.pdf>
[150] J.W. Lund, The use of down-hole heat exchangers, Geothermics 32 (2003) 535—543.
[151] M. Rosen, S. Koohi-Fayegh, Geothermal Energy: Sustainable Heating and Cooling Using the Ground, John Wiley & Sons, Inc, Chichester, West Sussex, United Kingdom, 2017.
[152] P. Bayer, D. Saner, S. Bolay, L. Rybach, P. Blum, Greenhouse gas emission savings of ground source heat pump systems in Europe: a review, Renew. Sustain. Energy Rev. 16 (2012) 1256—1267.

[153] A. Capozza, M. De Carli, A. Zarrella, Design of borehole heat exchangers for ground-source heat pumps: a literature review, methodology comparison and analysis on the penalty temperature, Energy Build. 55 (2012) 369–379.

[154] R. Curtis, J. Lund, B. Sanner, L. Rybach, G. Hellström, Ground source heat pumps—geothermal energy for anyone, anywhere: current worldwide activity, in: Proceedings of World Geothermal Congress, Antalya, Turkey, 24–29 April 2005.

[155] I. Sarbu, C. Sebarchievi, Ground-Source Heat Pumps Fundamentals, Experiments and Applications, Elsevier, 2016.

[156] ASHRAE, Commercial/Institutional Ground-Source Heat Pump Engineering Manual, American Society of Heating, Refrigerating and Air-Conditioning Engineers, Atlanta, GA, 1995.

[157] J.W. Lund, H.H. Freeston, T.L. Boyd, Direct utilization of geothermal energy Worldwide Review, in: Proceedings of World Geothermal Congress, No. 0007, Bali, Indonesia, 26–30 April 2010.

[158] S.J. Self, B.V. Reddy, M.A. Rosen, Geothermal heat pump systems: status review and comparison with other heating options, Applied Energy, Elsevier, 2013, pp. 341–348. vol. 101(C).

[159] Y. Bi, X. Wang, Y. Liu, H. Zhang, L. Chen, Comprehensive exergy analysis of aground-source heat pump system for both building heating and cooling modes, Appl. Energy 86 (2009) 2560–2565.

[160] K. Hellevang, Ground-source heat pumps, NSDU Extension, AE1483, 2018.

[161] Oklahoma State University, Division of Engineering Technology, Closed-loop/ground-source heat pump system—installation guide, International Ground Source Heat Pump Association, Stillwater, OK, 1988.

[162] C.K. Lee, Effects of multiple ground layers on thermal response test analysis and ground-source heat pump simulation, Appl. Energy 88 (2011) 4405–4410.

[163] I.W. Johnston, G.A. Narsilio, S. Colls, Emerging geothermal energy technologies, Korean society of civil engineers, J. Civ. Eng. 15 (4) (2011) 643–653.

[164] P.F. Healy, V.I. Ugursal, Performance and economic feasibility of ground source heat pumps in cold climates, Int. J. Energy Res. 21 (1997) 857–870.

[165] A.,D. Chiasson, Geothermal Heat Pump and Heat Engine Systems: Theory and Practice, first ed, John Wiley & Sons, Ltd, 2016.

[166] Office of Energy Efficiency, Heating and Cooling With a Heat Pump—Groundsource Heat Pumps (Earth-Energy Systems), Natural Resources Canada, Ottawa, CA, 2009.

[167] C.S. Smith, P.F. Ellis, Addendum to material selection guidelines for geothermal energy-utilization systems. Part I. Extension of the field experience data base. Part II. Proceedings of the Geothermal Engineering and Materials (GEM) Program Conference, San Diego, CA, 6-8 October 1982, 1983.

[168] G. Stryi-Hipp, Renewable Heating and Cooling Technologies and Applications, Elsevier, 2016.

[169] A. Kilicarslan, N. Müller, A comparative study of water as a refrigerant with some current refrigerantsWiley InterScience, July 18 Int. J. Energy Res. (2005) 947–959.

[170] P. Bahadorani, G.F. Naterer, S.B. Nokleby, Optimization of heat exchangers for geothermal district heating, Trans. Can. Soc. Mech. Eng. 33 (2009) 2.

[171] J.W. Lund, Design of closed-loop geothermal heat exchangers in the U.S. International Course on Geothermal Heat Pumps, 2002.

[172] A. Manzella, Geothermal energy, EPJ Web Conf. 148 (2017) 00012.

[173] M. Schultz, Geothermal Heating & Cooling Systems' Loop Fields [Blog post] (2020). Retrieved from https://earthrivergeothermal.com/geothermal-heating-cooling-systems-loop-fields/

[174] S.J. Rees, Advances in Ground Source Heat Pump Systems, Elsevier, Woodhead Publishing Series in Energy: Number 100, 2016.

[175] S. Hackel, G. Nellis, S. Klein, Optimization of cooling-dominated hybrid ground coupled heat pump systems, ASHRAE Trans. 115 (2009) 565–580.

[176] S. Choi, Enhancing thermal conductivity of fluids with nanoparticles, in: D.A. Siginer, H.P. Wang (Eds.), Developments applications of nonnewtonian flows, ASME, New York, 1995, p. 99. FED-vol 231/MD-vol 66.

[177] S.U.S. Choi, Z.G. Zhang, P. Keblinski, Nanofluids encyclopedia of nanoscience and nanotechnology, J. Nanosci. Nanotechnol. (2004) 727.

[178] J.A. Eastman, S.U.S. Choi, S. Li, W. Yu, L.J. Thompson, Anomalously increased effective thermal conductivities of ethylene glycol-based nanofluids containing copper nanoparticles, Appl. Phys. Lett. 78 (2001) 718–720.
[179] G. Huminic, A. Huminic, Application of nanofluids in heat exchangers: a review, Renew. Sustain. Energy Rev. 16 (8) (2012) 5625–5638.
[180] S.M.S. Murshed, K.C. Leong, C. Yang, Thermophysical and electrokinetic properties of nanofluids–A critical review, Appl. Therm. Eng. 28 (17–18) (2008) 2109–2125.
[181] B.C. Pak, Y.I. Cho, Hydrodynamic and heat transfer study of dispersed fluids with submicron metallic oxide particles, Exp. Heat. Transf. 11 (1998) 151–170.
[182] D. Wen, G. Lin, S. Vafaei, K. Zhang, Review of nanofluids for heat transfer applications, Particuology 7 (2) (2009) 141–150.
[183] M.H. Esfe, S. Saedodin, W. Yan, M. Afrand, M. Sina, Study on thermal conductivity of water-based nanofluids with hybrid suspensions of CNTs/Al2O3 nanoparticles, J. Therm. Anal. Calorim. 124 (2016) 455–460.
[184] G. Diglioa, C. Rosellia, M. Sassoa, U.J. Channabasappa, Borehole heat exchanger with nanofluids as heat carrier, Geothermics 72 (2018) 112–123.
[185] J.M. Olson, Nanofluids and a method of making nanofluids for ground source heat pumps and other applications, US Patent, Patent No.: US 8,580,138 B2, November 12 2013.
[186] I.E. Mouromtseff, Water and forced air cooling of vacuum tubes, Proc. IRE 30 (4) (1942) 190–205.
[187] L.B. Mapa, M. Sana, Heat Transfer in Mini Heat Exchanger Using Nanofluids, American Society for Engineering Education, Northern Illinois, University, DeKalb, IL, 2005.
[188] S. Mishra, S. Samantaray, K. Debendra, D.K. Das, Theoretical analysis on application of nanofluids in ground source heat pumps for building cooling, Int. J. Recent. Technol. Mech. Electr. Eng. (IJRMEE) 5 (6) (2018). ISSN: 2349-7947.
[189] I.M. Mahbubula, R. Saidura, M.A. Amalinaa, Heat transfer and pressure drop characteristics of Al_2O_3-R141b nanorefrigerant in horizontal smooth circular tube, Proc Eng. 56 (2013) 323–329.
[190] H. Li, W. Yang, Z. Yu, Z. Li, The performance of a heat pump using nanofluid (R22 + $TiO_{2)}$ as the working fluid–an experimental study, Energy Proc 75 (2015) 1838–1843.
[191] R. DiPippo, Geothermal Power Plants Principles, Applications, Case Studies and Environmental Impact, fourth ed, Elsevier, 2016.
[192] H. Moon, J. Sadiq, S.J. Zarrouk, Efficiency of geothermal power plants: a worldwide review, New Zealand Geothermal Workshop 2012 Proceedings, Auckland, New Zealand, 19–21 November 2012.
[193] J.W. Lund, Characteristics, development and utilization of geothermal resources, Ghc Bulletin, June 2007.
[194] S. Quoilin, M.,V.,D. Broek, S. Declaye, P. Dewallef, V. Lemort, Techno-economic survey of organic Rankine cycle (ORC) systems, Renew. Sustain. Energy Rev. 22 (2013) (2013) 168–186.
[195] T. Bornert, Organic Rankine cycle based power plant to utilize lowgrade waste heat sources, in: Cement Industry Technical Conference, 2011. pp. 1–10.
[196] G. Angelino, C. Invernizzi, G. Molteni, The potential role of organic bottoming Rankine cycles in steam power stations, Proc. Inst. Mech. Eng. Part. A: J. Power Energy 213 (2) (1999) 75–81.
[197] A. Allouhi, A. Boharb, A. Jamil, A. Ait Msaad, A. Benbassou, T. Kousksou, Simulation of a thermoelectric heating system for small-size office buildings in cold climates, IEEE 978-1-4673-7894-9, 2015.
[198] J. Siviter, A. Montecucco, A.R. Knox, Rankine cycle efficiency gain using thermoelectric heat pumps, Appl. Energy 140 (2015) 161–170.
[199] Z. Chen, G. Han, L. Yang, L. Cheng, J. Zou, Nanostructured thermoelectric materials: current research and future challenge, Prog. Nat. Sci Mater. Int. 22 (6) (2012) 535–549.
[200] X. Zhang, L. Zhao, Thermoelectric materials: energy conversion between heat and electricity, J. Mater 1 (2015) 92–105.
[201] G.J. Snyder, E.S. Toberer, Complex thermoelectric materials, Nat. Mater. 7 (2008) 105e14.
[202] T.M. Tritt, Thermoelectric phenomena, materials, and applications, Annu. Rev. Mater. 590 Res. 41 (2011) 433–448.

[203] J.H. Yang, Special section papers on thermoelectric materials and applications-foreword, J. Electron. Mater. 36 (2007) 703.

[204] I. Neya, J. Ramousse, M. Perier-Muzet, Thermodynamic analysis of water/water thermoelectric heat pumps: design considerations, in: Proceedings of ECOS 2015—The 28th International Conference on Efficiency, Cost, Optimization, Simulation and Environmental Impact of Energy Systems, 30 June-3 July, Pau, France, 2015.

[205] J.H. Li, Q. Tan, J.F. Li, Synthesis and property evaluation of CuFeS2-x as earth-abundant and environmentally-friendly thermoelectric materials, J. Alloy. Compd. 551 (2013) 143—149.

[206] T. Kyono, R. Suzuki, K. Ono, Conversion of unused heat energy to electricity by means of thermoelectric generation in condenser, IEEE Trans. Energy Convers. 18 (2) (2003) 330—334.

[207] A.R. Knox, E. McCulloch, J. Siviter, Method and apparatus for improvement of efficiency of thermal cycles, WO 2012 085551 A2, 2011.

CHAPTER THREE

Functional green-based nanomaterials towards sustainable carbon capture and sequestration

H.M. Yurdacan[1] and Mufrettin Murat Sari[2]
[1]University of Southern California, Los Angeles, CA, United States
[2]Texas A&M University, Commerce, TX, United States

3.1 Introduction

Increasing levels of Carbon dioxide (CO_2) emissions generated from the combustion of fossil fuels such as coal, oil or natural gas is the largest contributor of the global warming. CO_2 capture and sequestration (CCS) technologies have received great interest over the last years to reduce CO_2 emissions under various conditions such as precombustion, postcombustion, and oxy-fuel combustion from power plants and direct air capture (Fig. 3.1). CCS technology have three steps: capture of CO_2, transportation of carbon dioxide and sequestration via long term storage of CO_2 in plants, underground geological formations and oceans or conversion of CO_2 into value added products using chemical, biological, and physical methods [2].

There are two major challenges in CO_2 capture. The carbon dioxide concentration is low whereas N_2 gas composition is predominantly high in postcombustion gas mixture or in ambient air (Table 3.1). Second, high selectivity of CO_2 over other components in gas mixtures needed to ensure high CO_2 gas purity from the capture process. Carbon capture and storage technologies mainly rely on CO_2/N_2 separation for postcombustion, CO_2/H_2 seperation for precombustion, CO_2/CH_4 seperation for natural gas purification, and CO_2/CO separation for oxy-combustion. Selectivity of CO_2 capture process depends on both absorption and diffusion. As inverse relationship exists between adsorption and diffusion, it is crucial to device materials that can tune both parameters independently [3].

It is important to develop CO_2 capture technologies with the following properties:

- Low heat capacity;
- Low energy requirement for regeneration;
- High CO_2 capture capacity;
- High CO_2 selectivity;
- Thermal, chemical, and mechanical stability;
- Use of environmentally friendly, reusable and low-cost, green adsorbents.

Sustainable Materials for Transitional and Alternative Energy.
DOI: https://doi.org/10.1016/B978-0-12-824379-4.00004-5

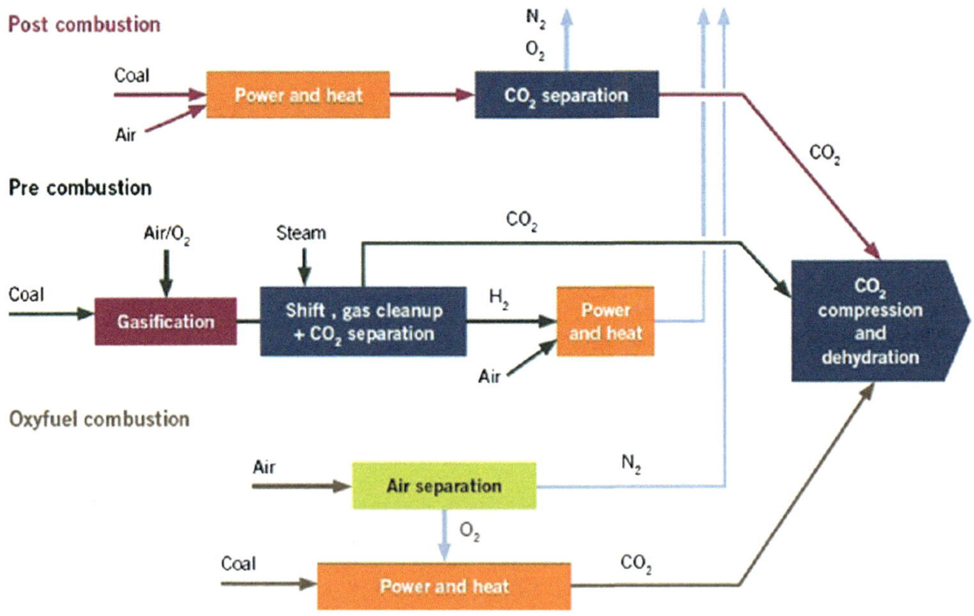

Figure 3.1 Types of CO_2 capture technologies from power plants [1].

Table 3.1 Parameters for typical gas compositions in pre and postcombustion processes [3].

Comp.	Postcombustion	Precombustion (after water−gas shift reaction)	Kinetic diameter	Polarizability	Quadropole moment
CO_2	15%−16%	35.5%	3.30	26.3×10^{-25}	13.4×10^{-40}
H_2O	5%−7%	0.2%	2.65		
N_2	70%−75%	0.25%	3.64	17.6×10^{-25}	4.7×10^{-40}
H_2		61.5%	2.89		
O_2	3%−4%		3.45		
CO	20 ppm	1.1%	3.75		
SO_x	<800 ppm				
NO_x	500 ppm				
H_2S		1.1%			

Several methods have been developed for CO_2 capture technologies including chemical absorption, physical adsorption, chemical adsorption and membrane systems [1]. These methods are summarized in Fig. 3.2 [4].

In physical adsorption, CO_2 is adsorbed under low temperature and high-pressure conditions whereas it is desorbed when the adsorbent is subjected to elevated temperatures and low pressure through pressure swing adsorption (PSA) in accordance with Henry's Law. This method is mainly based on the differences in physical and electronic properties of gases such as polarizability and quadrupole

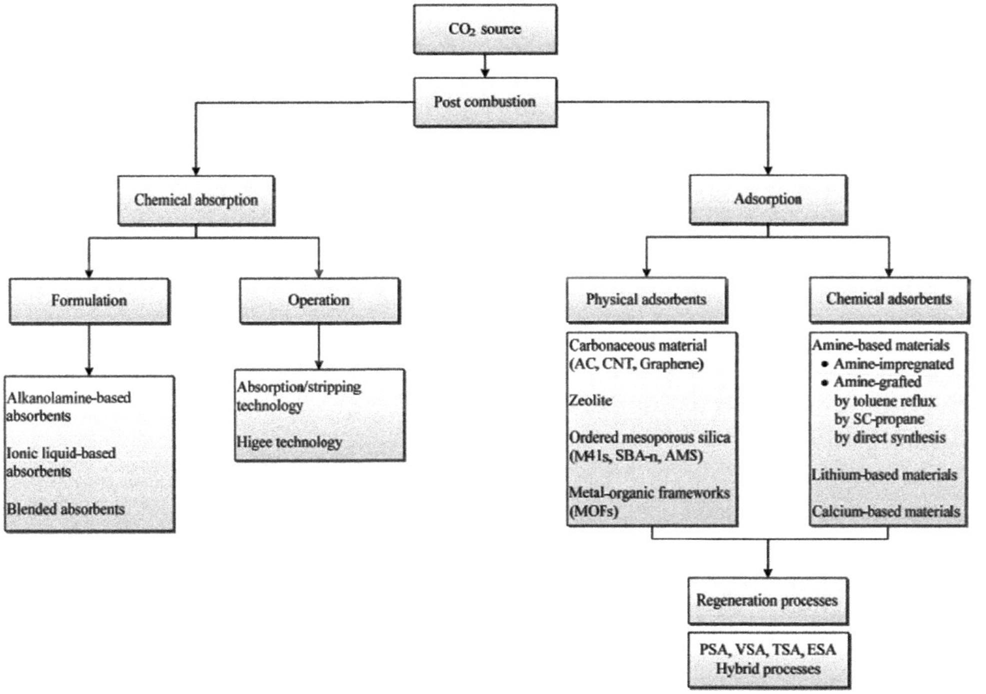

Figure 3.2 CO_2 capture technologies [4].

moment. It is notable for CO_2/N_2 or CO_2/H_2 separation, kinetic diameters of these gases are relatively similar. Luckily, CO_2 has a large quadrupolar moment and higher polarizability compared with N_2 resulting in higher affinity of materials surface for CO_2 (Table 3.1). Physical adsorbents such as Rectisol, Selexol, and ionic liquids have been used for decades.

Chemical absorption process, also called "Wet Scrubbing," is the most widely used technology that employs alkanolamine solvents and has been considered as state-of-art for CO_2 capture for decades. This method generally demonstrates high CO_2 selectivity as it involves chemical reaction between absorbent and CO_2. Primary, secondary, and tertiary structures of alkanolamines such as monoethanolamide (MEA), diethanolamine (DEA), and triethanolamine (TEA) have been extensively used for CO_2 capture so far. In typical chemical absorption process, primary and secondary amines react with CO_2 in order to form zwitterion first and then proceeded with the formation of carbamate, whereas reaction of tertiary amines results in the generation of bicarbonate (Fig. 3.3). Despite of high CO_2 absorption efficiency and selectivity, this technology suffers from high-energy requirement during regeneration step, solvent degradation, low absorption/desorption rate, equipment corrosion and environmental issues [5]. Moreover,

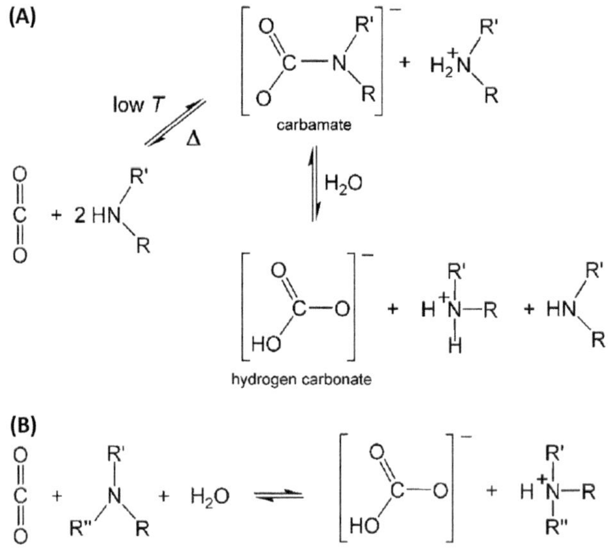

Figure 3.3 Reaction of CO_2 with (A) primary and secondary, (B) tertiary amine structures [3].

CO_2 capture process from conventional amine scrubbing methods constitutes 75% of the total cost of CO_2 capture, sequestration, and storage [3].

Due to these drawbacks of amine-based solvent absorption processes, solid adsorption methods have gained popularity in order to achieve efficient CO_2 capture with lower energy penalty during regeneration step and reusability. Thus, solids adsorbents offer a more environmentally friendly route for CO_2 capture compared with amine solvent methods. Solid sorbents having large surface area and high porosity are usually desirable as porous structure enables penetration of the gas stream from which CO_2 will be captured whereas large surface area allows better access to the active sites for adsorption. Zeolites [3], metal-organic frameworks (MOFs) [6], silica-based nanosorbents [7,8], amine-grafted or impregnated solid supports have been extensively studied for CO_2 capture [3,6,9].

For both CCS processes, the development of reusable, low cost and green-based adsorbents are particularly important for efficient and environmentally friendly removal of CO_2 from the atmosphere. In this chapter, we briefly discussed the functional materials derived from biomass, natural products or via bioinspired methods for CCS. This chapter includes latest research studies, reviews and examples for the development and application of new sustainable and environmentally friendly materials for efficient CO_2 capture processes. Synthesis and characterization of sustainable CO_2 adsorbents, functionalization of these materials for improved CO_2 adsorption and their overall CO_2 adsorption/desorption performance are reviewed in detail.

3.2 Chemically modified halloysite nanotubes for CO_2 capture

3.2.1 Introduction to halloysite nanotubes

HNTs are naturally occurring materials with unique properties. HNTs are the dominant form of naturally formed halloysite. Halloysite was first described by Berthier as a clay mineral of the Kaolin group [10]. It is chemically similar to kaolinite in terms of having high surface area and chemical stability except that halloysite has high water content due to having unit layers that are spaced by water molecules.

Halloysite is naturally mined in USA, Turkey, New Zealand, Korea, and China [11–13]. The typical geological occurrence of Halloysite is by weathering, pedagenesis and hydrothermal alteration of rocks and soils [14]. Halloysites may be formed in various morphologies such as platy, spheroidal and tubular depending on the crystallization conditions and geological formation. Aforementioned, tubular structure is the dominant form, also called HNTs.

HNT is a double-layered aluminosilicate composed of aluminum, oxygen, hydrogen and silicon. (Fig. 3.4). Halloysite has structural formula of $Al_2Si_2O_5(OH)_4.nH_2O$. Internal surface of Halloysites consists of Al-OH groups, whereas outer surface is mainly composed of Si – O – Si groups [11,15].

They are composed of ultra-small hollow nanotubular structures in the submicron sizes having lengths of 0.5–2 μm and diameters smaller than 200 nm shown by scanning electron microscopy (SEM) and transmission electron microscopy (TEM) images (Fig. 3.5) [16–18]. Nanotubular structure is primarily due to the strain as a result of lattice mismatch between oxygen-sharing aluminum oxide and silicone dioxide sheets [18]. In recent years, HNTs attracted significant interest in Nanotechnology applications due to their unique properties. They are low-cost and environmentally friendly

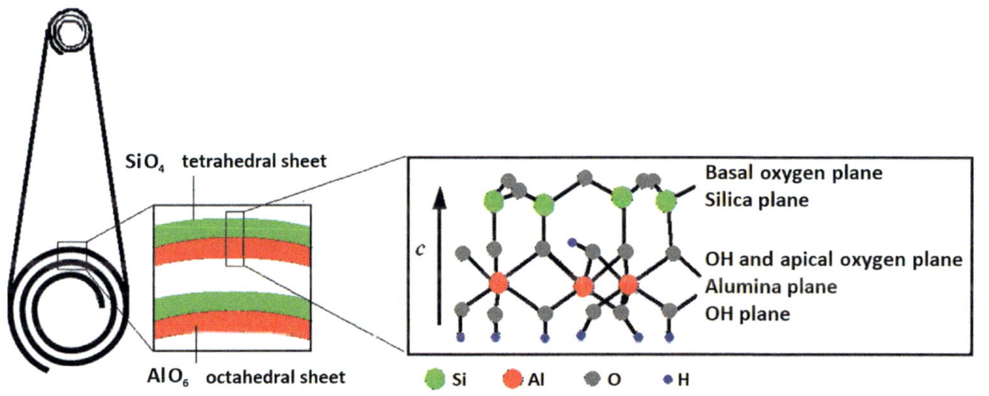

Figure 3.4 Representation of HNTs structure [11].

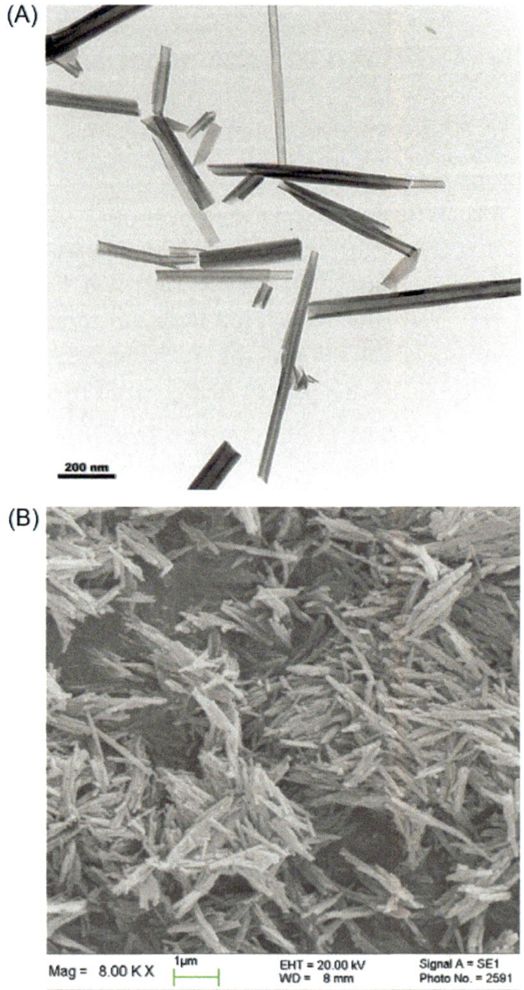

Figure 3.5 Morphology of the HNTs. (A) TEM photo; (B) SEM photo [16].

materials that possess excellent physicochemical properties, high surface area, high temperature resistance and several excellent mechanical properties [19–21].

They are increasingly considered as a green alternative in many applications including as nanoadditives [22], catalysts supports [23,24], and nanocarriers for controlled release of agents, drugs and proteins [25,26] due to being natural, nontoxic and biocompatible.

3.2.2 Modification of halloysite nanotubes for CO_2 capture applications

In addition to many fore mentioned exciting applications, chemically modified HNTs have recently been considered as a green solid CO_2 adsorbent primarily due to their

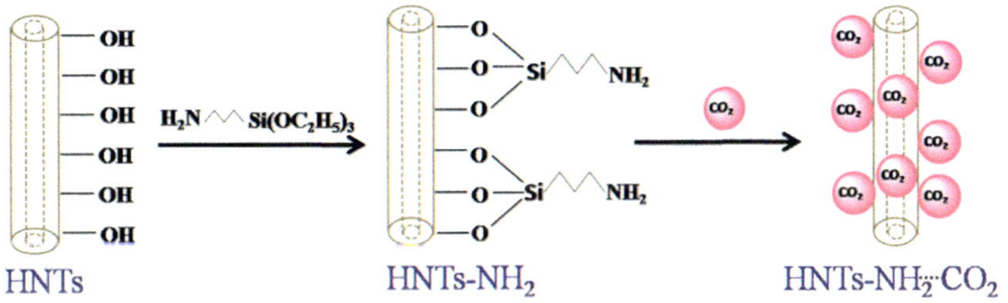

Figure 3.6 Schematic representation of aminosilane surface modification of HNTs and CO_2 capture at ambient temperature [15].

fine particle size, high thermal resistance, high surface area, regeneration ability and ease of chemical tunability. It is notable bare HNTs does not reveal any significant CO_2 adsorption even after 12 hours, therefore surface modifications are needed to develop adsorption properties [15]. HNTs can be modified through selective functionalization of the inner and outer surfaces. The outer surface of HNTs consists of Si-O-Si groups and has electrochemical properties similar to SiO_2 and negatively charged at pH of 6–7 whereas the inner surface composed of Al-OH groups and is positively charged [27].

Jana et al. modified the surface of the HNTs with 3-aminopropyl triethoxysilane in order to study the CO_2 adsorption efficiency from the atmosphere (Fig. 3.6) [15]. The chemical modification was characterized by FTIR spectroscopy. The three peaks at 1556 cm^{-1} were attributed to N—H deformation, whereas peaks at 2932 and 3453 cm^{-1} were assigned to the stretching vibration of C—H and N—H.

FESEM images demonstrated tubular structure with the length between 1.0 and 1.5 μm and the inner diameter of 15–20 nm (Fig. 3.7A–B). In addition, TEM images showed single HNT consisted of a multilayer structure post surface modification (Fig. 3.7D) [28].

Kim et al. prepared two different types of HNTs via amine grafting using (3-aminopropyl) triethoxysilane (APTES) and amino-containing polymer impregnation methods with polyethylenediamine (PEI). FTIR spectra of pure PEI and HNTs were used as reference for comparison with amine-modified HNTs (Fig. 3.8A). FTIR spectra of APTES-grafted HNTs showed amide bend at 1651 cm^{-1}. In addition, the peaks from PEI-impregnated HNTs were well in correlation with characteristics peaks of pure PEI showing indication of successful functionalization of amine groups by both grafting and impregnation. Zeta potential comparison of bare HNTs versus APTES-functionalized HNTs also indicates the grafting was successful as the outer surface of bare HNTs are negatively charged whereas postamine functionalization the zeta potential was measured as positive (Fig. 3.8B) [29]. SEM and TEM analysis showed some morphological changes due to surface functionalization (Fig. 3.9). amine grafting onto HNTs surfaces did not have a significant effect on the morphology shown by

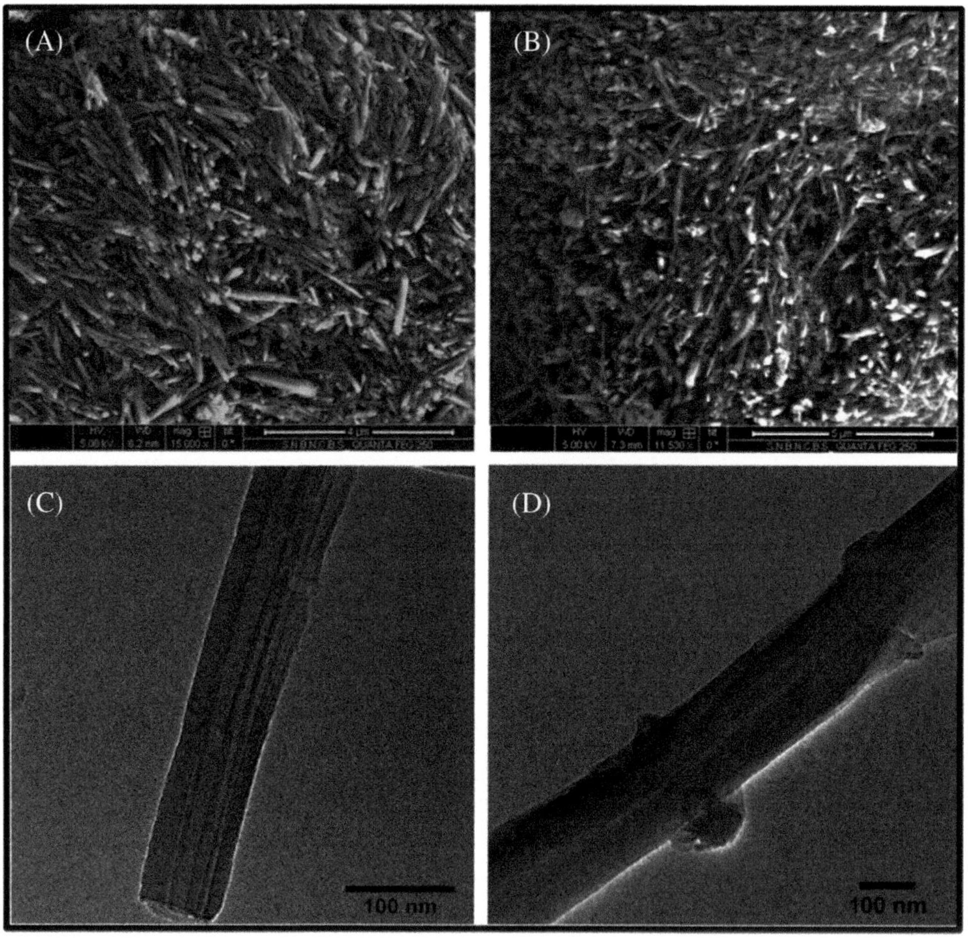

Figure 3.7 FESEM images of HNTs (A − C) preamine functionalization and (B − D) postamine functionalization [28].

SEM images. However, PEI-impregnated HNTs showed significant decrease in the inner diameter of HNTs indicating the association of PEI onto HNTs.

PEI loading to the HNTs was also performed by Cai et al. via similar methods described in the literature [30]. Pore volumes and the surface areas of PEI-HNTs with different PEI loadings via N_2 adsorption/desorption isotherms were studied. Once PEI was impregnated into the HNTs, surface area decreased as expected (Fig. 3.10). There was a slight increase in pore volume once the PEI loading was increased to 20% from 10%. This was attributed to the agglomeration of PEI-HNTs resulting in development of further large macropores. The proposed model was shown in Fig. 3.11. More than 20% PEI loading promoted reduction in pore volume as the inner space was almost fully loaded with PEI, thus restricting the access to the pores.

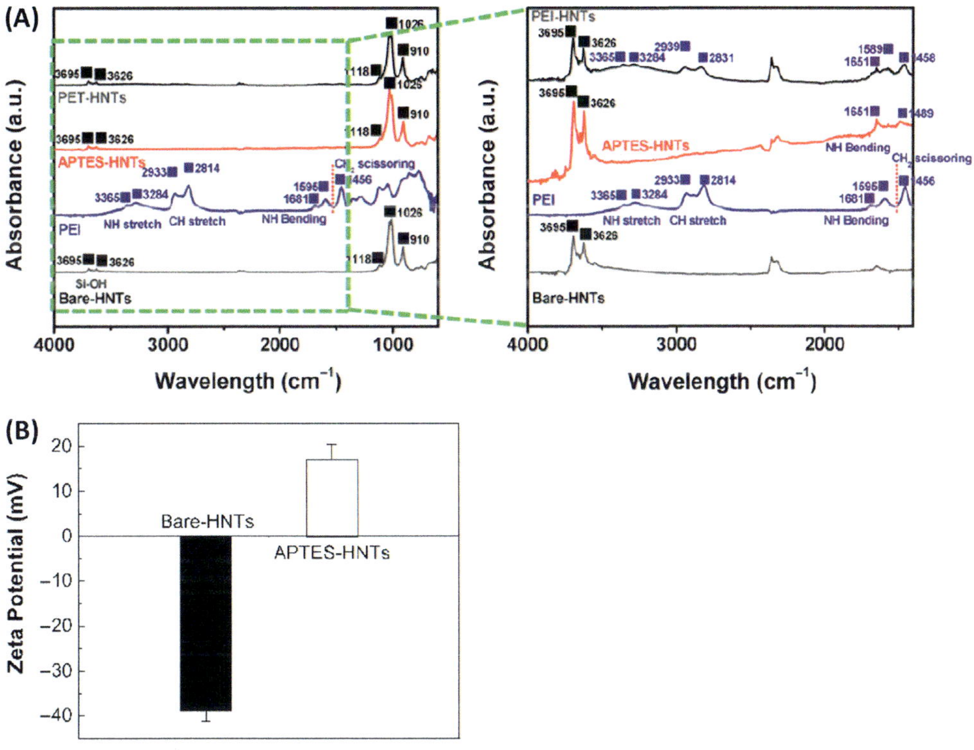

Figure 3.8 (A) FTIR Spectra of HNTs, PEI, APTES-HNTs, and PEI-HNTs. (B) Zeta potential of bare HNTs and APTES-functionalized HNTs [29].

In another study, HNTs were used to produce mesoporous silica nanotubes (MSiNTs) that may potentially be used as a substrate for CO_2 capture [31]. The HNTs were acid treated at 80°C in order to activate aluminosilicate sites and remove the alumina selectively in order to form MSiNTs. MSiNTs were then impregnated with PEI at different loading levels to facilitate CO_2 capture (Fig. 3.12).

MSiNTs showed large surface area and total pore volume; surface area was 6 times larger and the total pore volume was about 2 times larger than that of HNTs indicating more potential accessible sites for CO_2 capture (Table 3.2). Average pore size for MSiNTs decreased compared to that of HNTs even though the total volume doubled. This was attributed to formation of new micropores by the space from the removal of alumina component. More interestingly, with 30% PEI loadings on MSiNTs the average pore size almost tripled compared to bare MSiNTs. Moreover, as the content of PEI increased, both the surface area and pore volume decreased as expected since the surface of MSiNTs was loaded with PEI. It was suggested in the study that PEI impregnation occurred on the outer surface, instead of the inside of the nanotubes.

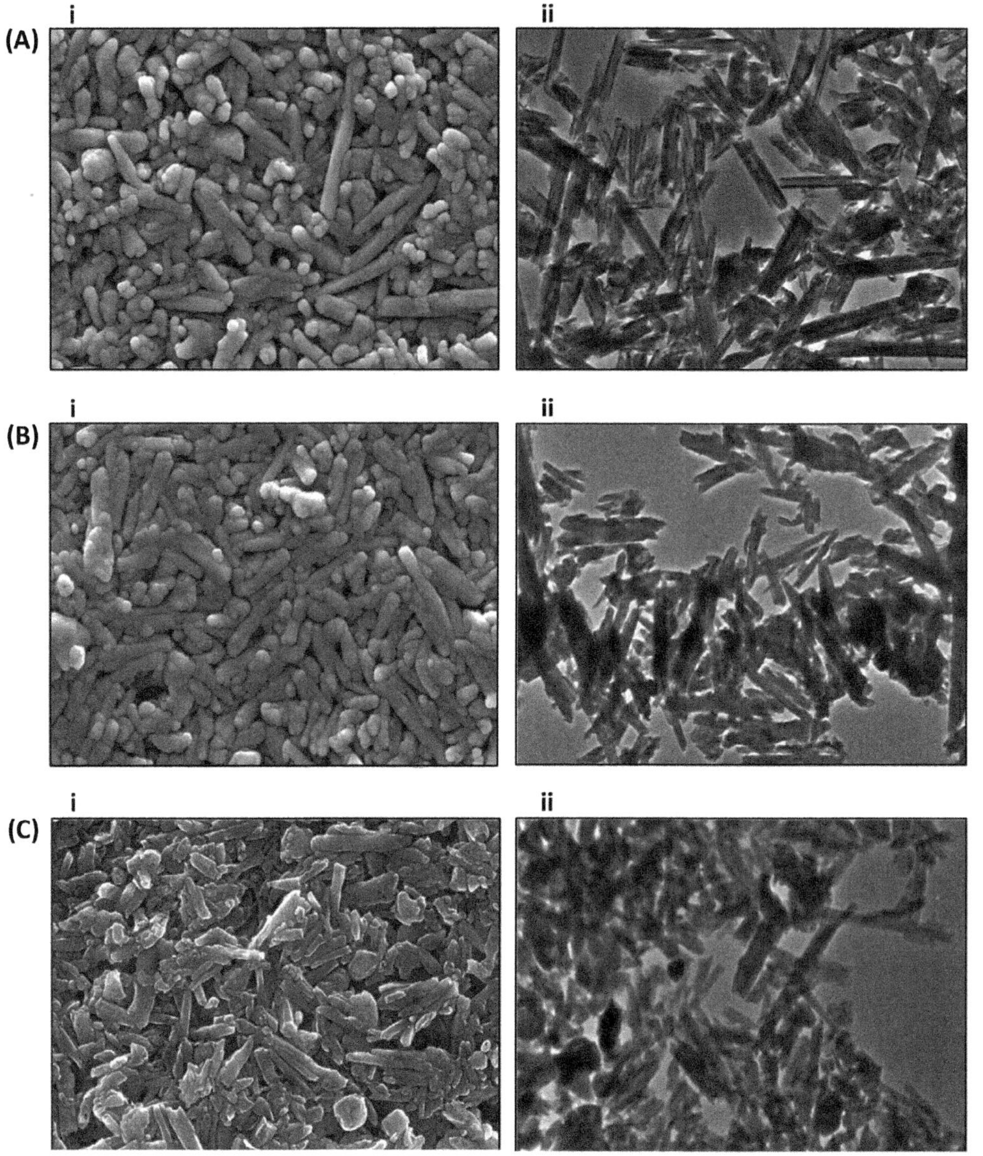

Figure 3.9 SEM (i) and TEM (ii) images of (A) HNTs, (B) APTES-HNTs, and (C) PEI-HNTs [29].

3.2.3 CO_2 adsorption/desorption studies

There have been many proposed ways to measure CO_2 adsorption and desorption of solid adsorbents. Some of the common methods include packed bed measurements, thermo gravimetric analysis (TGA) and laser based ICOS technique [15,29]. In order to measure CO_2 capture and release performance, air, pure CO_2 gas, or

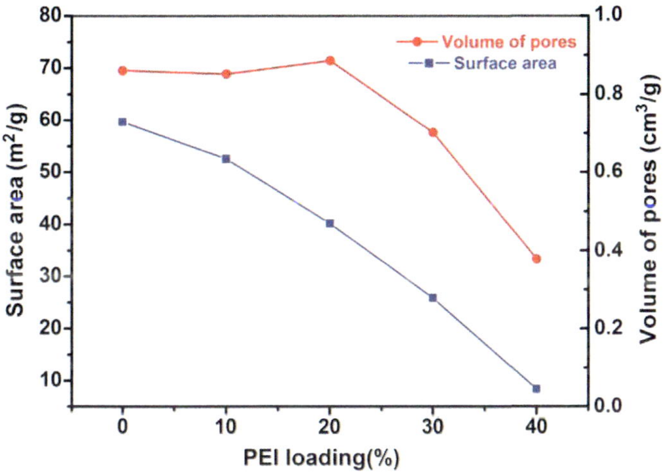

Figure 3.10 Surface area and Volume of pores of PEI-HNTs at different PEI loadings [30].

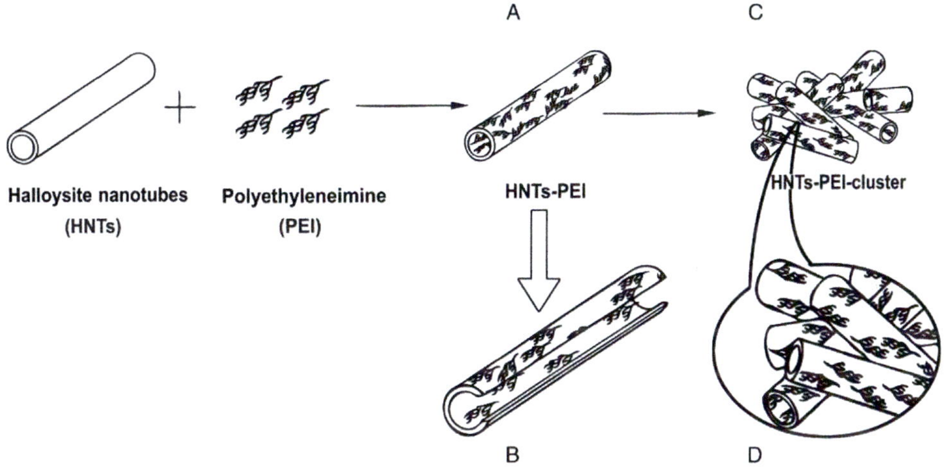

Figure 3.11 Schematic representation of PEI loading in the HNTs [30].

gas mixtures representing pre and postcombustion are usually introduced to the adsorbent medium.

There are many factors affecting the CO_2 adsorption/desorption performance of functionalized HNTs including CO_2 feed concentration, surface area of the HNT substrate, amine content, structure of amine groups grafted (primary, secondary, or tertiary), length of the functionalized amines, pressure, temperature, and reaction time. It was reported amine-grafted HNTs shows strong dependence on the concentration of CO_2 in the feed [15]. As concentration of CO_2 in the feed increases, the amount

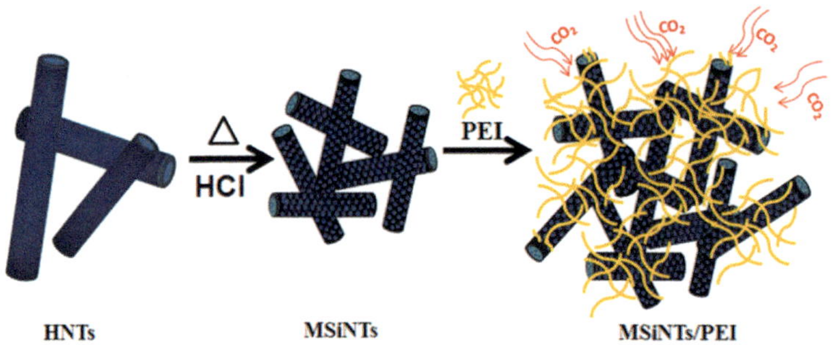

Figure 3.12 Schematic representation of proposed synthesis of PEI-MsiNTs [31].

Table 3.2 Surface area, pore size and pore volume of HNTs, MSiNTs, and PEI-MsiNTs [31].

Sample	Surface area, S_{BET} (m^2/g)	Total pore volume (cm^3/g)	Average pore size (nm)
HNTs	63.4	0.24	15.3
MSiNTs	366.4	0.55	6.0
MP-30	76.1	0.35	18.2
MP-40	52.5	0.17	12.9
MP-50	18.3	0.04	8.4
MP-60	23.5	0.03	4.5

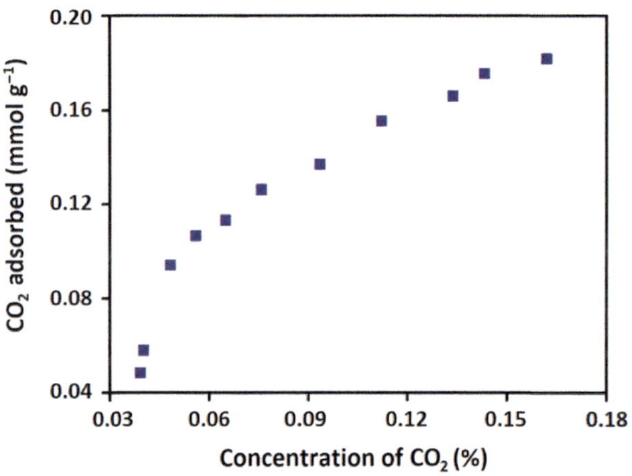

Figure 3.13 The effect of CO_2 concentration in the feed on the CO_2 adsorption of amine-functionalized HNTs [15].

of CO_2 adsorbed also increases (Fig. 3.13). This can be explained by the fact that as CO_2 concentration increases, more CO_2 will be available to the active sites on the large surface of the HNTs until those sites become saturated with CO_2.

Moreover, CO_2 adsorption shows strong dependence on the amine content as the interaction and reaction of amine groups with CO_2 promotes the adsorption [30].

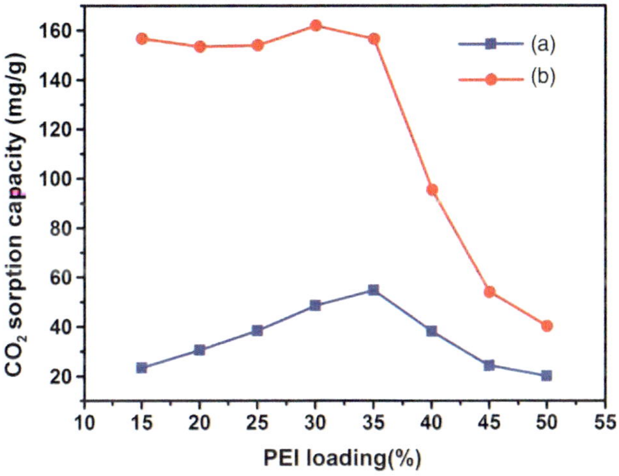

Figure 3.14 CO_2 adsorption performance with different PEI loadings in dry air at 25°C [30].

Chai et al. demonstrated CO_2 adsorption capacity at different PEI loadings (Fig. 3.14A) as the adsorption of CO_2 per gram adsorbent and PEI (Fig. 3.14B) as the adsorption of CO_2 per gram PEI. At low PEI content, CO_2 adsorption per gram adsorbent increased linearly with increasing PEI loading. Once polyethyleneimine (PEI) loading reached about 35%, adsorption started decreasing rapidly. This may be because maximum adsorption capacity was reached at this level as HNT lumen was almost completely filled with PEI inhibiting the access of the carbon dioxide molecules to the inner active sites of PEI.

Mesoporous Silica nanotubes (MSiNTs) derived from HNTs also showed similar behavior. CO_2 adsorption increased with increased PEI loadings up to 50%. However, once the PEI loading was increased from 50% to 60%, CO_2 sorption decreased substantially. It should be noted that the maximum CO_2 adsorption can be achieved at higher PEI loadings for MSiNTs compared to that of HNTs, probably due to the higher surface area and total pore volume. Moreover, when the CO_2 capture performances are compared between PEI-impregnated MsiNTs [31] and PEI-impregnated HNTs [29], CO_2 uptake capacity of PEI-MSiNTs is about 10 times larger than NH_2-HNTs at ambient temperature (Figs. 3.15B and 3.18). Given that CO_2 concentration in the feed has a strong effect on the adsorption, this comparison is not comprehensive, but still indicates that mesoporous silica (PEI-MSiNTs) derived from HNTs show promising CO_2 adsorption performance.

The structure and length of the grafted amine groups also have an effect on the CO_2 adsorption. Sankar et al. prepared HNTs functionalized with primary and secondary amines with different lengths [32] (Fig. 3.16). Secondary amine (D-HNTs and T-HNTs) functionalized HNTs possessed higher CO_2 adsorption, almost twice as primary amine-functionalized HNTs. As the amine length gets longer, an increase in adsorption was observed [32]. In addition, time dependence of CO_2 adsorption of functionalized HNTs

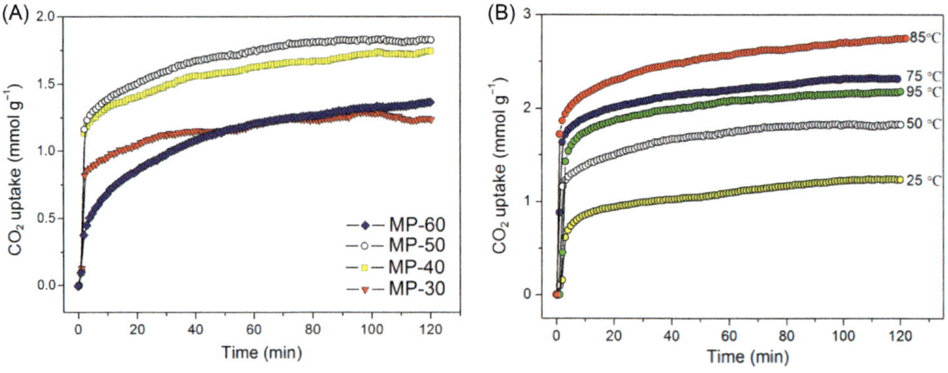

Figure 3.15 CO_2 adsorption of MSiNTs (A) with different PEI loadings at 50°C (B) at different temperatures [31].

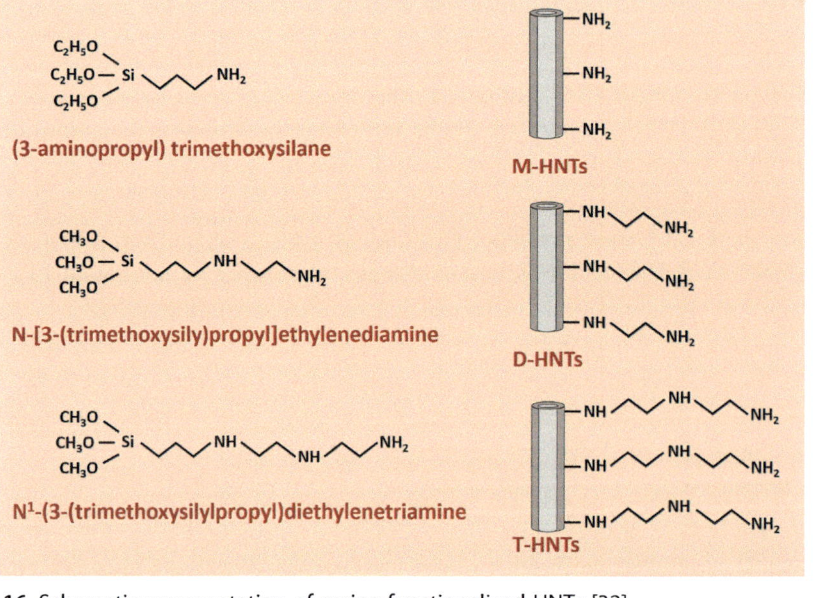

Figure 3.16 Schematic representation of amine-functionalized HNTs [32].

followed a nonlinear trend. Thus, for instance, rapid CO_2 uptake for amine-grafted HNTs was demonstrated in the first 2 hours at ambient temperatures (Fig. 3.17). After 2 hours, CO_2 adsorption did not change significantly and finally leveled off [15]. Similar time-dependent CO_2 adsorption behavior was also observed in PEI modified HNTs and Mesoporous Silica Nanotube prepared from HNTs [29–31].

Temperature is another factor affecting CO_2 uptake. Kim et al. demonstrated CO_2 adsorption performance of amine-grafted HNTs (APTES-HNTs) and PEI-impregnated HNTs (PEI-HNTs) at different temperatures (Fig. 3.18) [29]. CO_2 adsorption capacity of

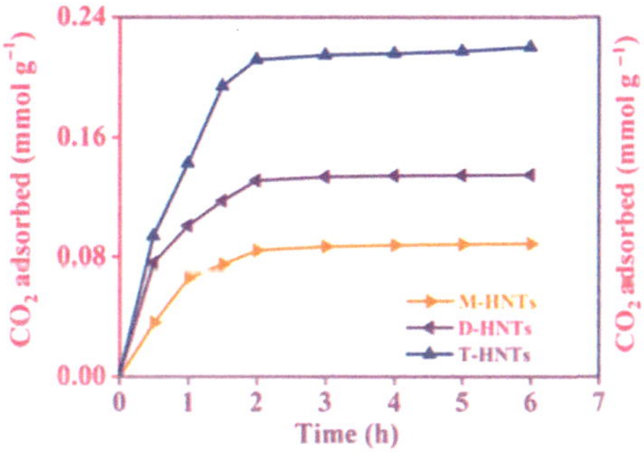

Figure 3.17 CO_2 uptake of amine-functionalized HNTs at ambient temperatures [32].

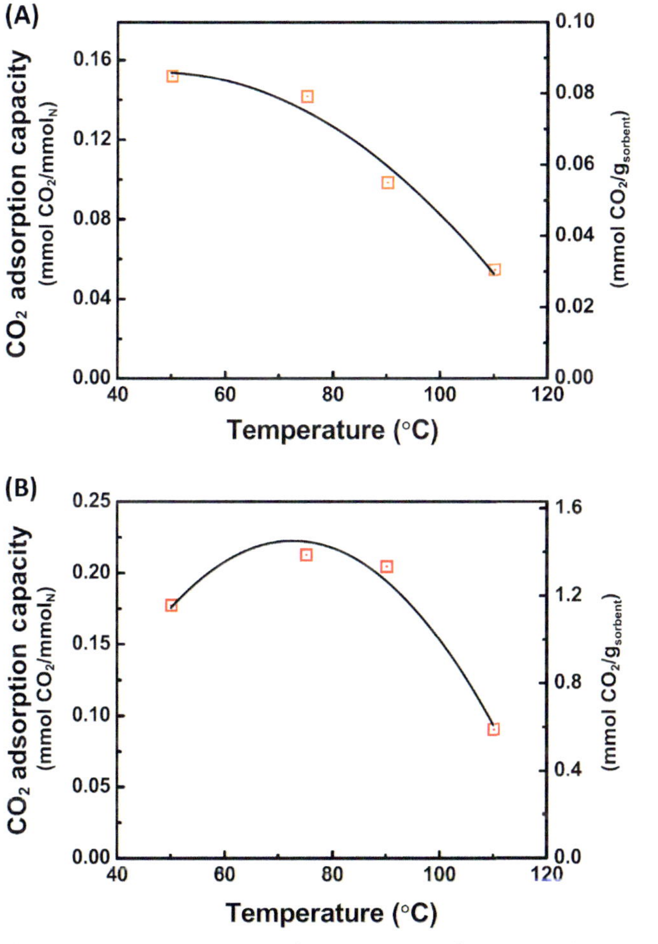

Figure 3.18 CO_2 adsorption versus temperature for (A) amine-grafted HNTs (B) PEI modified HNTs [29].

amine-functionalized HNTs decreased gradually with increasing temperatures whereas the CO_2 adsorption of PEI-impregnated HNTs increased and reached maximum uptake at 75°C, which proceeded by a decrease with increasing temperatures.

Desorption performance is as essential as adsorption capacity in CO_2 capture as it represents regeneration process performance. In most of the studies N_2 gas, reduced pressures and elevated temperatures are used to promote desorption. Desorption performance of amine-grafted and PEI-impregnated HNTs were studied at different temperatures in the literature [15,29,30]. It was shown that rapid increase of adsorption and desorption was followed by slower increase with time for both adsorbents. Moreover, CO_2 sorption of amine-grafted HNTs occurred very rapidly: 90% of adsorption and desorption was achieved in 1.2 and 1.7 min, respectively, whereas same amount of CO_2 sorption was achieved much slowly when PEI-HNTs are used as adsorbent. This was attributed to the larger amine content impregnated to PEI-HNTs [29].

Multiple cycles of adsorption/desorption was investigated to demonstrate reusability of adsorbent. Amine-functionalized adsorbents were recycled few times at various temperatures. It was shown that the adsorption performance of amine-grafted HNTs and PEI-impregnated HNTs only slightly degraded with extended cycles (up to 10 cycles) indicating easy regeneration and reusability of the solid adsorbent HNTs (Fig. 3.19) [15,29]. Moreover, elevated temperatures did not have significant impact on the regeneration process. Similarly, PEI-impregnated mesoporous silica did not show significant degradation in regeneration process up to 10 cycles.

3.3 Functionalized nanofibrillated cellulose

3.3.1 Classification and characterization of nanocellulose

There has been a great interest in nanofibrillated cellulose (NFC) due to its biodegradability, lox toxicity and its formation from renewable resources. NFC is mainly derived from cellulose, which is the main constituent of plant cell walls. Cellulose is the most abundant, biodegradable and renewable biopolymer [33]. It consists of a linear homopolysaccharide that forms highly ordered crystalline structure due to ability of glucose monomers to establish hydrogen bonds (Fig. 3.20). In most cases, cellulose does not occur in pure form and exists in form of fibers embedded in polymer matrices such as lignin and hemicelluloses [35]. Cellulose is used in potential applications such as paper production [36], nanocomposites, pharmaceuticals [34], membranes [37] and biomedical engineering [38].

Nanocellulose is defined as the cellulosic material having at least one dimension such as fibril particle diameters or width between 1–100 nm. Different levels of

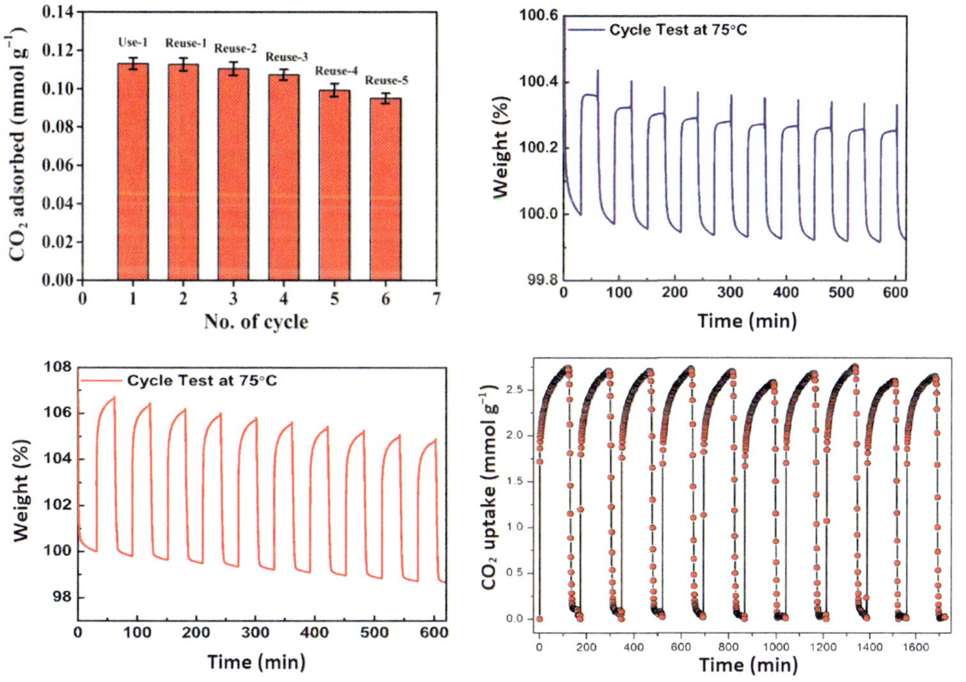

Figure 3.19 CO_2 capture during multicycling of amine-grafted HNTs at ambient temperature with 5 cycles (Top left) [10], amine-grafted HNTs at 75°C with 10 cycles (Top right) [24], PEI-HNTs at 75°C with 10 cycles (Bottom left) [24], PEI-impregnated mesoporous silica with 50% PEI loading at 85°C [31].

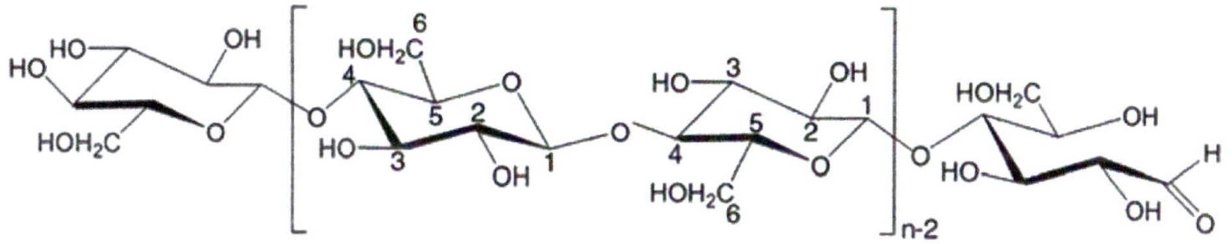

Figure 3.20 Molecular structure of cellulose [34].

wood-based cellulose are shown in Fig. 3.21. Among these subcategories, two main families have been identified as cellulose nanocrystals (CNC) and NFC.

NFCs have gained attention due to its promising mechanical properties such as high young modulus [40], high toughness and tensile strength [41], flexibility, renewability and biodegradability. In addition, due to availability of hydroxyl groups in cellulose, properties can easily be tuned via chemical modification (Fig. 3.22). NFC consists of stretched cellulose chain molecules with long, flexible and entangled cellulose nanofibers approximately 1–100 nm size, made of crystalline and amorphous regions. NFC can be manufactured from various sources. Wood is the main source

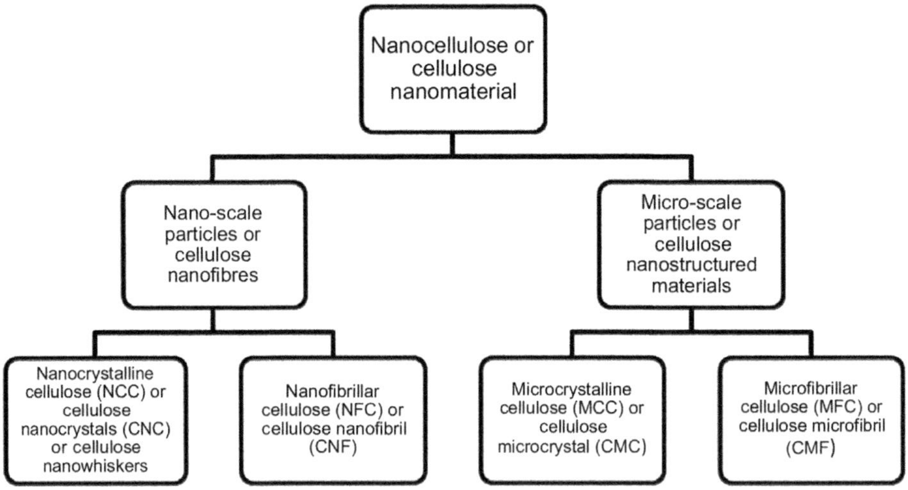

Figure 3.21 Substructures of wood-based cellulose [39].

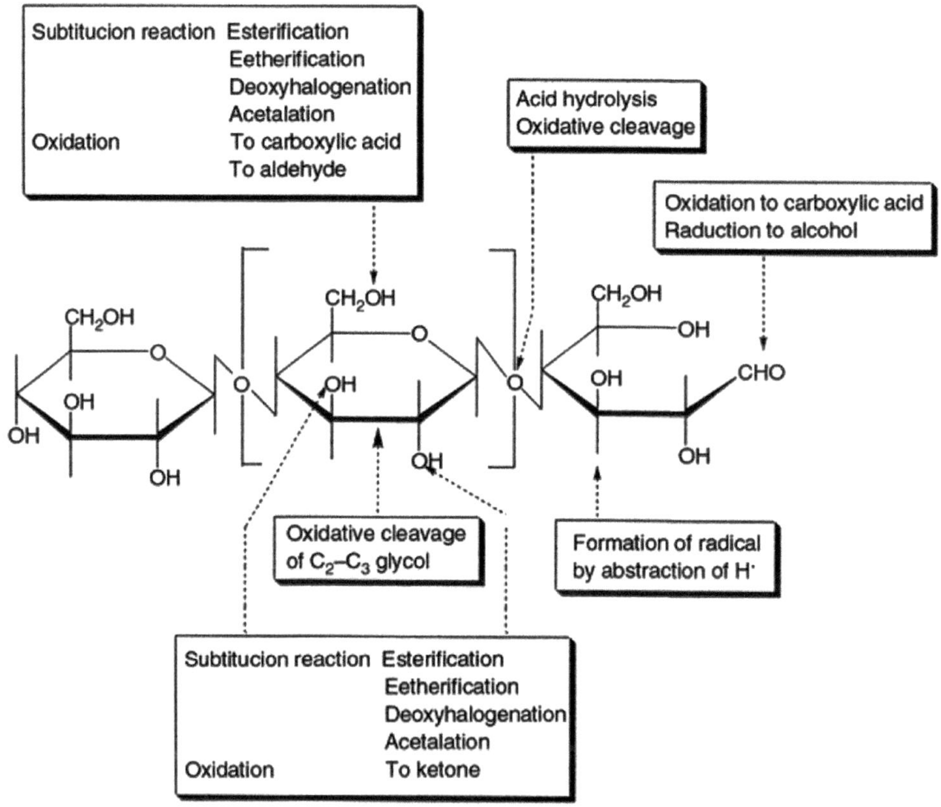

Figure 3.22 Schematic representation of cellulose structure for chemical modifications [34].

used to derive NFCs. Other materials include agricultural residues, plants and grass. It can be extracted from hemp, abaca, sisal, wheat straw, palm trees, sugar beet pulp, and carrots [42]. Compared to wood, nonwoody cellulose resources offer many advantages including short growth time and low energy consumption for production of nanocellulose. In addition, chemical treatment of the wood in order to remove noncellulose content such as lignin is essential for the manufacturing of NFCs whereas nonwoody materials generally have lower lignin composition [43]. On the other hand, nonwoody resources suffer from higher labor cost [44].

The morphology of NFC depends on the cellulose source demonstrated by FE-SEM images [45]. Fig. 3.23 shows mainly two different structures: NFCs originated from Abaca and Sisal shows thin and long fibres in widths about 20 nm and length between 200 nm to 1 um whereas Flax and Hemp fibres form thicker fibrils in widths around 30–50 nm. All of the fibrils formed highly entangled networks. In addition to the cellulose source, chemical and mechanical treatment conditions for nanofibrillation are an important determining factor for the morphology of NFCs.

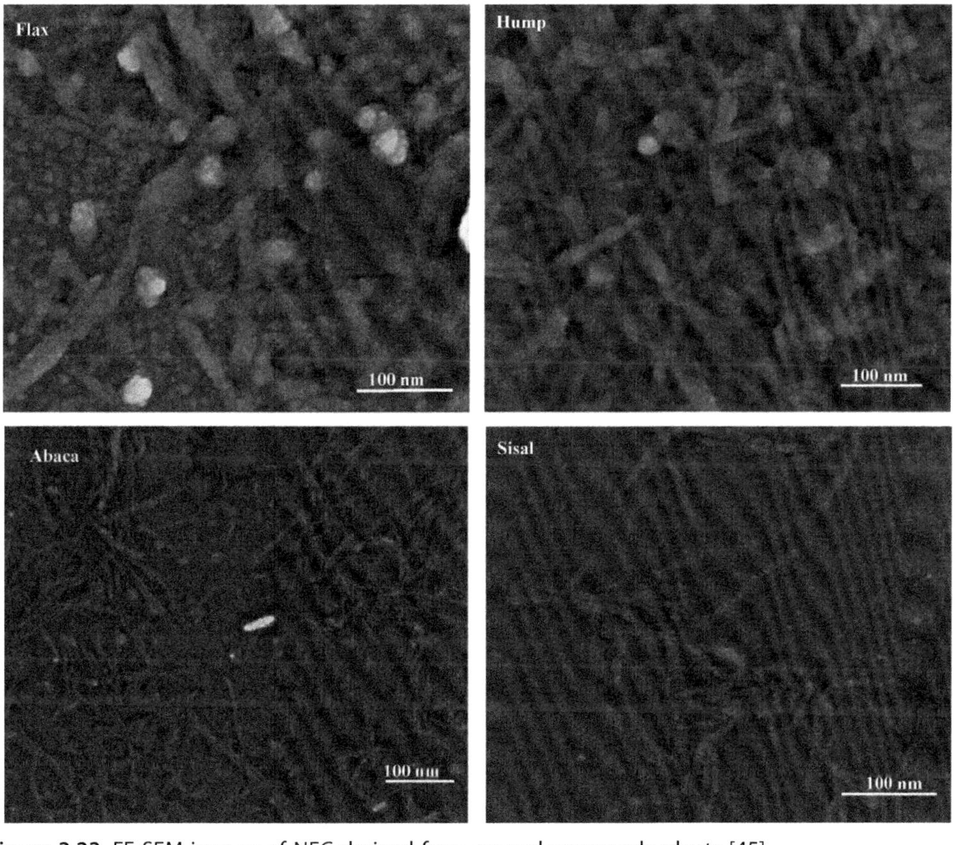

Figure 3.23 FE-SEM images of NFC derived from several nonwoody plants [45].

3.3.2 Mechanical processing of nanofibrillated celluloses

NFCs are generally formed by stratification of wood or other cellulose fibres using mechanical pressure in combination with chemical or enzymatic treatment. While several methods were reported in the literature, high-pressure homogenization, microfluidization, and grinding remain as the most widely used techniques that will be discussed in this chapter.

3.3.2.1 High-pressure homogenization

High-pressure homogenization is the most widely used method for both small and large scale NFC production due to its high efficiency, simplicity and lower cost compared to other alternatives. It involves passing the cellulose fiber-water suspension through a very narrow channel under high pressure. The delamination of fibres is advanced by extremely high shearing forces, high velocity, rapid pressure drop and frictional forces [44,46,47]. The degree of cellulose fibrillation depends on the applied pressure and extent of homogenization cycles [48]. NFCs produced by high-pressure homogenization have broad fibril diameter distribution, fibril length and high surface area [39].

There are two main disadvantages of this method. First, due to small nozzle pore size clogging is a problem [42]. Thus, it is essential to decrease the fiber size through various pretreatments before passing it through the system. In addition, the process is energy extensive and therefore scaling up the homogenization process is challenging [39].

3.3.2.2 Microfluidization

Microfluidization is a method used for production of micro and nanoscale size materials. It is commonly used in pharmaceutical industry to make liposomal products, emulsion [49] and in food industry to produce dairy products. Microfluidizers are also widely used to produce NFCs. A microfluidizer uses a pump to create high pressure to disintegrate fibers using shear forces. Fiber suspension is fed into the inlet and then forced through a Y-type or Z-type narrow channel under high pressure. This results in acceleration of the suspension that creates high shear rate and eventually breaks up the fibers. It was reported that increasing the pressure and number of cycles through the microfludizer or decreasing the size of the chamber increases the degree of fibrillation [39].

One main advantage of microfluidization over high-pressure homogenization is that it is less prone to clogging as it functions at a constant shear rate [50]. In addition, this method provides better reproducibility due to fixed geometry [44].

3.3.2.3 Grinding

Grinding is another commonly used technique to disintegrate the cellulose into NFCs. In grinding process, dilute cellulose suspension is passed between two stones: One static and other is rotating. Shearing force from the counter-rotating grinding stones and contact of the fibres with the hard surfaces of stones repeatedly result in fibrillation of cellulosic fibers. The degree of fibrillation depends on the structure of the stone channels and number of cycles through the grinder [51]. Main advantage of this technique is no requirement for pretreatments whereas this method potentially suffers from wearing of grinders that requires replacement and maintenance [44].

3.3.2.4 Pretreatment of fibers

There are two main problems of NFC production process: Clogging and high-energy requirement to disintegrate the cellulosic fibers. Pretreatment methods have been developed in order to alleviate these problems. Enzymatic and Chemical pretreatments remain to be most widely used techniques to enable environmentally friendly and energy efficient NFC manufacturing processes.

Enzymatic pretreatments are used to modify and degrade the cellulose by disabling the interactions between macro-fibrils [52,53]. This method enables production of NFCs with high aspect ratio, small diameter size with narrow distribution and with significant reduction in energy requirement and minimal clogging problem. Zhu et al demonstrated the degradation of wood pulp cellulose into NFC by enzymatic hydrolysis using cellulase enzyme. They achieved high yields in the range of 60–90% by adjusting hydrolysis duration. It was reported that Cellulase enzyme facilitated significant decrease in degree of polymerization of cellulose chain and reduction in the fiber length [54]. Janardhnan et al. reported high aspect ratio and small diameter size of cellulose nanofibers were achieved via Enzymatic treatment [55].

Chemical pretreatments are also widely used to improve fibrillation efficiency. The more commonly used chemical treatment strategy is 2, 2, 6, 6-tetramethylpiperidine-1-oxyl (TEMPO)-mediated oxidation. It facilitates disintegration via increasing the charge density of the cellulosic fibers under relatively mild conditions [51]. TEMPO-mediated oxidation pretreatment remarkably lessens the energy needed for delamination process [44] while improving the fibrillation efficiency. TEMPO-mediated oxidation of cellulose is shown in Fig. 3.24. The C6 hydroxyl groups of cellulose are selectively oxidized in the presence of TEMPO radical, NaBr, and NaClO at pH 9–11 [57]. Isogai et al. developed new type of environmentally friendly TEMPO-method that works at about pH 7 to prepare nanocellulose nanofibers of 3–4 nm wide [56].

3.3.3 Chemical modification and characterization of nanofibrillated celluloses for CO_2 capture

In addition to aforementioned applications above, NFCs have recently gained much attention as adsorbents for CO_2 capture due to its desirable properties such that they are low-toxic and environmentally friendly materials derived from biomaterials with high surface area, good mechanical strength and regenerability. Thanks to its reactive

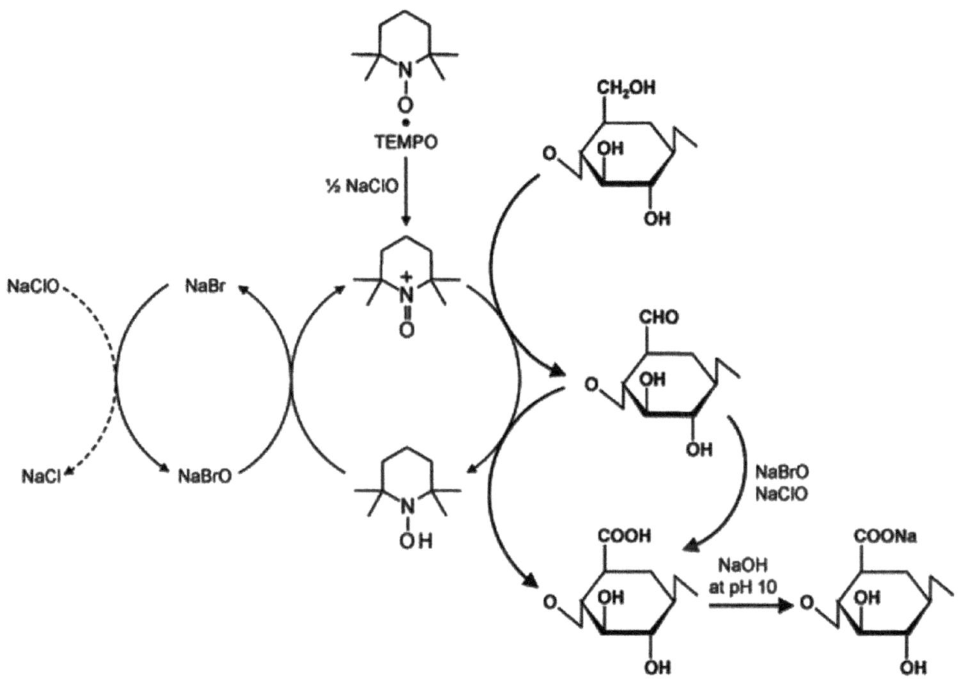

Figure 3.24 Schematic representation of regioselective oxidation of cellulose via TEMPO process [56].

hydroxyl groups, it can be chemically modified to achieve desired surface properties for CO_2 capture. PEI-impregnated and amine-functionalized NFCs were studied and tested for Direct air capture [58–60].

Gebald et al. studied CO_2 capture efficiency of amine-functionalized NFCs [58,59]. N-(2-aminoethyl)-3-aminopropylmethyldimethoxysilane was used to introduce amine functional groups to the NFC in order to promote CO_2 adsorption. SEM images showed a reduction in pore size, thus transformation from microporous to more like nanoporous structure after freeze-drying process (Fig. 3.25). It is notable that freeze-drying process is known to produce highly porous and flexible NFC sheets [61]. As ice crystals are formed during freezing process, the fibrils are squeezed around the crystals resulting in formation of cellulose sheets. The ice crystals are then removed by sublimation step and porous structure is created around thin cellulose sheets [60].

A decrease in surface area was seen postamine modification while retaining its nanofibrillar and hierarchically porous structure. It was also reported BET surface area decreased from 26.8 m^2/g to 7.1 m^2/g whereas density increased from 26 kg/m^3 to 61 kg/m^3 for freeze-dried samples postamine functionalization.

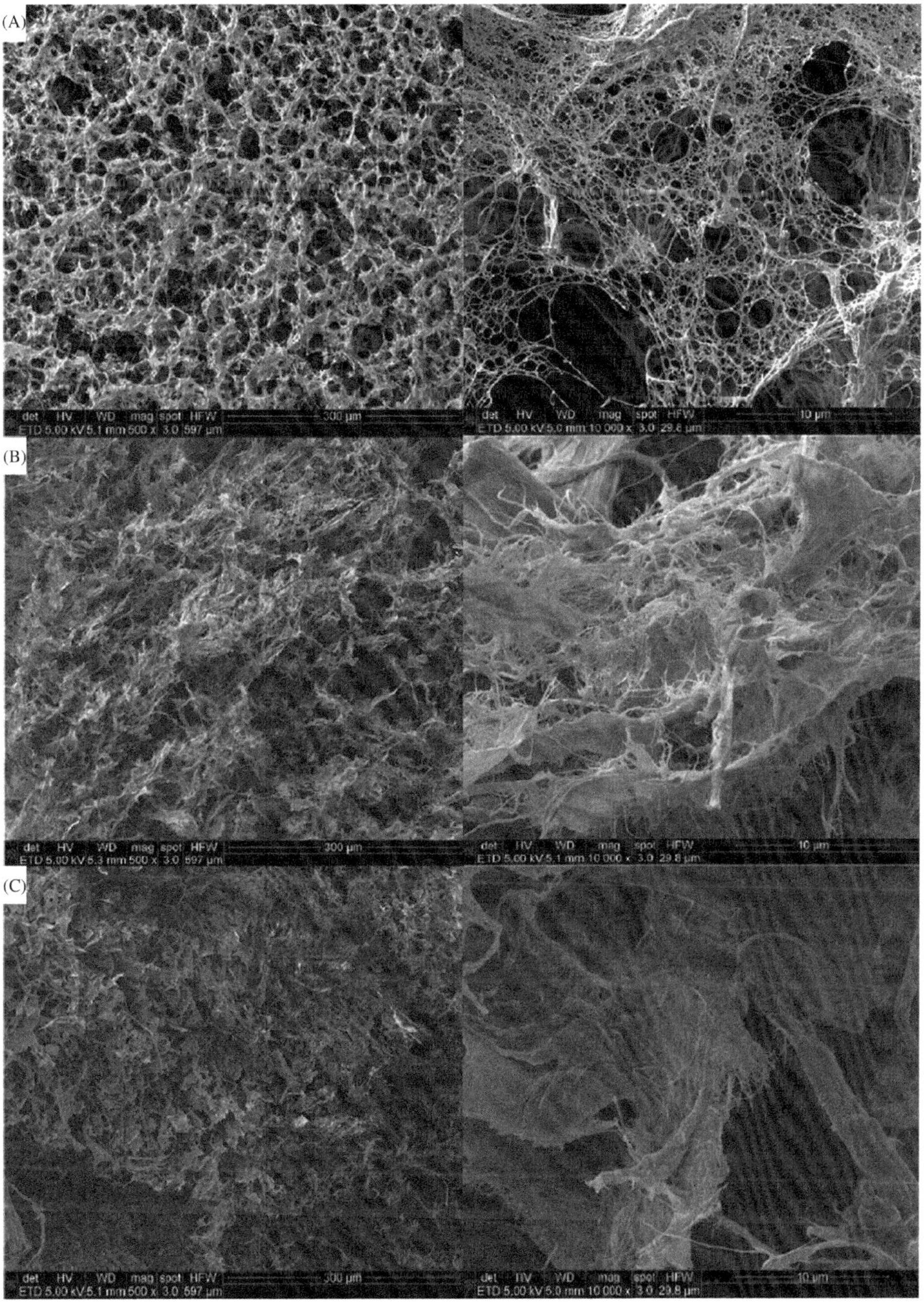

Figure 3.25 SEM images of (A) bare NFCs (B) freeze-dried bare NFC (C) freeze-dried amine-functionalized NFC [58].

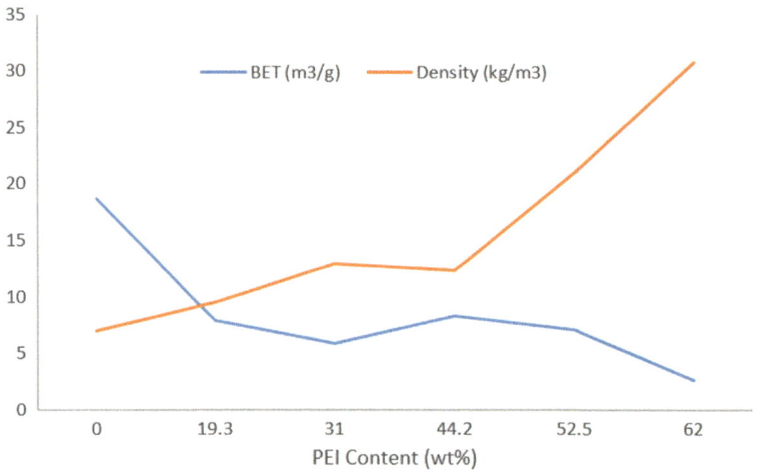

Figure 3.26 Surface area and density of NFC/PEI nanocomposites versus PEI content. *Illustrated from H. Sehaqui, et al. Fast and reversible direct CO_2 capture from air onto all-polymer nanofibrillated cellulose–polyethylenimine foams. Environ. Sci. Technol. 49 (2015) 3167–3174.*

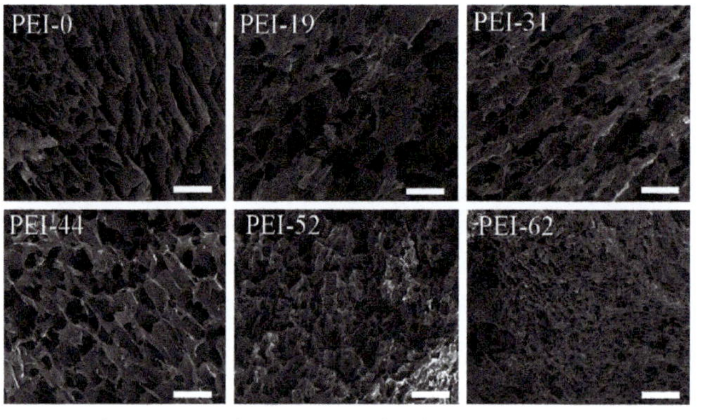

Figure 3.27 SEM images of NFC/PEI with various PEI loading [60].

Sehaqui et al. prepared fast and reversible CO_2 adsorbents from polymeric NFC/PEI foams with various PEI compositions at pH 11. These nanomaterials showed high porosity above 97% regardless of PEI content. The surface area of the NFC/PEI sorbents decreased from 18.7 to 2.7 m^2/g whereas density increased gradually with increased PEI content similar to the previously mentioned work by Gebald et al (Fig. 3.26) [60].

SEM images demonstrated all bare NFC and PEI-impregnated NFCs showed sheet like structures with high porosity. As PEI loading increases, pore diameter seems to be reduced (Fig. 3.27).

Thermal stability of adsorbents is also crucial especially for desorption and regeneration process as it generally occurs at elevated temperatures. It was shown that NFC

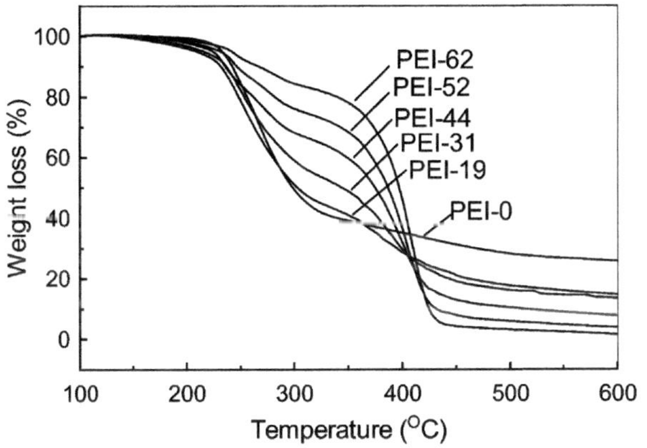

Figure 3.28 TGA thermograms of bare NFC and PEI-impregnated NFCs [60].

backbone starts degrading at about 200°C. However, PEI loading shifted both NFC and PEI degradation to higher temperatures (Fig. 3.28).

3.3.4 CO_2 adsorption and desorption studies

It was shown pure NFC did not exhibit any remarkable CO_2 adsorption [60]. Therefore, surface modification of NFCs is needed to promote adsorption. Amine-grafted NFCs showed CO_2 adsorption around 1.4 mmol/g and adsorption half time of 100 min for maximum adsorption at 25°C and 40% RH [58]. PEI-impregnated NFCs showed a linear increase in CO_2 adsorption capacity up and reached maximum adsorption amount at 44% PEI loading. PEI-44 showed a high adsorption capacity around 2.2 mmol/g at 25°C and 80% RH. When CO_2 adsorption half-time of various sorbents are compared between NFC and nonNFC based adsorbents, both amine-grafted and PEI-impregnated NFCs show promising adsorption efficiency (Fig. 3.29). At higher PEI loadings, reduction in adsorption capacity was observed. This was attributed to the agglomeration of components that reduced the accessibility of amine groups to the CO_2. In addition, increased PEI content may potentially block the pores, thus resulting in less permeability of air molecules. Therefore, both amine content and availability of the amine moieties on the surface play important role in CO_2 adsorption. Adsorption half time also greatly affected by increased PEI content after 44%. The half time was significantly low in the range of 5–11 min for PEI loadings up to 44% and then rapidly increased up to 50 min with further increased PEI loadings [60].

Furthermore, CO_2 adsorption capacity of NFC-based adsorbents reveals strong dependency on relative humidity. For instance, the adsorption of PEI/NFC nanosorbents was relatively low at 20% RH and it improved remarkably with increasing RH achieving the maximum adsorption capacity at 80% RH (Fig. 3.30). This is favorable given that Direct air capture is common at 40%–80% RH range.

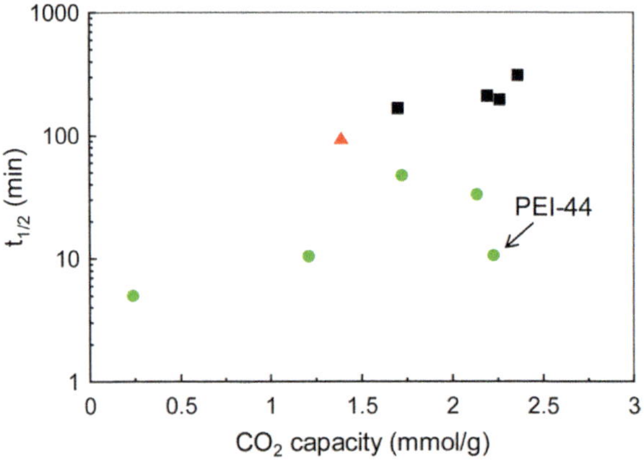

Figure 3.29 Adsorption half time for NFC/PEI sorbents (green circles), amine-grafted NFCs (red triangle) [58] and other various PEI/Silica sorbents (Black squares) [62].

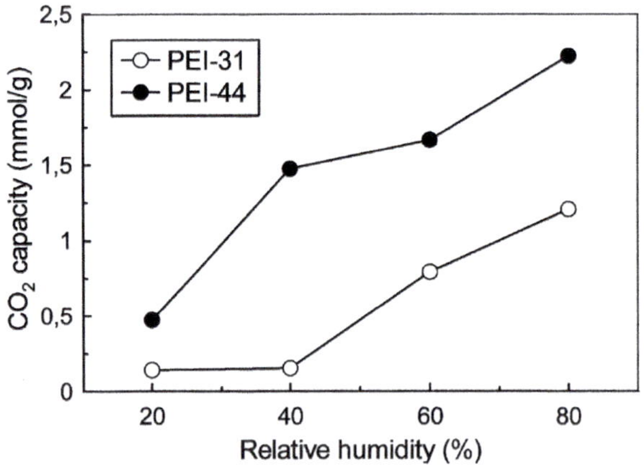

Figure 3.30 CO_2 adsorption capacity of PEI-31 and PEI-44 sorbents versus relative humidity [60].

Gebald et al. studied the stability of amine-functionalized NFCs over 100 temperature-vacuum swing (TSV) cycles (Fig. 3.31). Studies were conducted at desorption temperature of 90°C in the presence of O_2, H_2O, and CO_2 as these gases are the components of ambient air that can potentially affect degradation during Direct Air capture whereas N_2 gas is known to be inert. Prolonged adsorption/desorption cycles caused only up to 5% decrease in CO_2 adsorption capacity after 100 cycles showing high stability and reusability of the NFC adsorbent. More interestingly, prolonged TSV cycles did not cause any degradation in the morphology of the adsorbent demonstrated by SEM images (Fig. 3.32). Amine-functionalized NFCs retained their highly porous nanofibrillar sheet structure [59].

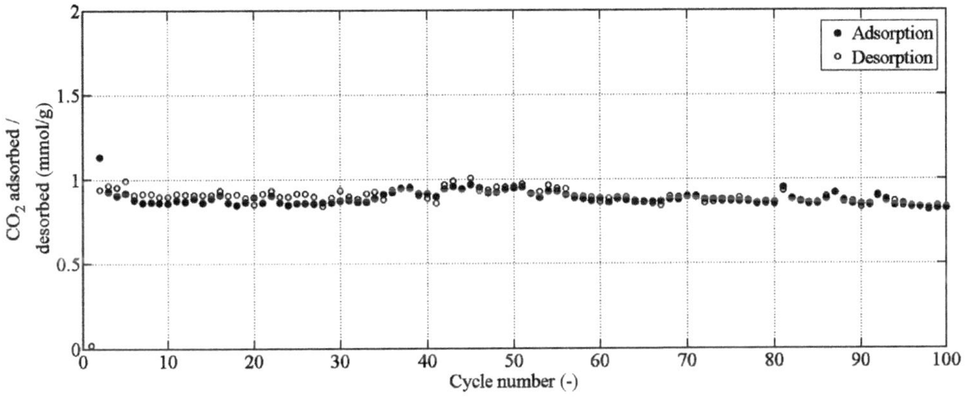

Figure 3.31 CO_2 adsorption/desorption behavior of amine-functionalized NFC over 100 consecutive TSV cycles [59].

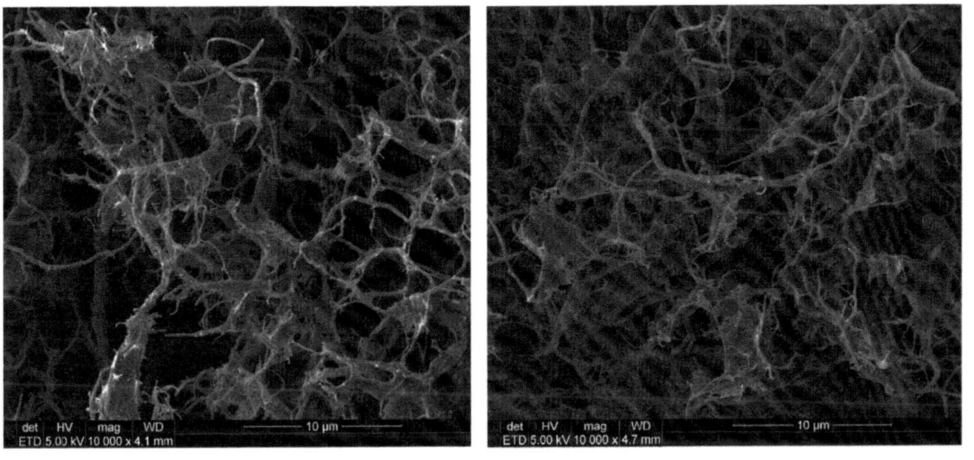

Figure 3.32 SEM of pristine amine-functionalized NFC (left), amine-functionalized NFC after 100 TVS cycles (right) [59].

3.4 Enzyme immobilized on bioinspired nanosorbents

3.4.1 Application of carbonic anhydrase in CO_2 sequestration

Mineralization-based methods have been widely used for CO_2 sequestration. Although mineral carbonation is a safe and viable process for CO_2 sequestration, this process is extremely slow under room temperature and pressure. Enzymatic conversion of CO_2 into mineral carbonates is an economically feasible and environmentally friendly approach for CO_2 sequestration.

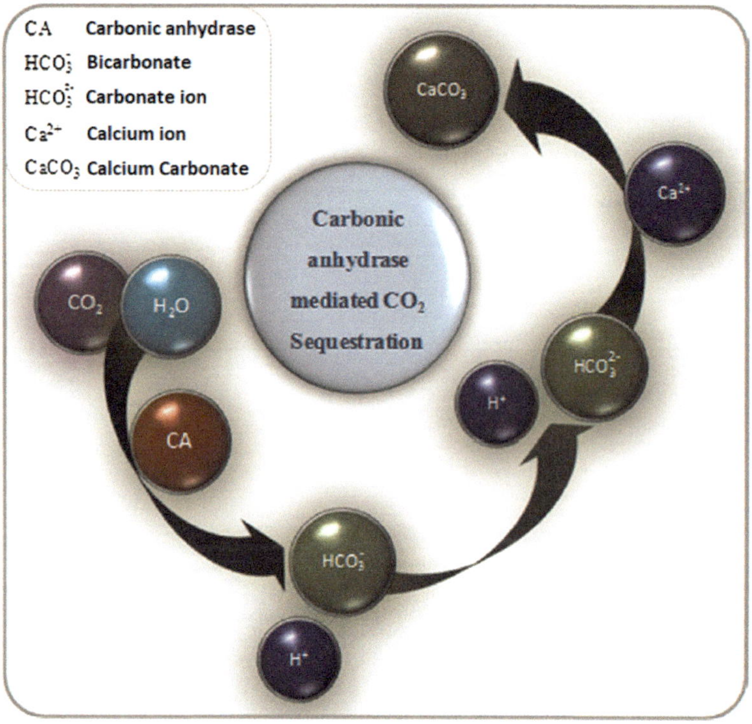

Figure 3.33 CA catalyzed CO_2 sequestration [63].

CA enzyme has gained much attention in this regard as it provides safe, stable and eco-friendly process. It is a zinc-containing metalloenzyme that is a key substance for all living forms. It facilitates stabilization of CO_2 present in the environment as well as in animal and plant cells. It mainly catalyzes hydration of carbon dioxide to bicarbonate and hydrogen ion in aqueous media as shown below for subsequent production of calcium carbonate via the reaction with calcium ion (Fig. 3.33). The final product $CaCO_3$ has wide range of applications including paints [64], drug delivery [65], personal care and biomedical applications [66], and polymer composites [67]. The typical catalytic rates are ranging from 10^4 to $10^6\ s^{-1}$ [68]. By utilizing CA enzyme-catalyzed CO_2 hydration approach (3.1), a fast and efficient CO_2 sequestration system can be facilitated to reduce CO_2 emissions.

$$CO_{2(aq)} + H_2O \leftrightarrows HCO_3^- + H^+ \quad (3.1)$$

3.4.2 Enzyme immobilization on bioinspired silica

Despite of crucial role of CA in sequestration process aforementioned above, CA suffers from low stability at high temperatures, poor reusability and recovery from the reaction due to high solubility that limits its use for CO_2 sequestration. Moreover, high cost of CA restricts its usage in pure form [69]. Therefore, enzyme immobilization onto a support or encapsulation techniques are quite important to enable CA mediated carbon sequestration to promote reusability, robustness and stability of CA.

A number of supports for have been studied so far including silica [69], hydrogels [70], chitosan [71], and polyurethane foam [72]. However, immobilization onto these supports involves techniques that suffers from long preparation times, complexity, and use of harsh conditions and more importantly reduction in enzyme activity.

Bioinspired silicification has been shown as a green method for the efficient enzyme immobilization and encapsulation under mild conditions [73,74]. Biosilicification is originally mediated by living organisms including plants, diatoms, and sponges in vivo in the presence of proteins, peptides and enzymes and biosilica is formed as a result of the process [75]. Biosilica is a nanostructured material that demonstrates highly hierarchical structure. It contains micro and nano pores as shown in Fig. 3.34 [76]. Unlike manmade silica production that requires harsh conditions such as high temperature, high pressure, strong acidic or alkaline conditions with long preparation times; Biosilica formation is an environmentally friendly process that occurs at mild pH, ambient temperatures [78] and in aqueous medium.

Inspired by the biosilicification process occurs in vivo, bio-inspired methods in vitro were developed to immobilize enzymes by the synthesis of bioinspired silica through sol-gel silica polycondensation and precipitation under mild conditions. Sol-gel polycondensation of the silicon derivatives were explained elsewhere [79]. Bioinspired silica materials have applications in several areas including biocatalysts [80], protein, and drug delivery [81,82].

In addition to these applications, bioinspired silica nanocomposite with immobilized CA enzyme has gained considerable attention as a next generation green nanomaterial for CCS though biosilicification process. As mentioned before, CA is known as a promising catalyst for CO_2 hydration and conversion into value added substances such as $CaCO_3$. As the enzyme is encapsulated within bioinspired silica, outstanding immobilization efficiency and better enzyme stability are achieved compared to other immobilization methods.

SEM images of enzyme encapsulated bioinspired silica revealed spherical structures having diameter size ranging from 300 to 400 nm (Fig. 3.35). The surface of the spheres was remarkably rough shown by TEM images. Rough surface is desirable in catalytic processes as it implies larger surface area than the smooth surfaces. It was reported that bioinspired silica with encapsulated CA synthesized from Neisseria gonorrhoeae has much smaller surface area ($\sim 7.7\ m^2/g$) compared with other silica materials utilized for enzyme fixation [83,84].

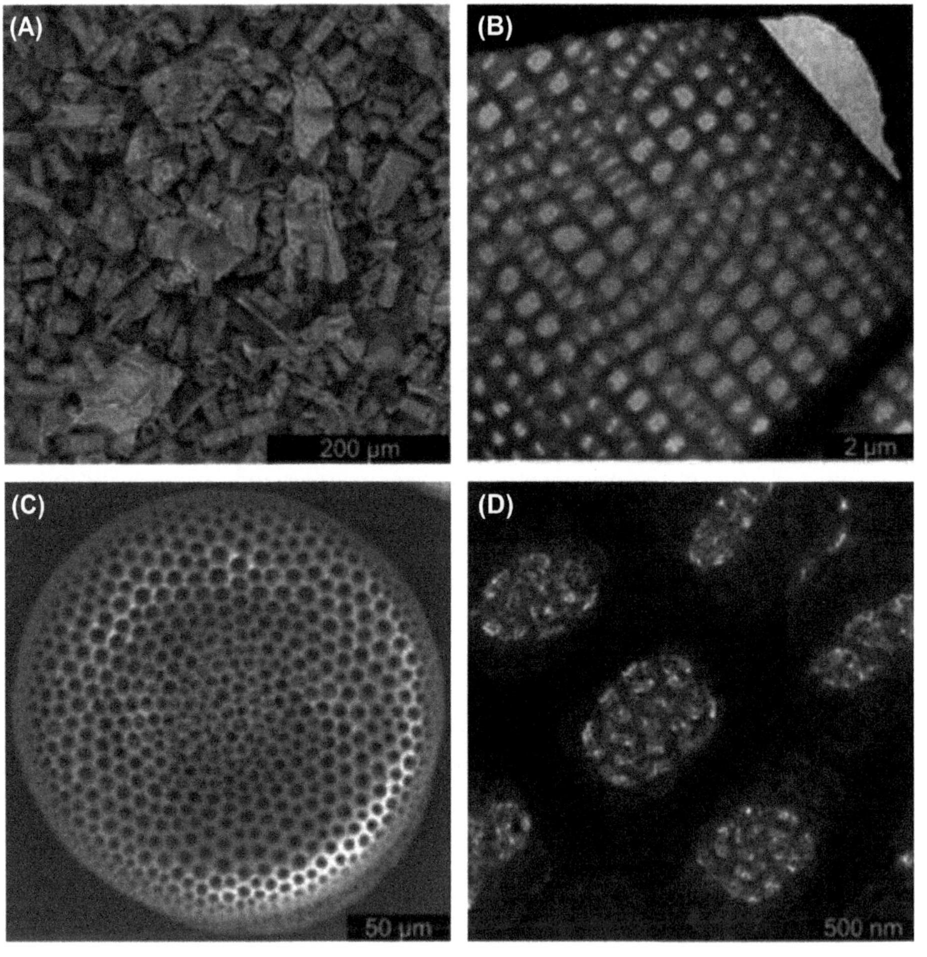

Figure 3.34 TEM images of different types of biosilica [76,77].

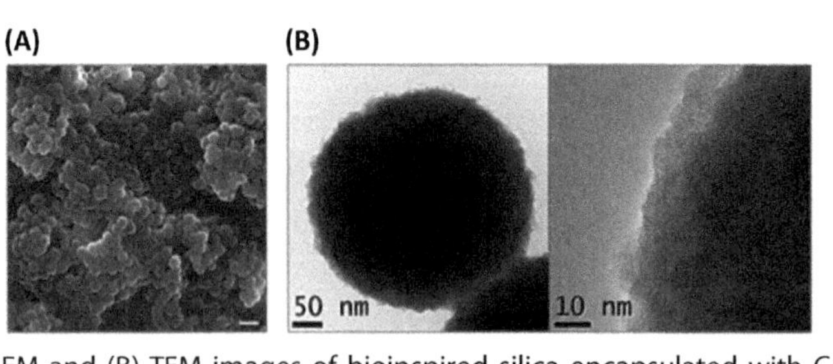

Figure 3.35 (A) SEM and (B) TEM images of bioinspired silica encapsulated with CA from Neisseria gonorrhoeae [83].

3.4.2.1 Enzyme activity, retention and immobilization efficiency

In the literature, several properties of the CA immobilized onto bioinspired silica were studied including such as thermal and pH stability, enzyme retention, enzyme activity and reusability that can potentially affect CO_2 sequestration performance.

Retention of enzyme in the nanocomposite matrix is desirable to ensure no leaching occurs from the solid support. Hence, immobilization efficiency, enzyme stability, and retention of enzyme activity are crucial factors for enzyme mediated CO_2 sequestration.

Jo et al. synthesized bioinspired silica nanocomposites with autoencapsulated CA from Neisseria gonorrhoeae [83]. Encapsulation efficiency reached about 90% using 5 mg/ml enzyme concentration. This auto-encapsulation method was found to be very effective showing almost 100% enzyme retention even after 50 hours. Relative enzyme activity of the nanomaterial to the free enzyme was reported as 56% at ambient temperature and pH 7.

Min et al. prepared highly stable CA formed from Hahella chejuensis (HCA) that was immobilized via biosilicification of amine (spermine) and monosilicate. They utilized alkaline-active CA through removal of signal peptide. It was found that signal peptide removed HCA showed almost 11 times higher activity and better solubility compared HCA. These enzymes showed high stability in alkaline conditions. Bioinspired silicification mediated by spermine occurred under ambient conditions with high enzyme immobilization efficiencies between 80% and 100%. Enzyme activity was about 58% which was comparable to covalently immobilized CA onto several other supports [85]. Enzyme encapsulated bioinspired silica showed improved enzyme activity and reusability [73].

3.4.2.2 Thermal and pH stability

As the adsorption/desorption processes are accomplished at high temperatures, enzyme activity needs to be retained at elevated temperatures. Forsyth et al. reported no significant change in enzyme activity up to 50°C for immobilized enzyme whereas the free enzyme lost majority of its activity between 40°C and 50°C (Fig. 3.36) [85]. In another study, it was shown that although enzyme activity declined at elevated temperatures, immobilized enzymes onto bioinspired silica has retained its activity even at elevated temperatures (Fig. 3.37A). For instance, the residual activity of amine-mediated Immobilized CA formed from HCA was 40% and 15% at 80°C−90°C, respectively. It worth mentioning amine-mediated enzyme encapsulation seems to improve the activity of immobilized CA especially at elevated temperatures (Figs. 3.36B, 3.37A).

Enzyme activity across a range of pH values were shown in Fig. 3.37B for the enzyme immobilized bioinspired silica. About 85% of residual activity was retained over 5−11 pH range. The activity reduced about ~10% at pH 11. On the other hand, Jo et al. reported a strong pH dependence of the enzyme activity: an increase in enzyme activity with

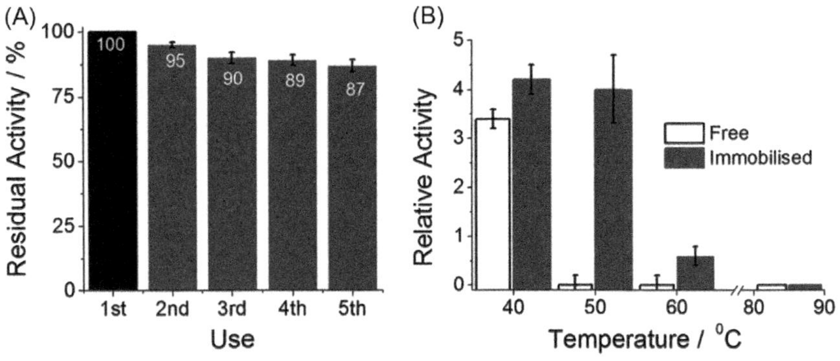

Figure 3.36 (A) Reusability (B) thermal stability relative to that at 20°C of immobilized enzyme [85].

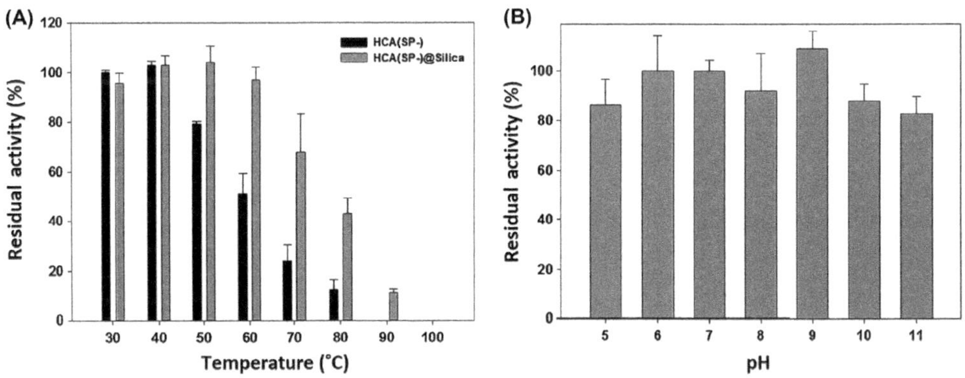

Figure 3.37 Immobilized CA from Hahella chejuensis enzyme activity versus temperature and pH post silica encapsulation (black-free enzyme, gray-bioinspired silica immobilized enzyme) [73].

increasing pH values for the bioinspired silica with autoencapsulated CA. Thus, chemical nature of immobilization affects pH behavior of the nanocomposites [83].

3.4.2.3 Reusability of enzyme immobilized bioinspired silica

Materials used for CCS needs to have high reusability to ensure an efficient and sustainable CCS system. Forsyth et al. reported almost 90% of enzyme activity was retained after five cycles after enzyme immobilization onto bioinspired silica (Fig. 3.36A). In other study, enzyme-immobilized biosilica was cycled even for prolonged time (10 cycles) and only showed ~10% loss of the activity [73].

3.4.2.4 CO_2 sequestration

Carbon sequestration efficiencies of both free and immobilized CA were reported in the literature. Jo et al. examined the relative rates of $CaCO_3$ formation from CO_2.

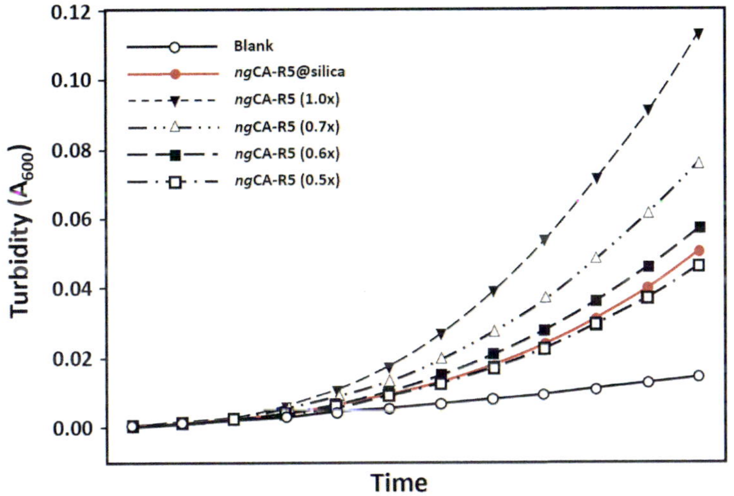

Figure 3.38 CO_2 sequestration performance determined by the relative initial rates of $CaCO_3$ precipitation for free CA and autoencapsulated CA [83].

Figure 3.39 SEM images of $CaCO_3$ precipitates (scale bar is 10 μm) [83].

When bioinspired silica with autoencapsulated CA was used, the precipitation rate of $CaCO_3$ was accelerated by about fivefold compared to noncatalyzed reaction. Compared to free enzyme, encapsulated CA had about 40% less CO_2 sequestration rate (Fig. 3.38). Overall, CA-bioinspired silica nanocomposite demonstrated promising CO_2 sequestration performance with good reusability, temperature and pH stability without affecting the morphology of the final product $CaCO_3$ (Fig. 3.39) [83]. Similarly, amine-based immobilized CA was shown to have 86% CO_2 sequestration efficiency whereas efficiency of free enzyme 90% [85].

3.4.3 Bioinspired silk protein hydrogels with encapsulated carbonic anhydrase

In addition to bioinspired silica-CA nanocomposites, silk protein hydrogels show promising CO_2 sequestration performance. Kim et al. synthesized highly stable dual

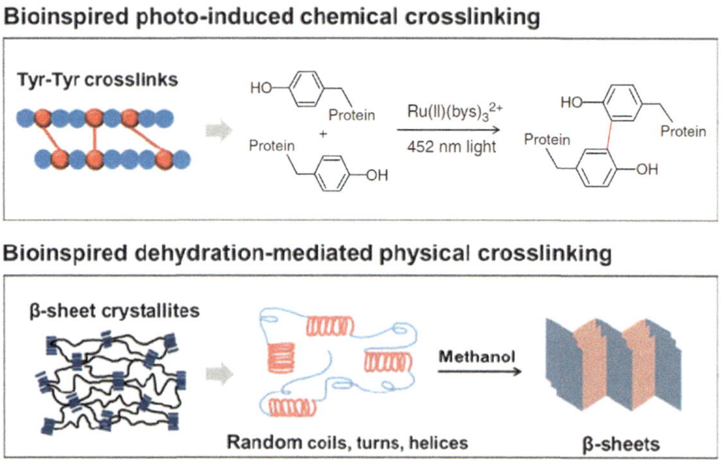

Figure 3.40 Schematic representation of the preparation of bioinspired silk protein hydrogels with encapsulated CA [86].

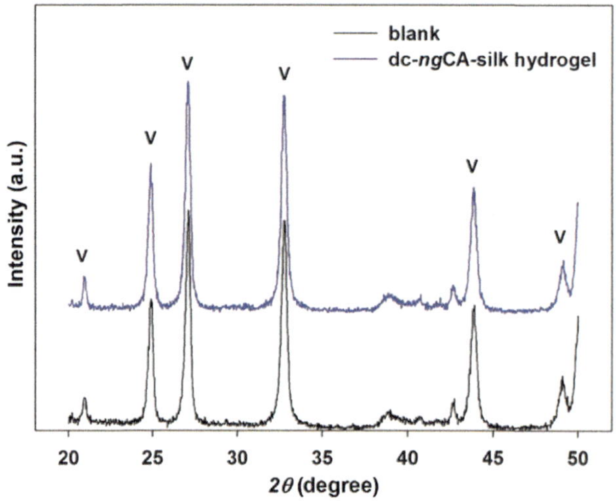

Figure 3.41 CO_2 sequestration efficiency of double crosslinked hydrogel with encapsulated CA into $CaCO_3$ [86].

chemical and physical crosslinked CA-encapsulating fibroin based protein hydrogel as shown in Fig. 3.40 [86].

Residual activity of free CA decreased rapidly only after 1 hour, whereas fibroin protein-CA hydrogel retain about 90% and 40% activity after 1 and 12 hours, respectively. Furthermore, hydrogel retained 97% enzyme activity after six cycles. Bioinspired hydrogels also demonstrated promising CO_2 sequestration performance. The CO_2 removing efficiency doubled by employing the CA-silk hydrogel compared to noncatalyzed reaction. It exhibited about 60% carbon sequestration performance of the free enzyme (Fig. 3.41).

3.5 Green metal-organic frameworks

3.5.1 Introduction to metal-organic frameworks

MOFs are considered as new type of materials that could be employed for capture and storage of various gases, including hydrogen [87], methane [88] and carbon dioxide [89] due to their large gas adsorption capacity. MOFs are prepared by combination of metal nodes and organic ligands using strong bonds to form crystalline meso- and micro- porous materials. MOFs consist of well-characterized highly-ordered porous structures with ultra-high surface area and porosity. For instance, Furukawa et al. reported MOFs with ultra-high surface area and porosity, ranging between 4530–6240 m^2/g and 2.16–3.60 cm^3/g, respectively.

MOFs can be tuned for various chemical behaviors into organized structures with a fine control over the structural arrangement (Fig. 3.42). The ease of chemical modification and control over the pore surface of MOFs allow for optimization to be utilized in applications such as drug delivery [91], fuel cells [92], gas separation [93] and storage [94]. In addition to these applications, MOFs are considered as next generation materials for the development of novel CO_2 adsorbents by taking the advantage of high capacity for gas adsorption and tunable pores resulting in highly selective CO_2 capture.

3.5.2 Thermal, chemical, and mechanical properties of metal-organic frameworks

Physical properties of MOFs play an important role for employing them as CO_2 adsorbents. It was reported that the chemical and thermal stability of Metal Organic Frameworks are usually lower than similar structures such as Zeolites.

One of the major concerns with the chemical stability of MOFs in aqueous conditions either in liquid and vapor state as many MOFs are air and moisture sensitive that can lead to the rapid degradation of the structure and loss of surface area. Especially for postcombustion CO_2 capture or direct air capture processes where high water or

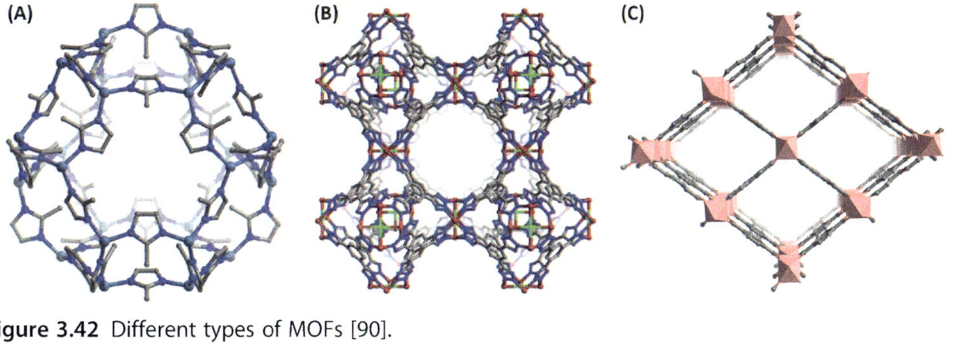

Figure 3.42 Different types of MOFs [90].

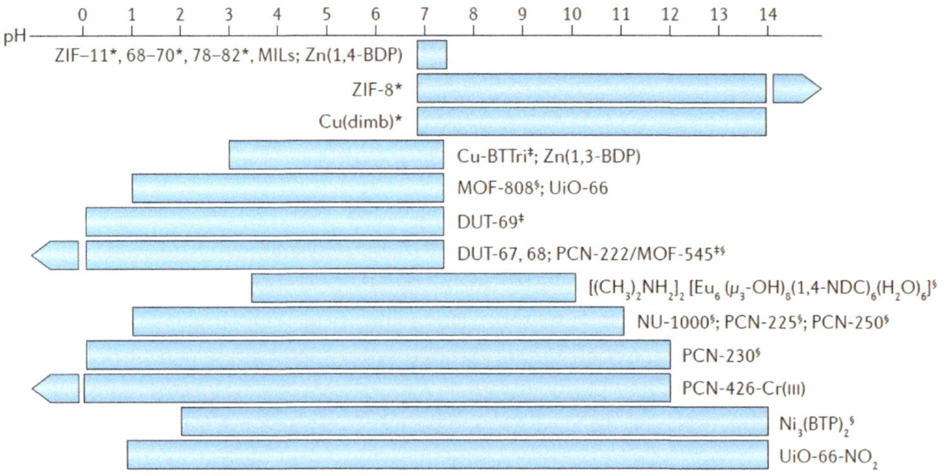

Figure 3.43 Chemical stability of different MOFs at a broad pH range [90].

air composition is present, it poses a significant problem [6]. Chemical stability of various MOFs at a broad pH range is shown in Fig. 3.43. There are many MOFs that are chemically stable at acidic, basic or neutral pH conditions.

The CO_2 adsorbents should be mechanically stable to facilitate a densely packed and highly ordered structures under high pressures and temperatures. Bundschuh et al. reported the young modulus of HKUST-1 type metal organic framework films as 10 GPa which is higher than that of most polymers [95]. However, a significant degradation of mechanical properties of MOFs was reported under large pressures. For instance, Cu_3 (1,3,5-benzenetricarboxylate)$_2$(H2O)$_3$ MOF showed a significant decrease from about 118 to 30 GPa under high pressure [96].

Thermal stability of MOFs mainly depends on number of linkers and the strength of node-linker as thermal degradation usually occurs from breaking of node-linker bonds. The other potential routes of thermal degradation for MOFs include crystalline-to-amorphous transformation, melting and dehydration of nodes [90].

There has been a great effort to synthesize MOFs with good chemical, thermal and mechanical stability. Demessence et al. reported ethylene diamine functionalized trizolate MOFs possessing high thermal stability up to 270°C and outstanding chemical stability in air, water and acidic media [97]. Wu et al. reported highly porous Zirconium MOFs with excellent mechanical stability under shear stress having shear modulus of 13.7 Gpa [98].

3.5.3 Functionalization of metal-organic frameworks for improved CO_2 capture and storage

In addition to physical properties of MOFs that has strong influence on CO_2 adsorption performance, there are other factors need to be considered such as heat capacity, enthalpy, adsorption capacity and CO_2 selectivity to tailor highly efficient MOF adsorbents.

Heat capacity of MOF is remarkably lower than that of liquid amine-based adsorbents. This is quite important given that it is associated to the energy requirement for desorption in temperature swing adsorption processes [6]. Low heat-capacity adsorbent is expected to have lower energy penalty during regeneration process.

Additionally, enthalpy of CO_2 uptake has a strong effect on the CO_2 capture performance of the material. Enthalpy describes the adsorbent selectivity and the energy needed during desorption process. If there is a strong interaction between CO_2 and the adsorbent surface, the enthalpy of CO_2 adsorption and the selectivity of CO_2 will be high, however energy penalty to break the interaction for regeneration will also be high. Likely, if the enthalpy is too low, smaller quantities of energy will be needed for regeneration but CO_2 selectivity will decline. Therefore, it is important to have control over the strength of interaction between $CO_2 - $ MOFs.

High CO_2 selectivity of MOFs over other gases such as N_2 (postcombustion CO_2 capture), CH_4 (from natural gas purification) needs to be achieved to ensure high purity of captured CO_2 from various gas mixtures. Several MOFs were reported in the literature that possess high CO_2/N_2 and CO_2/CH_4 selectivity (Table 3.3) [99–101]. As seen in Table 3.3, selectivity is highly dependent on nature of metal nodes and organic linkers. Overall, high selectivity of CO_2 over CH_4, and N_2 was reported.

Adsorption capacity is another important metric to evaluate CO_2 performance of MOFs. Both gravimetric and volumetric capacity need to be taken into account as gravimetric capacity defines the CO_2 uptake per unit mass whereas volumetric capacity characterizes density of CO_2 storage within the adsorbent volume [6]. Metal Organic Frameworks typically have large surface area and high porosity being advantageous for CO_2 capture as it provides densely packing and close interaction of CO_2 molecules on the surface of pores.

In order to facilitate efficient CO_2 adsorption, surface properties of MOFs can be modified by tuning organic ligands, metal nodes or by postmodification of existing MOFs. The use of functional groups having interactions with CO_2 has been utilized. Amine or nitrogen-based functionalized ligands have been widely used for this purpose due to their high affinity for CO_2 molecules. Couck et al. prepared amine-functionalized MIL-53 MOFs having large CO_2 capture capacity with good CO_2/CH_4 selectivity [103]. An alternative strategy for promoting CO_2 adsorption is to employ exposed metal cation sites onto pore surfaces. These sites are usually formed

Table 3.3 Selectivity of CO_2 over N_2, and CH_4 in selected MOFs [99–101].

MOF	CO_2/CH_4	CO_2/N_2	Conditions	References
$[Zn(L) \cdot H_2O] \cdot DMA$	52	134	273K, 1 bar	[99]
$[Cu(bcppm)H_2O]$	-	590	273K, 1 bar	[2]
[HEMIM][DCA]/ZIF-8	30	-	298K, 1 bar	[101]
$CO_2(ad)_2(CO_2CH_3)_2$ 2DMF $0.5H_2O$	-	75	298K, 1 bar	[102]

by the removal of the solvent molecules from the coordination site of the metal center. This method allows highly selective interactions with the gas molecules that can facilitate efficient CO_2 capture and storage [6,104]. An alternative method employs polar moieties into MOFs for physical adsorption of CO_2. It is worth mentioning that most of MOFs adsorbents capture CO_2 via dipole-quadruple interactions or through active open metal sites instead of chemical adsorption.

3.5.4 Green metal-organic frameworks

While MOFs are considered as promising materials for the development of next generation CO_2 adsorbents due to their high capacity for adsorption and storage, majority of MOFs are made from nonrenewable petrochemical materials using toxic solvents. Therefore, there is a need for development of green and sustainable MOFs that can efficiently capture carbon dioxide. MOFs derived from biomolecules such as peptides, amino acids proteins [105], nucleobases, and carbohydrates [106–108] have been reported in the literature. One challenge is to form highly porous and crystalline bio-MOFs due to the low symmetry of the building blocks [106]. An et al. synthesized highly porous crystalline zinc-adeninate MOFs using adenine nucleobase building blocks having very large surface area (4300 m^2/g) and pore volume (4.3 cm^3/g) [109,110]. Smaldone and Gassensmith et al. prepared porous symmetrical MOFs under mild conditions using γ-cyclodextrin derived from starch and coordinated by K^+ and Rb^+ cations having large surface area 1220 m^2/g and pore density of 0.47 g/cm^3. Its structure is shown in Fig. 3.44 where eight monosaccharides incorporated to form round shape assembly with inner diameter of 0.9 nm [106,107].

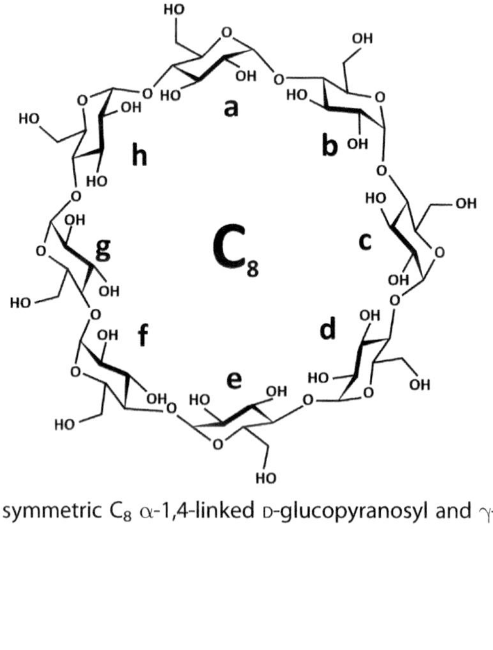

Figure 3.44 Structure of symmetric C_8 α-1,4-linked D-glucopyranosyl and γ-cyclodextrin [106].

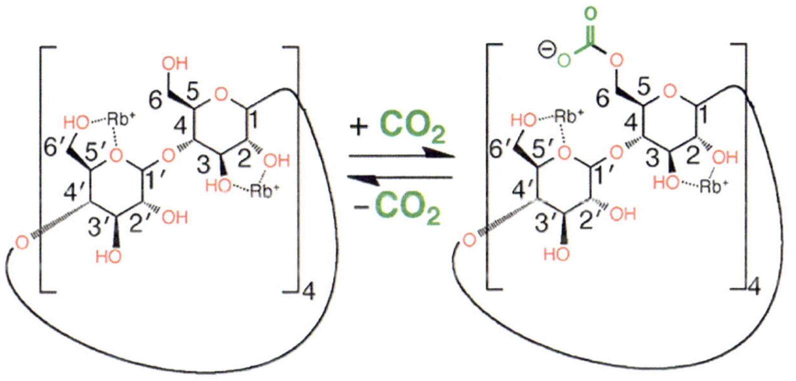

Figure 3.45 CO_2 capture mechanism of γ-cyclodextrin-based MOFs [107].

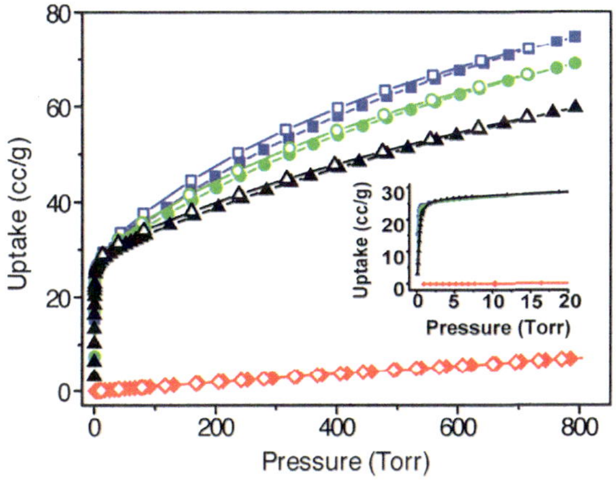

Figure 3.46 CO_2 uptake of γ-cyclodextrin derived MOFs versus pressure at 273K (blue squares), 283K (green circles), 298K (black triangles); and CH_4 uptake at 298K (red diamonds) [107].

Gassensmith et al. synthesized inexpensive and green MOFs from γ-cyclodextrin (CD-MOF) linked to each other by Rb^+ cations building a three-dimensional body-centered-cubic structure and studied CO_2 adsorption properties. CO_2 was fixed reversibly by chemical adsorption using nucleophilic groups (Fig. 3.45).

CO_2 uptake of γ-cyclodextrin derived MOFs at different temperatures and pressures was shown in Fig. 3.46. Even at lower pressures (<1 torr) carbon dioxide uptake was ~25 cm^3/g between 273K and 298K. CO_2 adsorption exhibited a decrease with increasing temperatures whereas it improved up to ~70 cm^3/g at higher pressures. Moreover, CO_2 uptake become more dependent on temperature with increasing pressure. CH_4 adsorption was measured as a comparison and found to be much smaller compared to CO_2 under same temperature/pressure conditions. This indicates γ-cyclodextrin derived MOFs possess very high CO_2/CH_4 selectivity [107].

Table 3.4 Differential enthalpies of CO_2 adsorption of MOFs derived from γ-cyclodextrin [108].

Coverage of CO_2/nm^2	Differential enthalpy of adsorption (kJ/mol CO_2)		
	Round 1	**Round 2**	**Round 3**
~0	−113.5 ± 0.9	−77.7 ± 1.1	−71.1 ± 0.8
0.1−0.3	−73.0 ± 0.8	−68.8 ± 1.5	−65.9 ± 1.2
>0.4	−67.5 ± 1.1	−66.0 ± 0.8	−63.7 ± 1.2

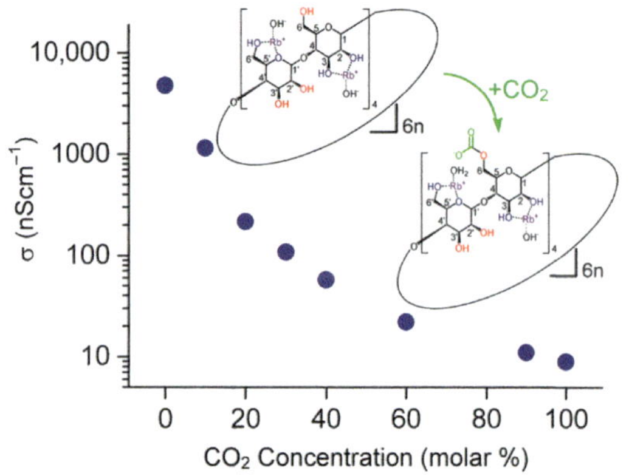

Figure 3.47 Average conductivity values in γ-cyclodextrin derived MOFs versus CO_2 concentration [111].

Differential enthalpies of CO_2 adsorption were measured via direct calorimetric measurement and shown in Table 3.4. After the each round of experiment, sample was subjected to vacuum treatment to initiate desorption in order to study regeneration performance of the adsorbent. Large enthalpy adsorption of CO_2 at zero coverage shows strong binding between CO_2 and γ-cyclodextrin MOF. As the CO_2 coverage on the adsorbent surface increases, the enthalpy of adsorption becomes less exothermic. Moreover, the adsorption was much less exothermic on the second round compared to first round and it did not change much on the third round. This concludes that CO_2 cannot be desorbed readily when it interacts with the most reactive hydroxyl groups whereas adsorption is regenerable on the less reactive hydroxyls. Wu et al. reported that CO_2 adsorption occurs mainly on γ-cyclodextrin's structural alcohol groups rather than OH^- present in the pores [108].

Furthermore, Gassensmith et al. demonstrated the electrochemical sensing ability of γ-cyclodextrin MOFs. They exhibit large reversible ion conductivity especially at low CO_2 concentrations. Significant reduction in conductivity was observed with increase in CO_2 concentration, potentially due to the hindrance of protons movement (Fig. 3.47).

3.6 Bio-derived porous carbons

3.6.1 Introduction

Biomass has been utilized to form porous carbonaceous materials for decades. Bioderived porous carbons can be derived from various biomass resources such as palm shell [112], coconut shell, carbohydrates [113], and biomass feedstocks [114]. They are abundant, cost-effective, environmentally friendly, easily regenerable and reusable materials with excellent thermal, chemical, and mechanical stability. Due to their hydrophobic properties, they are insensitive to water and can operate well under humid conditions. Moreover, these materials have large surface area and well-defined and tunable pores in micro and macro size range with large pore volume. Potential applications of bio-derived microporous carbon materials include their use supercapacitors [115], hydrogen storage materials [116], and CO_2 adsorbents [117–120].

3.6.2 Synthesis of biomass derived porous carbons

3.6.2.1 Carbonization

Carbonization is defined as the formation of carbon materials from organic matter. The carbonization methods for the formation of porous carbons from Biomass sources are discussed in detail in the literature [121]. Typical methods are listed below:

Hydrothermal carbonization (HTC);

Template-directed synthesis;

Direct synthesis from biomass precursors.

HTC is the most widely used technique for the synthesis of porous carbon materials. Carbohydrates and biomass feedstock are usually used as biomass sources. HTC can be performed at low or high temperature depending on the purpose of the end material. High temperature HTC processes typically occur temperatures between 300°C and 800°C and under high pressures. This process is mainly utilized to produce carbonaceous materials with high surface area and porosity such as carbon nanotubes, graphite and activated carbons [121]. (Fig. 3.48).

Low temperature HTCs are considered as more environmentally friendly method that is performed temperatures up to 250°C for the production of monodispersed colloidal carbon spheres mainly from carbohydrate [123] and cellulosic precursors [124]. This method involves utilization of subsequent processes such as dehydration, polymerization and aromatization.

Sevilla et al. demonstrated the size of microspheres can be tuned by modifying operation conditions such as precursor concentration, temperature and reaction time (Fig. 3.49; Table 3.5).

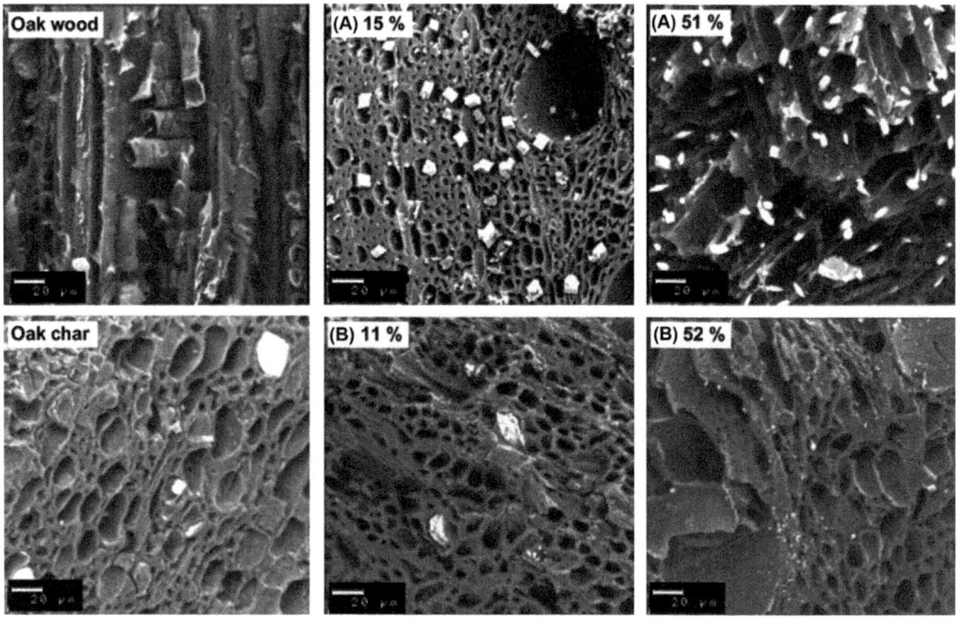

Figure 3.48 SEM images of oak char and oak wood with supercritical water and steam [122].

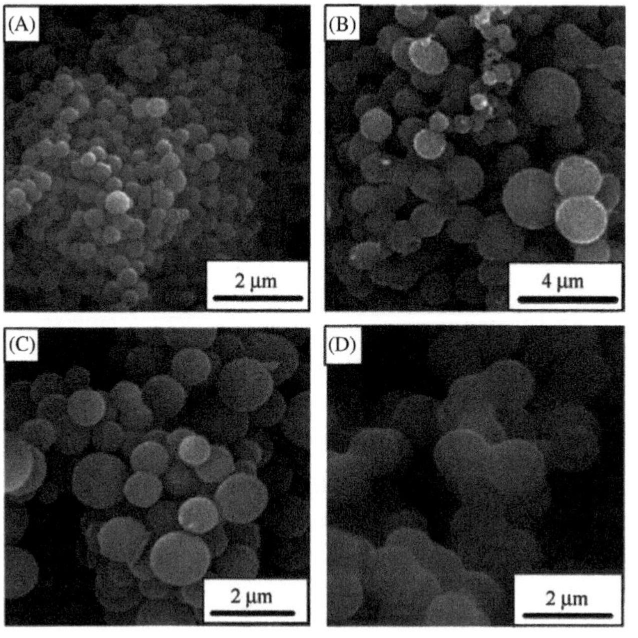

Figure 3.49 SEM micrographs of microspheres by hydrothermal carbonization of glucose [123].

Table 3.5 Summary of conditions for hydrothermal carbonization of glucose shown in Fig. 3.49 [123].

SEM micrograph	Precursor concentration (M)	Hydrothermal carbonization temperature (°C)	Reaction time (hour)
A	0.5	170	4.5
B	0.5	230	4.5
C	0.5	170	15
D	1	240	0.5

Template-directed synthesis is another method to synthesize well-defined porous carbonaceous materials with controlled architecture and pore size distribution. Hard-template, soft-template, and dual-template techniques are reported in the literature. Hard-template method uses MOFs, silicas, and zeolites as templates. The carbonization occurs in multiple steps as described below [121]:

Template preparation;

Addition of carbon precursors to the pores of the template;

Cross-linking and carbonization;

Template removal.

Although carbonaceous materials with controlled porosity can be formed by hard template synthesis, this method suffers from disintegration of pore structure and tough removal of the template at the end of the process. Soft-template synthesis utilizes assembly between carbon precursors and agents such as amphiphilic molecular and surfactants to minimize pore collapse problem. Dual-template synthesis combines hard and soft template synthesis to generate macro-porous and mesoporous carbon materials [121].

3.6.2.2 Activation of porous carbons

In carbonization process, char, defined as porous carbon skeleton, is formed. The tar being formed during thermal process can block off the pores of char lowering the adsorption performance. Therefore, the activation of carbon materials are essential in order to form ordered structures with increased porosity and better control over the pore size distribution for CO_2 capture. There are two main activation methods: physical (thermal activation) and chemical activation.

Physical activation is a two-step process. The disorganized part of the material is first removed by pyrolysis at 400°C–900°C, with the subsequent partial gasification with oxidizing gas such as steam and air at 350°C–1000°C [125,126].

The activation performance strongly depends on oxidizing gas and activation temperature. As the activation temperature increases, porosity of the materials get larger but with broader pore size distribution [125]. Bouchelta et al. prepared porous carbons having large surface area ~ 635 m^2/g and microporous volume ~ 0.716 $cm^3\ g^{-1}$ with physical activation [127]. Chemical activation process is a one-step method that involves heating the carbon precursor and the activating agent at temperatures

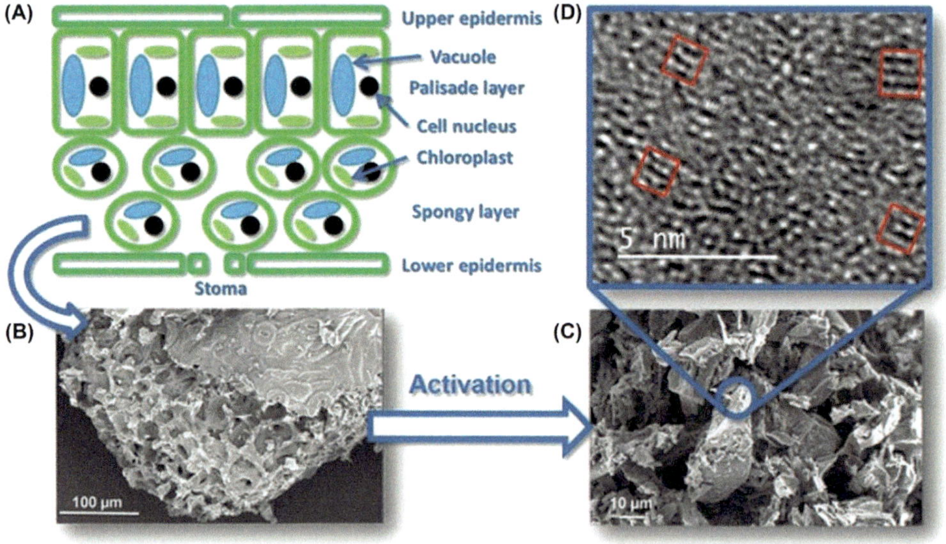

Figure 3.50 (A) Schematic representation of cellular structure of leaf (B) SEM image of carbonized leaf (C) SEM image of leaf-derived porous carbon after KOH activation (D) TEM image of leaf-derived porous carbon after KOH activation [130].

between ~450°C and 900°C. While $ZnCl_2$, H_3PO_4, $MgCl_2$, $AlCl_3$, NaOH, and KOH are used as activating agents [125], KOH activation is the most commonly used to form porous carbons as CO_2 adsorbents [128,129].

Compared with physical activation, chemical activation owns the following advantages [125]:

Higher carbon yield;

Porous carbons with high surface area;

Well-defined, controlled and narrowly distributed microporosity;

Lower pyrolysis temperatures.

SEM and TEM images of porous carbon derived from leaves showed randomly aligned nanoporous structure embedded in micropores post KOH activation (Fig. 3.50C – D) [130].

3.6.3 Carbon dioxide adsorption studies

Bio-derived microporous carbon has gained great attention in CO_2 capture applications due to their large surface area with high porosity, superior thermal and chemical stability, insensitivity to humid conditions, fast adsorption-desorption kinetics and low-cost. In addition, these materials are renewable, eco-friendly, reusable and easily regenerable. Several studies has been carried out to investigate CO_2 capture properties of bioderived porous carbon materials [118,128–131].

Table 3.6 Porous texture properties, compositions, CO_2 uptakes of carbons derived from coconut shell and wheat under different conditions [128].

Sample	S_{BET} (m^2/g)	V (cm^3/g)	N (wt%)	CO_2 uptake at 25°C (mmol/g)	CO_2 uptake at 0°C (mmol/g)	Ref
Carbonized coconut shell	21	0.02	0.26	1.5	2.0	[128]
NC	N/A	N/A	4.47	1.4	1.9	[128]
NC-600-2	1023	0.38	1.35	4.1	5.6	[128]
NC-600-3	1383	0.52	1.08	4.0	6.0	[128]
NC-600-4	1604	0.60	0.81	4.3	6.7	[128]
NC-700-2	1550	0.57	0.86	4.1	6.1	[128]
NC-700-3	1596	0.60	0.73	4.7	7.0	[128]
NC-700-4	1937	0.72	0.61	4.4	6.8	[128]
Wheat derived MCC	648	0.299	N/A	2.3	3.4	[129]
MCC-K1	916	0.432	N/A	2.4	3.8	[129]
MCC-K2	1057	0.581	N/A	2.8	4.4	[129]
MCC-K3	1438	0.654	N/A	3.5	5.7	[129]
MCC-K4	1801	0.840	N/A	3.1	5.3	[129]
MCC-K5	2192	1.076	N/A	2.6	4.4	[129]

Chen et al. synthesized nitrogen-doped microporous carbons from Coconut shells and urea (Table 3.6). The prepared samples were shown as NC-X-Y whereas X is the KOH activation temperature and Y is KOH/N ratio for doped carbonized coconut shell (NC). KOH-activated microporous carbons have remarkably larger surface area ranging between 1023 and 1937 m^2/g compared with nonactivated carbonized coconut shell (21 m^2/g). Total pore volume of coconut shell increased from 0.02 to 0.72 post KOH activation, therefore, carbon activation plays a significant role to form microporous carbons. As KOH penetrates deeper layers of nitrogen-doped carbon, it reacts with nitrogen groups resulting in formation of more porous structure [128]. When activation temperature was increased from 600°C to 700°C, an improvement was seen in both surface area and total pore volume. KOH-activated carbon samples showed almost 4 times larger CO_2 adsorption compared with nonactivated samples at ambient temperature. As CO_2 adsorption measurement temperatures was reduced from 25°C to 0°C, an overall increase in CO_2 uptake was seen for all carbon samples. Son et al. prepared metal doped coconut char with surface area up to 519 m^2/g having high CO_2/N_2 and CO_2/CH_4 selectivity [131].

Table 3.6 also shows the surface area, porosity and CO_2 uptake performance of Wheat derived microporous carbons (MCC). Carbon samples were marked as MCC-KX whereas X is the KOH/C ratio [129]. Similar to the nitrogen-doped porous carbons from Coconut shells, the surface area and total porosity of wheat-derived microporous carbons increased with larger amount of KOH used during activation. Morever, wheat-derived porous carbon showed relatively less CO_2 adsorption compared to nitrogen-doped coconut shell-derived porous carbon materials. Wheat-derived KOH-activated

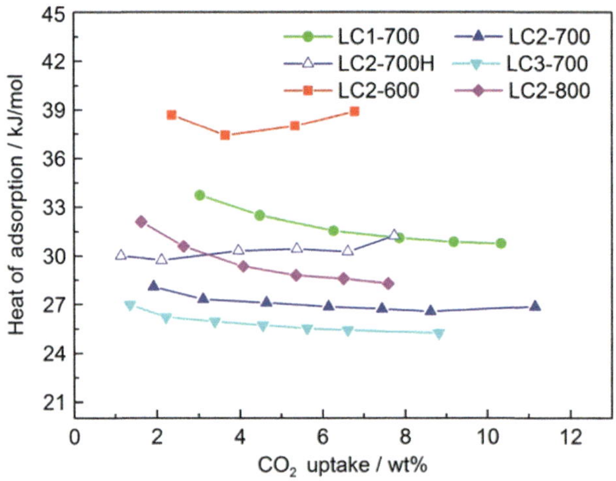

Figure 3.51 CO_2 adsorption isotherms of leaf-derived carbon adsorbent versus heat of adsorption [130].

microporous carbon lost only 1% of CO_2 adsorption capacity after 10 cycles that can be attributed to reusability of bio-derived porous carbon adsorbents. In addition to these porous carbons, carbon molecular sieves were synthesized from palm shell and peach stones having high CO_2/N_2 selectivity [132–134].

Adsorption heats of leaf-derived (LC) microporous carbons doped with metals and nitrogen were reported by Zhu et al. [130]. Prepared samples were designated as LX-Y, where X is KOH/C ratio and Y is the thermal treatment temperature. LC2–700H sample was further washed with HCl acid post KOH activation. Carbon sample prepared with KOH/C ratio of 3:1 and at 700°C had the lowest heat of adsorption (Fig. 3.51). It can be concluded that it is possible to modify the heat of adsorption of microporous Carbons by tuning conditions such as KOH/C ratio and thermal treatment temperature. Furthermore, it was reported dopants and porosity also affect the heat of adsorption such that increasing Calcium dopant content on carbon materials increased the heat of adsorption [130].

Starbons are mesoporous carbons derived from starch or alginic acid. They are synthesized by addition of the acid catalyst to the starch and subsequently heating the substance to the temperatures more than 150°C under nonoxidizing conditions [135]. They can be carbonized at temperatures between 300°C and 1200°C. Starbons has large surface area up to 826 m^2/g and total pore volume of 1.3 cm^3/g depending on the carbonization temperature of polysaccharide. They demonstrated very rapid CO_2 capture with adsorption up to 3 mmol/g. Superior CO_2/N_2 selectivity of Starbon derived from starch and alginic acid at room temperature was reported as ~14 [118].

3.7 Conclusion and outlook

Given that huge amount of carbon dioxide needs to be removed from the atmosphere, chemical adsorption using solid adsorbents seems to be appropriate and viable solution. Although there has been huge number of published studies on the development of inexpensive, reusable adsorbents for efficient Carbon Dioxide Capture and Storage with minimum energy requirement for regeneration, only few of these involve materials derived from green and renewable sources. In this chapter, synthesis, functionalization, characterization and CO_2 adsorption/desorption performance of green adsorbents derived from natural, bio-based or bio-inspired nanomaterials were reviewed in detail.

HNTs are inexpensive, abundant, eco-friendly and natural nanomaterials with high surface area and porosity. They exhibit excellent mechanical, thermal and chemical stability. Although HNT did not adsorb any carbon dioxide in its pure state, its surface properties can be easily tuned to improve adsorption properties. Amine grafted and impregnated HNTs were studied in the literature and found to have promising carbon dioxide uptake and regeneration performance.

NFCs derived from cellulose are highly porous, low-cost, abundant, biodegradable nanomaterials. They have superior mechanical properties such as high young modulus and tensile strength. Similar to HNTs, pure NFCs possess no CO_2 adsorption. However, efficient CO_2 adsorbents can easily be designed from NFCs via surface modification due to the availability of OH groups in its structure. Amine modified NFCs showed superior CO_2 capture and lost only 5% of carbon dioxide adsorption/desorption capacity after 100 cycles.

Use of CA enzyme for large scale CO_2 sequestration seems to be very promising as it remarkably speeds up the CO_2 conversion. However, low stability and leaching problem of CA limits its use for this purpose. Biosilicification is a green and efficient method to immobilize and encapsulate CA to prevent leaching and improve adsorbent stability. This method yields to the formation of immobilized CA-bioinspired silica nanocomposites showing superior carbon sequestration performance compared to free CA enzyme with high stability and reusability.

MOFs possess ultra-high surface and well-defined porosity. In the recent years, there has been great effort to synthesize green MOFs from biomolecules. It was shown that bio-MOFs had very promising CO_2 capture performance and high CO_2 selectivity. However, there are some concerns related to chemical and mechanical stability of MOFs. Most of MOFs are prone to chemical degradation under humid conditions. Moreover, it was reported mechanical strength of MOFs can degrade significantly at high pressures. Therefore, there is a need to address these problems for the development of next generation CO_2 capture MOF-based nanomaterials.

Bio-derived activated porous carbons has also gained considerable attention as they are inexpensive, environmentally friendly and have high chemical, mechanical and thermal stability. As they have high surface area, porosity and are insensitive to humid conditions, they are considered as promising materials for CO_2 adsorption. They exhibited high CO_2 uptake even at ambient temperatures.

Overall, properties such as adsorption capacity, performance, adsorption rates, lifetime, and reusability of these materials still need to be improved in order to enhance efficient CCS. Although amine modification remains to be the most widely used surface functionalization technique to improve carbon dioxide adsorption of these nanomaterials, there is a need to consider other functional moieties with less energy penalty during regeneration. In addition, most of the research focuses on the adsorption characteristics, therefore there is a need to investigate CO_2 selectivity of nanomaterials over other gases such as N_2 and CH_4 in more detail.

References

[1] P. Kumar, K.-H. Kim, Recent progress and innovation in carbon capture and storage using bioinspired materials, Appl. Energy 172 (2016) 383–397.

[2] G.A. Olah, A. Goeppert, G.K.S. Prakash, Chemical recycling of carbon dioxide to methanol and dimethyl ether: from greenhouse gas to renewable, environmentally carbon neutral fuels and synthetic hydrocarbons, J. Org. Chem. 74 (2008) 487–498.

[3] D.M. D'Alessandro, B. Smit, J.R. Long, Carbon dioxide capture: prospects for new materials, Angew. Chem. Int. Ed. 49 (2010) 6058–6082.

[4] C.-H. Yu, C.-H. Huang, C.-S. Tan, A review of CO_2 capture by absorption and adsorption, Aerosol Air Qual. Res. 12 (2012) 745–769.

[5] X. Ma, X. Wang, C. Song, "Molecular basket" sorbents for separation of CO_2 and H_2S from various gas streams, J. Am. Chem. Soc. 131 (2009) 5777–5783.

[6] K. Sumida, et al., Carbon dioxide capture in metal–organic frameworks, Chem. Rev. 112 (2012) 724–781.

[7] M. Czaun, et al., Organoamines-grafted on nano-sized silica for carbon dioxide capture, J. CO_2 Util. 1 (2013) 1–7.

[8] A. Goeppert, et al., Carbon dioxide capture from the air using a polyamine based regenerable solid adsorbent, J. Am. Chem. Soc. 133 (2011) 20164–20167.

[9] Y. Bae, R.Q. Snurr, Development and evaluation of porous materials for carbon dioxide separation and capture, Angew. Chem. Int. Ed. 50 (2011) 11586–11596.

[10] P. Berthier, Analyse de l'halloysite, Annales de. Chimie et. de Phys. 32 (1826) 332–335.

[11] E. Abdullayev, Y. Lvov, Halloysite clay nanotubes for controlled release of protective agents, J. Nanosci. Nanotechnol. 11 (2011) 10007–10026.

[12] J.H. Kirkman, Morphology and structure of halloysite in New Zealand tephras, Clays Clay Min. 29 (1981) 1–9.

[13] O.I. Ece, P.A. Schroeder, Clay mineralogy and chemistry of halloysite and alunite deposits in the Turplu area, Balikesir, Turkey, Clays Clay Min. 55 (2007) 18–35.

[14] E. Joussein, et al. Halloysite clay minerals—a review, 2005.

[15] S. Jana, S. Das, C. Ghosh, A. Maity, M. Pradhan, Halloysite nanotubes capturing isotope selective atmospheric CO_2, Sci. Rep. 5 (2015) 8711.

[16] M. Liu, B. Guo, M. Du, X. Cai, D. Jia, Properties of halloysite nanotube–epoxy resin hybrids and the interfacial reactions in the systems, Nanotechnology 18 (2007) 455703.

[17] R. Price, B.P. Gaber, Y. Lvov, In-vitro release characteristics of tetracycline HCl, khellin and nicotinamide adenine dineculeotide from halloysite; a cylindrical mineral, J. Microencapsul. 18 (2001) 713–722.
[18] T.F. Bates, F.A. Hildebrand, A. Swineford, Morphology and structure of endellite and halloysite, Am. Mineral. J. Earth Planet. Mater. 35 (1950) 463–484.
[19] B. Lecouvet, M. Sclavons, S. Bourbigot, C. Bailly, Thermal and flammability properties of polyethersulfone/halloysite nanocomposites prepared by melt compounding, Polym. Degrad. Stab. 98 (2013) 1993–2004.
[20] D. Lu, H. Chen, J. Wu, C.M. Chan, Direct measurements of the Young's modulus of a single halloysite nanotube using a transmission electron microscope with a bending stage, J. Nanosci. Nanotechnol. 11 (2011) 7789–7793.
[21] P. Yuan, D. Tan, F. Annabi-Bergaya, Properties and applications of halloysite nanotubes: recent research advances and future prospects, Appl. Clay Sci. 112 (2015) 75–93.
[22] P. Pasbakhsh, R. De Silva, V. Vahedi, G.J. Churchman, Halloysite nanotubes: prospects and challenges of their use as additives and carriers—A focused review, Clay Min. 51 (2016) 479–487.
[23] M. Massaro, et al., Halloysite nanotubes as support for metal-based catalysts, J. Mater. Chem. A 5 (2017) 13276–13293.
[24] S. Barrientos-Ramírez, et al., Use of nanotubes of natural halloysite as catalyst support in the atom transfer radical polymerization of methyl methacrylate, Microporous Mesoporous Mater. 120 (2009) 132–140.
[25] G. Cavallaro, et al., Halloysite nanotubes: controlled access and release by smart gates, Nanomaterials 7 (2017) 199.
[26] R.F. Fakhrullin, Y.M. Lvov, Halloysite clay nanotubes for tissue engineering, 2016.
[27] R. Kamble, M. Ghag, S. Gaikawad, B.K. Panda, Halloysite nanotubes and applications: a review, J. Adv. Sci. Res. 3 (2012).
[28] S. Das, A. Maity, M. Pradhan, S. Jana, Assessing atmospheric CO_2 entrapped in clay nanotubes using residual gas analyzer, Anal. Chem. 88 (2016) 2205–2211.
[29] J. Kim, I. Rubino, J.-Y. Lee, H.-J. Choi, Application of halloysite nanotubes for carbon dioxide capture, Mater. Res. Express 3 (2016) 45019.
[30] H. Cai, et al., Preparation and characterization of novel carbon dioxide adsorbents based on polyethylenimine-modified Halloysite nanotubes, Environ. Technol. 36 (2015) 1273–1280.
[31] M. Niu, H. Yang, X. Zhang, Y. Wang, A. Tang, Amine-impregnated mesoporous silica nanotube as an emerging nanocomposite for CO_2 capture, ACS Appl. Mater. Interfaces 8 (2016) 17312–17320.
[32] S. Das, C. Ghosh, S. Jana, Moisture induced isotopic carbon dioxide trapping from ambient air, J. Mater. Chem. A 4 (2016) 7632–7640.
[33] D. Klemm, et al., Nanocelluloses as innovative polymers in research and application, Polysaccharides Ii, Springer, 2006, pp. 49–96.
[34] S. Kamel, N. Ali, K. Jahangir, S.M. Shah, A.A. El-Gendy, Pharmaceutical significance of cellulose: a review, Express Polym. Lett. 2 (2008) 758–778.
[35] L.R. Lynd, C.E. Wyman, T.U. Gerngross, Biocommodity engineering, Biotechnol. Prog. 15 (1999) 777–793.
[36] C. Ververis, K. Georghiou, N. Christodoulakis, P. Santas, R. Santas, Fiber dimensions, lignin and cellulose content of various plant materials and their suitability for paper production, Ind. Crop. Prod. 19 (2004) 245–254.
[37] E.M. Renkin, Filtration, diffusion, and molecular sieving through porous cellulose membranes, J. Gen. Physiol. 38 (1954) 225–243.
[38] W.K. Czaja, D.J. Young, M. Kawecki, R.M. Brown, The future prospects of microbial cellulose in biomedical applications, Biomacromolecules 8 (2007) 1–12.
[39] S.H. Osong, S. Norgren, P. Engstrand, Processing of wood-based microfibrillated cellulose and nanofibrillated cellulose, and applications relating to papermaking: a review, Cellulose 23 (2016) 93–123.
[40] S. Iwamoto, W. Kai, A. Isogai, T. Iwata, Elastic modulus of single cellulose microfibrils from tunicate measured by atomic force microscopy, Biomacromolecules 10 (2009) 2571–2576.

[41] M. Henriksson, L.A. Berglund, P. Isaksson, T. Lindstrom, T. Nishino, Cellulose nanopaper structures of high toughness, Biomacromolecules 9 (2008) 1579–1585.

[42] K. Missoum, M. Belgacem, J. Bras, Nanofibrillated cellulose surface modification: a review, Materials (Basel), 6, 2013, pp. 1745–1766.

[43] G. Marques, J. Rencoret, A. Gutiérrez Suárez, J.C. del Río Andrade, Evaluation of the chemical composition of different non-woody plant fibers used for pulp and paper manufacturing, 2010.

[44] S. Kalia, S. Boufi, A. Celli, S. Kango, Nanofibrillated cellulose: surface modification and potential applications, Colloid Polym. Sci. 292 (2014) 5–31.

[45] S. Alila, I. Besbes, M.R. Vilar, P. Mutjé, S. Boufi, Non-woody plants as raw materials for production of microfibrillated cellulose (MFC): a comparative study, Ind. Crop. Prod. 41 (2013) 250–259.

[46] A.N. Frone, D.M. Panaitescu, D. Donescu, Some aspects concerning the isolation of cellulose micro-and nano-fibers, UPB Bul. Stiint. Ser. B Chem. Mater. Sci. 73 (2011) 133–152.

[47] H.P.S.A. Khalil, et al., Production and modification of nanofibrillated cellulose using various mechanical processes: a review, Carbohydr. Polym. 99 (2014) 649–665.

[48] J. Li, et al., Homogeneous isolation of nanocellulose from sugarcane bagasse by high pressure homogenization, Carbohydr. Polym. 90 (2012) 1609–1613.

[49] Y.-F. Maa, C.C. Hsu, Performance of sonication and microfluidization for liquid–liquid emulsification, Pharm. Dev. Technol. 4 (1999) 233–240.

[50] K.L. Spence, R.A. Venditti, O.J. Rojas, Y. Habibi, J.J. Pawlak, A comparative study of energy consumption and physical properties of microfibrillated cellulose produced by different processing methods, Cellulose 18 (2011) 1097–1111.

[51] K. Abe, H. Yano, Comparison of the characteristics of cellulose microfibril aggregates of wood, rice straw and potato tuber, Cellulose 16 (2009) 1017.

[52] M. Henriksson, G. Henriksson, L.A. Berglund, T. Lindström, An environmentally friendly method for enzyme-assisted preparation of microfibrillated cellulose (MFC) nanofibers, Eur. Polym. J. 43 (2007) 3434–3441.

[53] S. Janardhnan, M.M. Sain, Isolation of cellulose microfibrils—an enzymatic approach, Bioresources 1 (2007) 176–188.

[54] J.Y. Zhu, R. Sabo, X. Luo, Integrated production of nano-fibrillated cellulose and cellulosic biofuel (ethanol) by enzymatic fractionation of wood fibers, Green. Chem. 13 (2011) 1339–1344.

[55] S. Janardhnan, M.M. Sain, Targeted disruption of hydroxyl chemistry and crystallinity in natural fibers for the isolation of cellulose nano-fibers via enzymatic treatment, BioResources 6 (2011) 1242–1250.

[56] A. Isogai, T. Saito, H. Fukuzumi, TEMPO-oxidized cellulose nanofibers, Nanoscale 3 (2011) 71–85.

[57] T. Saito, et al., Individualization of nano-sized plant cellulose fibrils by direct surface carboxylation using TEMPO catalyst under neutral conditions, Biomacromolecules 10 (2009) 1992–1996.

[58] C. Gebald, J.A. Wurzbacher, P. Tingaut, T. Zimmermann, A. Steinfeld, Amine-based nanofibrillated cellulose as adsorbent for CO_2 capture from air, Environ. Sci. Technol. 45 (2011) 9101–9108.

[59] C. Gebald, J.A. Wurzbacher, P. Tingaut, A. Steinfeld, Stability of amine-functionalized cellulose during temperature-vacuum-swing cycling for CO_2 capture from air, Environ. Sci. Technol. 47 (2013) 10063–10070.

[60] H. Sehaqui, et al., Fast and reversible direct CO_2 capture from air onto all-polymer nanofibrillated cellulose—polyethylenimine foams, Environ. Sci. Technol. 49 (2015) 3167–3174.

[61] M. Pääkkö, et al., Long and entangled native cellulose I nanofibers allow flexible aerogels and hierarchically porous templates for functionalities, Soft Matter 4 (2008) 2492–2499.

[62] S. Choi, M.L. Gray, C.W. Jones, Amine-tethered solid adsorbents coupling high adsorption capacity and regenerability for CO_2 capture from ambient air, ChemSusChem 4 (2011) 628–635.

[63] R.R. Yadav, et al., Carbonic anhydrase mediated carbon dioxide sequestration: Promises, challenges and future prospects, J. Basic. Microbiol. 54 (2014) 472–481.

[64] P.C. Sahoo, F. Kausar, J.H. Lee, J.I. Han, Facile fabrication of silver nanoparticle embedded $CaCO_3$ microspheres via microalgae-templated CO_2 biomineralization: application in antimicrobial paint development, RSC Adv. 4 (2014) 32562–32569.

[65] W. Wei, et al., Preparation of hierarchical hollow $CaCO_3$ particles and the application as anticancer drug carrier, J. Am. Chem. Soc. 130 (2008) 15808–15810.
[66] D.B. Trushina, T.V. Bukreeva, M.V. Kovalchuk, M.N. Antipina, $CaCO_3$ vaterite microparticles for biomedical and personal care applications, Mater. Sci. Eng. C. 45 (2014) 644–658.
[67] J.S. de C. Campos, A.A. Ribeiro, C.X. Cardoso, Preparation and characterization of PVDF/$CaCO_3$ composites, Mater. Sci. Eng. B 136 (2007) 123–128.
[68] I.G. Kim, et al., Biomineralization-based conversion of carbon dioxide to calcium carbonate using recombinant carbonic anhydrase, Chemosphere 87 (2012) 1091–1096.
[69] M. Vinoba, M. Bhagiyalakshmi, S.K. Jeong, Y.I.I. Yoon, S.C. Nam, Immobilization of carbonic anhydrase on spherical SBA-15 for hydration and sequestration of CO_2, Colloids Surf. B Biointerfaces 90 (2012) 91–96.
[70] Y.-T. Zhang, L. Zhang, H.-L. Chen, H.-M. Zhang, Selective separation of low concentration CO_2 using hydrogel immobilized CA enzyme based hollow fiber membrane reactors, Chem. Eng. Sci. 65 (2010) 3199–3207.
[71] R. Yadav, T. Satyanarayanan, S. Kotwal, S. Rayalu, Enhanced carbonation reaction using chitosan-based carbonic anhydrase nanoparticles, Curr. Sci. 100 (2011) 520–524.
[72] F. Migliardini, et al., Biomimetic CO_2 capture using a highly thermostable bacterial α-carbonic anhydrase immobilized on a polyurethane foam, J. Enzyme Inhib. Med. Chem. 29 (2014) 146–150.
[73] K.-H. Min, R.G. Son, M.-R. Ki, Y.S. Choi, S.P. Pack, High expression and biosilica encapsulation of alkaline-active carbonic anhydrase for CO_2 sequestration system development, Chemosphere 143 (2016) 128–134.
[74] W.D. Marner II, A.S. Shaikh, S.J. Muller, J.D. Keasling, Enzyme immobilization via silaffin-mediated autoencapsulation in a biosilica support, Biotechnol. Prog. 25 (2009) 417–423.
[75] K. Shimizu, et al., Exploration of genes associated with sponge silicon biomineralization in the whole genome sequence of the hexactinellid euplectella curvistellata BT - Biomineralization, in: K. Endo, T. Kogure, H. Nagasawa (Eds.), Springer Singapore, 2018, 147–153.
[76] S.V. Patwardhan, Biomimetic and bioinspired silica: recent developments and applications, Chem. Commun. 47 (2011) 7567–7582.
[77] M.J. Leng, et al., The potential use of silicon isotope composition of biogenic silica as a proxy for environmental change, Silicon 1 (2009) 65–77.
[78] D.G. Mann, F.E. Round, R.M. Crawford, The diatoms Biology and morphology of the genera, Cambridge University Press, 1990.
[79] P. Innocenzi, From the precursor to a sol BT - the sol to gel transition, in P. Innocenzi (Ed.), Springer International Publishing, 2016, 7–25. doi: https://doi.org/10.1007/978-3-319-39718-4_2.
[80] L. Betancor, H.R. Luckarift, Bioinspired enzyme encapsulation for biocatalysis, Trends Biotechnol. 26 (2008) 566–572.
[81] D. Shao, et al., Bioinspired diselenide-bridged mesoporous silica nanoparticles for dual-responsive protein delivery, Adv. Mater. 30 (2018) 1801198.
[82] K.E. Papathanasiou, M. Vassaki, A. Spinthaki, A. Moschona, K.D. Demadis, Silica-based polymeric gels as platforms for delivery of phosphonate pharmaceutics BT-polymer gels: synthesis and characterization, in: V. K. Thakur, M. K. Thakur (Eds.), Springer Singapore, 2018, 127–140. doi: https://doi.org/10.1007/978-981-10-6083-0_5.
[83] B.H. Jo, et al., Bioinspired silica nanocomposite with autoencapsulated carbonic anhydrase as a robust biocatalyst for CO_2 sequestration, ACS Catal. 4 (2014) 4332–4340.
[84] Y. Wang, F. Caruso, Mesoporous silica spheres as supports for enzyme immobilization and encapsulation, Chem. Mater. 17 (2005) 953–961.
[85] C. Forsyth, T.W.S. Yip, S.V. Patwardhan, CO_2 sequestration by enzyme immobilized onto bioinspired silica, Chem. Commun. 49 (2013) 3191–3193.
[86] C.S. Kim, Y.J. Yang, S.Y. Bahn, H.J. Cha, A bioinspired dual-crosslinked tough silk protein hydrogel as a protective biocatalytic matrix for carbon sequestration, Npg Asia Mater. 9 (2017) e391.
[87] A.R. Millward, O.M. Yaghi, Metal–organic frameworks with exceptionally high capacity for storage of carbon dioxide at room temperature, J. Am. Chem. Soc. 127 (2005) 17998–17999.

[88] P.D.C. Dietzel, V. Besikiotis, R. Blom, Application of metal–organic frameworks with coordinatively unsaturated metal sites in storage and separation of methane and carbon dioxide, J. Mater. Chem. 19 (2009) 7362–7370.
[89] Y. Lin, C. Kong, Q. Zhang, L. Chen, Metal-organic frameworks for carbon dioxide capture and methane storage, Adv. Energy Mater. 7 (2017) 1601296.
[90] A.J. Howarth, et al., Chemical, thermal and mechanical stabilities of metal–organic frameworks, Nat. Rev. Mater. 1 (2016) 15018.
[91] P. Horcajada, et al., Porous metal–organic-framework nanoscale carriers as a potential platform for drug delivery and imaging, Nat. Mater. 9 (2009) 172.
[92] H. Kitagawa, Transported into fuel cells, Nat. Chem. 1 (2009) 689.
[93] J.-R. Li, R.J. Kuppler, H.-C. Zhou, Selective gas adsorption and separation in metal–organic frameworks, Chem. Soc. Rev. 38 (2009) 1477–1504.
[94] N.L. Rosi, et al., Hydrogen storage in microporous metal-organic frameworks, Sci. (80-) 300 (2003) 1127–1129.
[95] S. Bundschuh, et al., Mechanical properties of metal-organic frameworks: an indentation study on epitaxial thin films, Appl. Phys. Lett. 101 (2012) 101910.
[96] K.W. Chapman, G.J. Halder, P.J. Chupas, Guest-dependent high pressure phenomena in a nanoporous metal–organic framework material, J. Am. Chem. Soc. 130 (2008) 10524–10526.
[97] A. Demessence, D.M. D'Alessandro, M.L. Foo, J.R. Long, Strong CO_2 binding in a water-stable, triazolate-bridged metal–organic framework functionalized with ethylenediamine, J. Am. Chem. Soc. 131 (2009) 8784–8786.
[98] H. Wu, T. Yildirim, W. Zhou, Exceptional mechanical stability of highly porous zirconium metal–organic framework UiO-66 and its important implications, J. Phys. Chem. Lett. 4 (2013) 925–930.
[99] X. Lv, L. Li, S. Tang, C. Wang, X. Zhao, High CO_2/N_2 and CO_2/CH_4 selectivity in a chiral metal–organic framework with contracted pores and multiple functionalities, Chem. Commun. 50 (2014) 6886–6889.
[100] W.M. Bloch, R. Babarao, M.R. Hill, C.J. Doonan, C.J. Sumby, Post-synthetic structural processing in a metal–organic framework material as a mechanism for exceptional CO_2/N_2 selectivity, J. Am. Chem. Soc. 135 (2013) 10441–10448.
[101] M. Zeeshan, et al., Core–shell type ionic liquid/metal organic framework composite: an exceptionally high CO_2/CH_4 selectivity, J. Am. Chem. Soc. 140 (2018) 10113–10116.
[102] J. An, S.J. Geib, N.L. Rosi, High and selective CO_2 uptake in a cobalt adeninate metal–organic framework exhibiting pyrimidine- and amino-decorated pores, J. Am. Chem. Soc. 132 (2010) 38–39.
[103] S. Couck, et al., An amine-functionalized MIL-53 metal–organic framework with large separation power for CO_2 and CH_4, J. Am. Chem. Soc. 131 (2009) 6326–6327.
[104] T.M. McDonald, et al., Capture of carbon dioxide from air and flue gas in the alkylamine-appended metal–organic framework mmen-Mg2(dobpdc), J. Am. Chem. Soc. 134 (2012) 7056–7065.
[105] Y. Xie, et al., Rational design of MOFs constructed from modified aromatic amino acids, Chem. Eur. J. 13 (2007) 9399–9405.
[106] R.A. Smaldone, et al., Metal–organic frameworks from edible natural products, Angew. Chem. Int. Ed. 49 (2010) 8630–8634.
[107] J.J. Gassensmith, et al., Strong and reversible binding of carbon dioxide in a green metal–organic framework, J. Am. Chem. Soc. 133 (2011) 15312–15315.
[108] D. Wu, J.J. Gassensmith, D. Gouvêa, S. Ushakov, J.F. Stoddart, A. Navrotsky, Direct calorimetric measurement of enthalpy of adsorption of carbon dioxide on CD-MOF-2, a green metal–organic framework, J. Am. Chem. Soc. 135 (18) (2013) 6790–6793.
[109] J. An, S.J. Geib, N.L. Rosi, Cation-triggered drug release from a porous zinc – adeninate metal–organic framework, J. Am. Chem. Soc. 131 (2009) 8376–8377.
[110] J. An, et al., Metal-adeninate vertices for the construction of an exceptionally porous metal-organic framework, Nat. Commun. 3 (2012) 604.
[111] J.J. Gassensmith, et al., A metal–organic framework-based material for electrochemical sensing of carbon dioxide, J. Am. Chem. Soc. 136 (2014) 8277–8282.

[112] W.M.A.W. Daud, W.S.W. Ali, Comparison on pore development of activated carbon produced from palm shell and coconut shell, Bioresour. Technol. 93 (2004) 63–69.

[113] S. Kubo, R.J. White, N. Yoshizawa, M. Antonietti, M.-M. Titirici, Ordered Carbohydrate-derived porous carbons, Chem. Mater. 23 (2011) 4882–4885.

[114] A. Jain, R. Balasubramanian, M.P. Srinivasan, Hydrothermal conversion of biomass waste to activated carbon with high porosity: a review, Chem. Eng. J. 283 (2016) 789–805.

[115] W. Tian, et al., Bio-inspired beehive-like hierarchical nanoporous carbon derived from bamboo-based industrial by-product as a high performance supercapacitor electrode material, J. Mater. Chem. A 3 (2015) 5656–5664.

[116] F. Cheng, J. Liang, J. Zhao, Z. Tao, J. Chen, Biomass waste-derived microporous carbons with controlled texture and enhanced hydrogen uptake, Chem. Mater. 20 (2008) 1889–1895.

[117] N.S. Nasri, U.D. Hamza, S.N. Ismail, M.M. Ahmed, R. Mohsin, Assessment of porous carbons derived from sustainable palm solid waste for carbon dioxide capture, J. Clean. Prod. 71 (2014) 148–157.

[118] G. Durá, et al., Importance of micropore–mesopore interfaces in carbon dioxide capture by carbon-based materials, Angew. Chem. Int. Ed. 55 (2016) 9173–9177.

[119] S. Shahkarami, R. Azargohar, A.K. Dalai, J. Soltan, Breakthrough CO_2 adsorption in bio-based activated carbons, J. Environ. Sci. 34 (2015) 68–76.

[120] M. Sevilla, P. Valle-Vigón, A.B. Fuertes, N-doped polypyrrole-based porous carbons for CO_2 capture, Adv. Funct. Mater. 21 (2011) 2781–2787.

[121] S. De, A.M. Balu, J.C. van der Waal, R. Luque, Biomass-derived porous carbon materials: synthesis and catalytic applications, ChemCatChem 7 (2015) 1608–1629.

[122] F. Salvador, M.J. Sánchez-Montero, C. Izquierdo, C/H_2O reaction under supercritical conditions and their repercussions in the preparation of activated carbon, J. Phys. Chem. C. 111 (2007) 14011–14020.

[123] M. Sevilla, A.B. Fuertes, Chemical and structural properties of carbonaceous products obtained by hydrothermal carbonization of saccharides, Chem. – A Eur. J. 15 (2009) 4195–4203.

[124] M. Sevilla, A.B. Fuertes, The production of carbon materials by hydrothermal carbonization of cellulose, Carbon N. Y. 47 (2009) 2281–2289.

[125] M. Sevilla, R. Mokaya, Energy storage applications of activated carbons: supercapacitors and hydrogen storage, Energy Environ. Sci. 7 (2014) 1250–1280.

[126] F. Rodríguez-Reinoso, M. Molina-Sabio, Activated carbons from lignocellulosic materials by chemical and/or physical activation: an overview, Carbon N. Y. 30 (1992) 1111–1118.

[127] C. Bouchelta, M.S. Medjram, O. Bertrand, J.-P. Bellat, Preparation and characterization of activated carbon from date stones by physical activation with steam, J. Anal. Appl. Pyrolysis 82 (2008) 70–77.

[128] J. Chen, J. Yang, G. Hu, X. Hu, Z. Li, S. Shen, et al., Enhanced CO_2 capture capacity of nitrogen-doped biomass-derived porous carbons, ACS Sustain. Chem. Eng. 4 (3) (2016) 1439–1445.

[129] S.-M. Hong, E. Jang, A.D. Dysart, V.G. Pol, K.B. Lee, CO_2 capture in the sustainable wheat-derived activated microporous carbon compartments, Sci. Rep. 6 (2016) 34590.

[130] B. Zhu, K. Qiu, C. Shang, Z. Guo, Naturally derived porous carbon with selective metal- and/or nitrogen-doping for efficient CO_2 capture and oxygen reduction, J. Mater. Chem. A 3 (2015) 5212–5222.

[131] S.-J. Son, et al., Development of carbon dioxide adsorbents using carbon materials prepared from coconut shell, Korean J. Chem. Eng. 22 (2005) 291–297.

[132] M.A. Ahmad, W.M.A. Wan Daud, M.K. Aroua, Adsorption kinetics of various gases in carbon molecular sieves (CMS) produced from palm shell, Colloids Surf. A Physicochem. Eng. Asp. 312 (2008) 131–135.

[133] C. Gómez de Salazar, A. Sepúlveda-Escribano, F. Rodriguez-Reinoso, Preparation of Carbon Molecular Sieves by Controlled Oxidation Treatments, Carbon 38 (2000).

[134] A.R. Mohamed, M. Mohammadi, G.N. Darzi, Preparation of carbon molecular sieve from lignocellulosic biomass: a review, Renew. Sustain. Energy Rev. 14 (2010) 1591–1599.

[135] M.-M. Titirici, et al., Sustainable carbon materials, Chem. Soc. Rev. 44 (2015) 250–290.

CHAPTER FOUR

Nanocatalysts and sensors in coal gasification process

Irfan Celal Engin[1] and Mufrettin Murat Sari[2]
[1]Mining Engineering Department, Faculty of Engineering, Afyon Kocatepe University, Afyonkarahisar, Turkey
[2]Texas A&M University, Commerce, TX, United States

4.1 Introduction

Today, global energy needs are increasing rapidly in parallel with industrialization and population growth. The expectation is that the energy demand will also increase in the coming years. Energy security, environmental problems and the cost of energy are becoming more important with increasing energy needs. The environmental problems arising from the use of fossil fuels as the primary energy source necessitate the more efficient use and the reduction of their environmental impact. Renewable energy sources such as solar energy, wind energy and geothermal energy, which are shown as alternatives to fossil fuels, have yet to reach the required energy demand. Nuclear energy allows for the production of high capacity power, but also radioactive waste resulting from nuclear reaction and storage of these wastes are the problem. Especially in recent years, researches have focused on the production of cleaner and environmentally friendly liquid fuels, various chemicals and electrical energy with advanced transformation technologies instead of conventional systems from fuels such as coal and biomass. The gasification process in this area stands out as an alternative technology that enables the production of clean gas products that can be used in many areas such as internal combustion engine, fuel cell, liquid fuel production. Increasing environmental sensitivity around the world and the limitations on carbon emission against global warming are the main factors that increase the interest in the gasification process.

Gasification is the process of converting hydrocarbon fuels such as coal, petroleum, biomass and solid wastes into gases, such as CO, H2, CO_2, and CH_4. The gasification process is carried out in the reactors called gasifier and the gas mixture formed as the product is called the synthesis gas. The synthesis gas (Syngas) generally consists of CO, CO_2, H_2. Synthesis gas takes its name because of it is synthetic natural gas intermediate product and is involved in the synthesis of different chemicals such as ammonia, fertilizer or methanol. The gasification process among solid fuel and solid waste based electricity generation technologies stands out as a clean and environmentally friendly

Sustainable Materials for Transitional and Alternative Energy.
DOI: https://doi.org/10.1016/B978-0-12-824379-4.00001-X

technology with the lowest emissions and solid and liquid waste values. In the gasification process, a smaller amount of CO_2, SO_2 and NOx is produced than combustion technologies.

Gasification technologies have been used commercially for the production of both liquid fuel and various chemicals worldwide for nearly a century. The gasification process can be carried out using an oxidant as air, oxygen, water vapor or a mixture in various proportions; in a pressurized or atmospheric fixed bed, fluidized bed, or entrained flow gasifiers. There are many gasifier systems known worldwide under the trade name [1].

Coal is a flammable sedimentary fossil fuel whose main component is carbon. In addition to carbon, it contains hydrogen, oxygen and a small amount of sulfur, nitrogen, and inorganic compounds. Coal is the flammable, fossilized solid organic masses in brown and black color which is occurred as a result of the change of the plant survivors against decaying in swamps in marsh conditions and the accumulation of them in time due to geographic, geological, biochemical and physical effects [2].

The difference of coal which is very similar to petroleum due to the raw materials that make up the structure is caused by high amount of carbon and low amount of hydrogen content. Because of its low hydrogen content, coal is not fluid in nature, such as oil, but in a solid state [3,4]. Coal is one of the most important raw materials of today.

In the classification of coal: coal type, constant carbon content, carbon / hydrogen ratio, volatile matter content, agglomeration, thermal value and petrographic properties are taken into consideration [5]. According to ASTM D 388 Standard, the classification from meta anthracite to high volatile A bituminous coal is based on the constant carbon content and volatile matter content of coal. The clumping property of coal also affects this classification [6]. In the classification of coal in Europe, thermal value, volatile matter content, fixed carbon content and caking characteristics is taken into consideration. Coal is divided into two as bituminous and brown coals. The bituminous coal is divided into subgroups as anthracite, bituminous coal and subbituminous coal, and the brown coal as semibituminous coal and lignite. According to the coal classification of the International Coal Classification Committee of the European Economic Commission (UNECE), coal with an upper thermal value of over 5700 kcal/kg is called bituminous (hard) coal. Coals with an upper thermal value of less than 5700 kcal/kg and a volatile content of more than 31% are called brown coal [7].

Coal has a large share in the world's electricity production. However, the most important problem in the use of coal is environmental pollution caused by fossil fuels. Gases released into the atmosphere cause global warming. Efforts to remove SO_2 caused by the combustion of coal started in England in 1850s. Since the 1970s, it has become necessary to implement environmental control technologies in order to stay below the emission values determined by international agreements, especially the European Union and the development of environmental legislation. With the emergence of the environmental impacts of coal, which has an important role in meeting the energy needs of the world, processes have been started to minimize waste and emissions in terms of environmental pollution [8].

The concept of clean coal technologies generally refers to a definition that encompasses efficiency and environmental dimensions in the production, preparation and use of coal. These technologies include; on the one hand, reduction of emissions and wastes, on the other hand increasing the energy to be obtained from unit coal. The research covers a wide range of applications such as increasing the efficiency of coal-based thermal power plants and reducing CO2 emissions from these plants, coal washing, gasification, carbon capture and storage [9].

Coal gasification technologies are used effectively in the world for both energy and chemical production. The synthesis gas obtained by gasification is utilized in power generation plants with high efficiency in Integrated Gasification Combined Cycle (IGCC) power generation plants, and by catalytic conversion, hydrogen, methanol, ammonia, methane, ethanol, propanol, di-methyl-ether, ethylene, propylene, acetic acid, acetone, methyl It is possible to produce liquid hydrocarbons such as ethyl ketone, synthetic gasoline, synthetic diesel, organic chemicals, aliphatic hydrocarbons and many other types of chemicals.

4.2 The importance of coal gasification in terms of fossil fuels

Fossil fuels, also known as mineral fuels, are natural energy sources such as coal, petroleum and natural gas, which contain hydrocarbons and high levels of carbon, which are made of rotting plant and animal wastes in past ages. When the general distribution of fossil fuels is examined, it is seen that oil and natural gas reserves are concentrated in certain geographical regions of the world, coal is distributed regularly and its production is realized in more than 50 countries.

Since it is the main source of many chemical raw materials, especially liquid fuels used in transportation, petroleum has an important place in fossil fuels. The oil, which was first used as fuel of gas lamps, has increased its importance with the production of automobiles and its use especially in the 1st World War. With the 20th century, the use of oil dominated chemical industries and processes. Today, fuels from crude oil meet 96–98% of the worldwide demand for transportation (cars, ships, airplanes), and more than 55% of the oil is refined to produce fuel. However, oil reserves are increasingly depleted and are estimated to have a life span of 40 to 60 years at the current consumption rate [10]. Since natural gas is a clean fuel with high energy efficiency and no sulfur and ash, it is a growing fossil fuel. It is also used in electricity generation from cycle power plants. It is the third energy source in fossil fuels after oil and coal.

Coal is the most abundant fossil fuel in the world and has been used for many years as a solid fuel in electricity production, iron and steel plants, chemical industry and houses. Its use is increasing especially in Asia—Pacific and developing countries. The long-term availability of resources to respond to consumption, the availability of resources in many countries, and the low production costs keep the interest in coal high.

In addition, it is seen as coal comes at the beginning of the fossil fuels responsible for the global warming that led to the release of the greenhouse gas carbon dioxide. Coal is the most dangerous energy source in terms of environmental impacts. Coal consists mainly of carbon and hydrogen and a small amount of sulfur and iron. When burned, it releases some harmful gases which are not environmentally friendly. By burning carbon, carbon is released, it mixes with oxygen and forms carbon dioxide. As a result of coal combustion, harmful substances such as sulfur dioxide, oxides and ash also appear. Acid rain and air pollution are some of the environmental problems that are caused. The increase in global warming is directly related to the increased coal consumption for our daily needs. While developments in the field of technology have enabled the capture of a few impurities before being released into the atmosphere, large-scale production and use of coal has badly affected the environment.

All these environmental pressures have led to restrictions on the use of fossil fuels or the development of cleaner use technologies. Coal has been the most exposed to these criticisms because it is the fossil fuel that causes the most carbon dioxide emissions. In order to achieve the same amount of energy, the greenhouse gas to be released into the atmosphere is twice as high as natural gas. As a solution to this, clean coal technologies have been developed and coal gasification applications have pioneered them.

In gasification, higher energy efficiency is achieved than direct combustion of coal. This is because less coal is used to produce the same amount of energy in coal gasification, which results in less carbon dioxide (CO_2) emissions. Instead of direct coal burning, gasification converts the entire carbon of the coal into electrical, hydrogen and other energy types through partial oxidation. Coal is combined with hot steam and controlled air or oxygen at high temperatures (up to 2600°F) and high pressures (up to 1200 psig). Depending on the conditions in the gasifier and the coal used, Syngas, a mixture of carbon dioxide, hydrogen, methane, nitrogen and carbon monoxide, is obtained. The gas is cooled and 99 percent of the impurities forming the pollutant are removed. Gasification is a more environmentally friendly technology than combustion technologies and is much more advantageous in terms of CO_2, SO_2, NOx emissions. The present sulfur is mostly converted into H_2S, which can be removed more easily than SO_2. NOx, dioxin and furan problems do not occur during gasification.

To further increase efficiency, the cleaned gas is burned in a conventional gas turbine to produce electrical energy and the steam generates steam for the steam turbine. It can be converted into portable gas in portable quality in the pipeline and can be sent directly to people's homes. It can also be used as a building block to produce more complex products in refined and petrochemical industries.

Coal gasification offers one of the cleanest ways of producing many chemicals with valuable energy sources such as coal, electricity, hydrogen and other liquid fuels. This situation is also very important in terms of reducing the dependence of countries on oil. It can be produced domestically and production facilities that produce these alternative energy sources contribute to the development of local economies. Gasification facilities will provide business, income and economic growth for the areas where the facilities are established [11].

Coal gasification power plants are currently operating commercially in the United States and other countries. According to experts, next generation clean coal technology will be at the center of its facilities.

4.3 Types of coal gasification process

Gasification; It is a thermochemical process that converts carbon-rich fuels such as coal, oil and biomass into carbon monoxide and hydrogen. While the basis of gasification is based on combustion, it is different from burning. In oxygenation processes, less oxygen is used than the stoichiometric oxygen amount which ensures complete combustion. The amount of gases obtained by gasification depends on factors such as the size of the coal, gasification gas used, pressure, reactor type, temperature. The gas content obtained as a result of gassing is generally CO, CH_4, H_2, CO_2, H_2O and N_2, although it contains tar and small amounts of contaminants such as H_2S, COS, NH_3 etc. The so-called syngas gas contains mainly CO and H_2 gases [12].

The first step in coal gasification is the drying of coal. This step can have a significant effect, especially when a high-humidity, low-grade coal is gasified. The effect of water evaporation on bituminous coals is relatively small. With the pyrolysis of coal at temperatures above about 320°C, the bonds between carbons, carbon and oxygen, nitrogens or sulfurs, which form the backbone of the organic compounds, break.

Pyrolysis is the process of chemical decomposition of organic materials by heating to a sufficient temperature in the absence of oxygen. Pyrolysis and devolatilization are often used interchangeably. The coal may be considered as a network of aromatic clusters polymers linked together by aliphatic bridges as shown in Fig. 4.1. Side chains of different functional groups, such as methyl, ethyl or carboxylic groups, may also be added to the batches. As the temperature of the coal increases, the

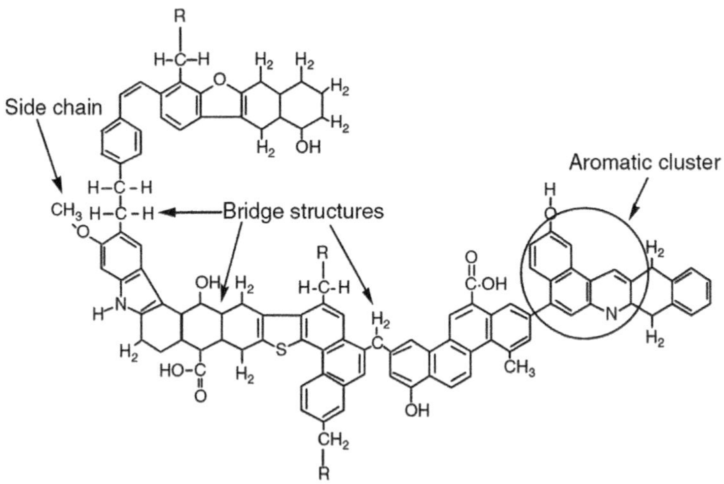

Figure 4.1 Hypothetical coal structural model, illustrating aromatic clusters, bridge structures, and side chains [13].

bridges between the clusters can be broken and as a result the coal polymer is divided into smaller pieces.

At the start of the reaction, unstable molecular particles are formed. These particles may also be pyrolyzed or reacted to form relatively stable compounds. When the pyrolytic products are cooled to ambient temperature, they may be composed of gases, coal-tary liquids or solids. With the increase in the amount of volatiles released, the gasification begins after this phase. However, if the heating rate is increased, the volatilization and gassing phases are carried out at the same time, and the outlines are composed of (fixed carbon), tar (heavy hydrocarbons) light gases (CH_4, CO, H_2, water vapor) [14]. Oxygen is supplied to the gasifier as a very pure oxygen or as normal air. When oxygen reacts with coal, carbon monoxide or carbon dioxide is produced. Oxygen may also react with gas phase substances.

The content of volatile gases in general; H_2O, H_2, N_2, O_2, CO, CH_4, H_2S, NH_3, C_2H_6 and a small amount of olefin, aromatic and unsaturated hydrocarbons such as tar [15]. Basic reactions are given in Table 4.1 [16].

For the temperature and pressure dependence of the composition of the produced gas, it is necessary to take a brief look at the thermodynamics and kinetics of the gasification event. From a thermodynamic point of view, the increase in pressure in gasification by hydrogen converts the balance in favor of methane, while the increase in temperature reduces the methane in the balance. In the methanization reaction, the increase in pressure increases the amount of water vapor and methane in the balance. Exothermic gasification reaction with hydrogen and endothermic heterogeneous water

Table 4.1 Basic reactions in coal gasification process [16].

Reaction	**Enthalpy**	**Definition**
$H_2O(l) \rightarrow H_2O(g)$	+ 43.99 MJ/kmole	Drying of coal
$Coal \rightarrow Coke(s) + Volatiles(g)$	+ variable	Pyrolysis of coal
$C(s) + O_2(g) \rightarrow CO_2(g)$	− 393.98 MJ/kmole	Oxidation of carbon
$2C(s) + O_2(g) \rightarrow 2CO(g)$	− 221.31 MJ/kmole	Partial oxidation of carbon
$2CO(g) + O_2(g) \rightarrow 2CO_2(g)$	− 566.65 MJ/kmole	Partial oxidation of carbon monoxide
$2H_2(g) + O_2(g) \rightarrow 2H_2O(g)$	− 484.23 MJ/kmole	Partial oxidation of hydrogen
$2CH_4(g) + O_2(g) \rightarrow 2CO(g) + 4H_2(g)$	− 71.44 MJ/kmole	Partial oxidation of methane
$C(s) + H_2O(g) \rightarrow CO(g) + H_2(g)$	+ 131.46 MJ/kmole	Char reaction (steam gasification reaction)
$C(s) + 2H_2(g) \rightarrow CH_4(g)$	− 74.94 MJ/kmole	Char reaction (methanation reaction)
$C(s) + CO_2(g) \rightarrow 2CO(g)$	+ 172.67 MJ/kmole	Char reaction (Boudouard reaction)
$CO(g) + H_2O(g) \rightarrow CO_2(g) + H_2(g)$	− 41.21 MJ/kmole	Additional Gas Phase Reactions (steam exchange reaction)
$CH_4(g) + H_2O(g) \rightarrow CO(g) + 3H_2(g)$	+ 206.2 MJ/kmole	Additional Gas Phase Reactions (steam methane reforming reaction)
$Ash(s) \rightarrow Slag(l)$	+ variable	Slagging (Melting of ash)

gas reaction and Boudouard reactions cause a decrease in the amount of water vapor, CO_2 and CH_4 at high temperatures [17].

The steps in a typical coal gasification process are shown schematically in Fig. 4.2.

It examines the reaction kinetics of transport events affecting the speed and speed of the chemical equilibrium. Three important transport steps in the gasification of coals are as follows:

1. Diffusion of the gasifier in the hydrodynamic boundary film formed on the coal surface.
2. Diffusion into pore.
3. Chemical reaction in pore surfaces.

The slowest of these steps in reaction conditions determines the reaction rate. The char gasification occurs at the same time as the drying and pyrolysis stages. The char is the remaining carbon from the coal and is the remaining carbon. In the char gasification, there are reactions that limit the carbon transformation. These reactions occur more slowly than the drying and volatile release phases. Heterogeneous char reactions

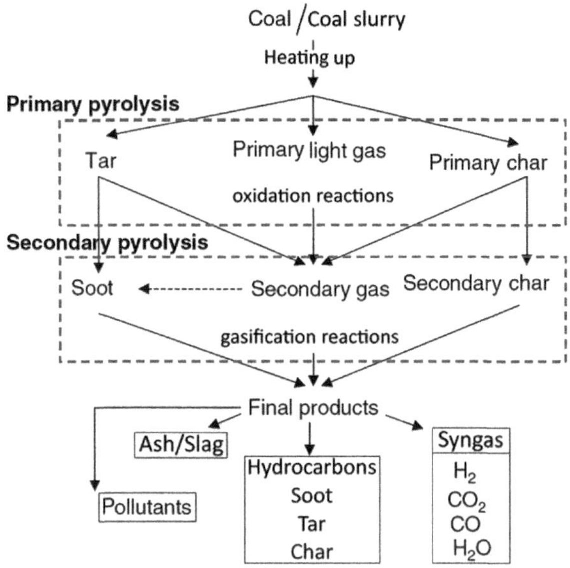

Figure 4.2 Flow sheet of coal gasification process [13,18].

occur during the char oxidation phase. Boudouard and heterogeneous water gas reaction occur faster than the methane reaction. The char gasification reactions are very slow reactions due to low temperature and abundant CO and H_2 in volatile release phase [19]. During the gas phase phase, homogenous gasification reactions occur between the gases after the char gasification and the first volatilization phase.

The way to increase the energy value of the syngas is to support the endothermic gas generation reactions as much as possible. These reactions do not occur spontaneously, and are reliable for exothermic reactions during gasification to increase the temperature of the mixture to the desired gasification temperature and to provide heat for endothermic gas forming reactions. Oxidation reactions occur very rapidly and consume all the oxygen contained in the gasifier. Reducing ambient conditions are thus achieved in the gasifier [20].

The heat required for endothermic vapor gasification reactions can be provided by an external heat source such as solar heat, nuclear heat and external coal or gas combustion. In the case of an external heat source, the design of heat transfer to withstand high gasification temperatures and corrosive atmospheres complicates the system.

Typical feed materials of a gasifier are coal, oxygen and water. Oxygen can be supplied as an almost pure oxygen stream or air from an air separation unit. Water enters the gasifier as coal moisture, coal slurry or steam. Oxygen and water supply quantities/ratios must be sufficient to completely gasify the feed. If the oxygen supply is too high, the reaction is becoming burning rather than gasification, and gases with a low heating value are produced. The temperature is controlled by changing the oxygen/water balance. All of the reactions with oxygen are exothermic, therefore oxygen tend to increase the gasification temperature. The vapor gasification reaction provides additional gas formation. Steam gasification is endothermic, therefore it tends to reduce the gasification temperature. In order to optimize the gasification process, the correct ratio of (O_2 H_2 H_2O)/coal and the ratio of O_2/H_2O must be chosen. Steam can be used instead of carbon dioxide, but CO_2 gasification reaction is slower than the steam gasification reaction. Most syngas applications support the higher H_2/CO ratio produced by the vapor gasification reaction. Oxygen/water/coal ratios depend on gasification configuration, operating conditions and coal selection.

In gasification, the synthesis gas is divided into three groups as low, medium and high thermal values in terms of energy values. The coals with a thermal value in the range of 3.35–7.53 MJ/Nm^3 are included in the group of low, the coals with a thermal value of 7.53–15.07 MJ/Nm^3 are included in the group of medium and the coals with a thermal value of 33.6 MJ/Nm^3 are included in the group of high energy synthesis gas.

Especially in coal gasification, it is aimed to obtain clean energy by removing the pollutants before burning the energy source. For this purpose, the syngas produced must be free of harmful H_2S, SO_2, COS, NO, NO_2, N_2O, HgS, $HgCl_2$ gases, and CO_2 be captured and stored. Various processes have been developed for this purpose.

Gasification applications are divided into two main groups as surface and underground coal gasification. While underground gasification applications are classified as Shaft UCG methods, Shaftless UCG methods; above ground gasification applications are divided into 3 main groups as fixed/moving bed type, fluidized bed type, and entrained flow type.

4.3.1 Above-ground coal gasification

Clean energy under short life of the plant above ground syngas production by partial oxidation process is called coal gasification surface. The history of coal gasification dates back to 1600 years, but the gasification of coal on a commercial scale was carried out in 1972 by Scottish engineer William Murdock. This gas, called a gas or air gas, was used to illuminate factories and streets in England [21].

In 1816, the Baltimore Gas Company in the United States established the first coal gasification facility. In the 1920s, with the use of oxygen, generated via the commercialization of Carl von Linde's air through the cooling cycle into its components, synthesis gas and hydrogen production were started to be made in the gasification process which was carried out by air until then. In the following period, Winkler Fluidized Bed Process in 1926, Lurgi-Pressure Gasifier Process in 1931,

Koppers—Totzek Drift Flow Gasification Process in 1940s was found and applied. In the mid-1920s, the US began to produce 20% of its gas supply from coal. Before World War 2, at least 20,000 gasifiers were operating in the US, and gas production from coal began to lose its importance in the 1940s due to the low cost of natural gas. In the 1950s and 1960s, the oil market was the leader and a relatively small amount of coal-producing facilities remained.

In the 1970s, the shortage of natural gas resources in the USA began to be noticed and the gas trade was limited. This situation has raised interest in research related to the production of gas and liquid fuel from coal by introducing coal as an important source [21]. Since the 1980s, researchers have begun to evaluate the environmental effects of gasification technologies. The commercialization of gasification processes has gained momentum with the encouragement of governments and industry stakeholders, which have forced to reduce the emissions and greenhouse gases, by the harsh and restrictive environmental standards particularly in power plants and industrial establishments [15].

In the coal gasification process, coal, can be used as fuel to generate electricity or as a basic chemical building block for many chemicals, is converted to a substantially Syngas comprising hydrogen and carbon monoxide. The dry or slurry coal feed reacts with steam and oxygen in the gasifier (or air) at a high temperature and pressure in a reduced atmosphere. Clean syngas can be obtained after removal of solid particles, sulfur and other impurities by subsequent purification [13].

According to the coal feed and oxidant flow through the gasifier, coal gasification technologies can be classified into three basic types: mobile or fixed bed, fluidized bed and entrained flow technology. The entrained flow gasification process among all these processes has the greatest purification capability and therefore the least environmental impact. Thus, entrained flow processes are applied in most commercial enterprises. Commercial gasifiers such as British Gas Lurgi (BGL), Multi Purpose Gasifier (MPG), Lurgi Mark IV Gasifier, Plasma Gazification are among the fixed/mobile bed gasifiers. U-GAS, High Temperature Winkler Gasifier's commercial gasifiers are among the fluidized bed gasifiers. At present, General Electric Corporation (GE), Shell, Siemens, Mitsubishi Heavy Industries, ConocoPhillips and GSP (Gas Schwarze Pumpe), are the leading supplier of entrained flow gasification technology. General structures of coal gasification systems are given in Fig. 4.3.

Moving bed gasifiers (fixed bed gasifiers) use lump coal of size 6—50 mm. Generally, the reagent is fed from the bottom while the coal is loaded from the top of the gasifier. The coal is slowly downwardly moved under gravity while being gasified by countercurrent. In a moving bed gasifier, the coal is preheated by the hot syngas rising from the gasification zone and the coal is pyrolyzed. Compared to other gasification processes, the oxygen consumption in this process is low, but the synthesis

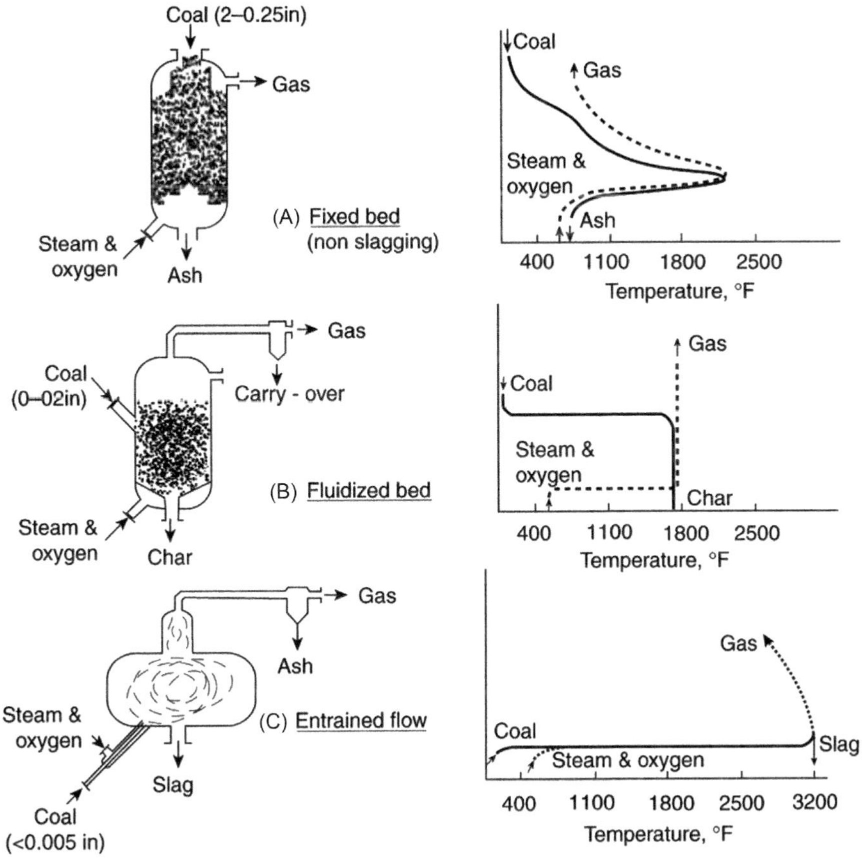

Figure 4.3 Gasifier types: (A) moving bed (dry ash), (B) fluidized bed, and (C) entrained flow and temperature changes in gasifiers [22,23].

product contains pyrolysis products. The temperature of the syngas at the outlet is generally low. The design and operation of moving bed gasifier devices is simple and has high reliability. Furthermore, the inverse current contact between the coal and the reactant gas results in a high coal conversion, hence high heat efficiency. 4 types are available: upstream gasifiers, downstream gasifiers, counter flow gasifiers, open flow gasifiers.

The fluidised-bed gasifier uses a bed of sand or other inert material as well as small coal pieces or ground coal. The coal particles are fluidized by the oxidant flow in the gasifier. Fluidized bed gasifiers offer a homogeneous temperature environment and support heat and mass transfer between gas and solid phases. As in other fluidized bed applications, the size of the coal particles in the feed is critical. Furthermore, the fluidized bed reactor may only operate at temperatures below the softening point of the ash. Low operating temperature also means that they are suitable for gasifying reactive feedstocks, such as low-grade coals. However, as a continuous stirred tank reactor (CSTR), the carbon conversion in the fluidized bed is generally lower than in other gasification methods.

The entrained flow reactors operate in simultaneous flow. The pulverized coal is contained in the oxidant (oxygen and vapor) and fed to the gasifier (sometimes using a coal water slurry [CWS] and pumped to the gasifier as in GE gasifiers). The reaction takes place at a high temperature (1500°C–1900°C) and produces CO, H_2, CO_2 and other gases. The ash is removed from the bottom of the reactor as molten slag. Due to the short residence time in the reactor, high temperatures are required to ensure a good coal conversion and therefore the operating temperature is high. In order to maintain the high operating temperature, the oxygen flow in the entrained flow gasifier is also high. Due to the high operating temperature, the entrained flow gasifiers do not have any technical limitation on the type of feed used. However, coals with high moisture or ash content require higher oxygen consumption and reduce heat efficiency compared to alternative processes. A comparison of all these technologies is given in Table 4.2.

Table 4.2 Comparison of different gasification technologies [24].

Gasifiers	Advantages	Disadvantages
– Fixed bed	– High thermal efficiency – Low evaporation temperature	– Separator required for tar removal – The feed size should be > 0.25 mm
– Entrained flow	– Simple to design – Easy raw material feeding – High C conversion efficiency – Thermal flexibility	– Large volume of process gas – Oxygen is required for operation – Large product container requirement – Cooling system required – Feed should be in powder form
– Fluidized bed	– Working with air – Processing various sizes of raw materials – Homogeneous temperature distribution – Reduction of tar and semi product quantities – High gasification rate – Bedding material can be changed according to fuel	– Remaining carbon – Limited conversion

4.3.2 Underground gasification

Gasification of coal at underground is a method that aims to obtain flammable gas from low quality and uneconomic coal deposits when extracted by traditional mining methods. In this method, two wells are drilled into the coal strata and air, pure oxygen and water vapor are sent in a controlled manner to obtain flammable artificial gas from the production well. The content of the artificial gas is mostly composed of CO_2, CO, H_2 and CH_4. The process is abbreviated as ISCG, representing the initials of the in-situ coal gasification words, or UCG representing the initials of underground coal gasification words (Fig. 4.4).

In this method, coal beds serve as reactors. Thus, gasification is not carried out on the ground, but underground. Process is not combustion is a partial oxidation and the resulting combustible gas (Syngas) can be used in the established IGCC (integrated gasification combined cycle) power plant to produce electricity as well as in particularly in the production of liquid fuel with the FTS (Fischer-Tropsch) method and chemicals.

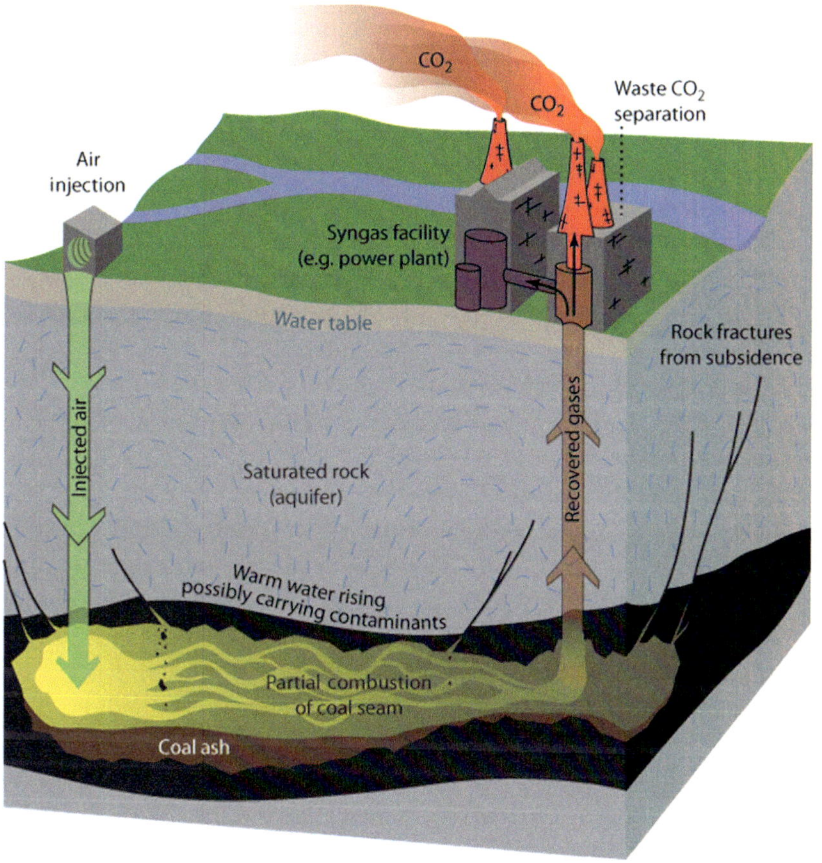

Figure 4.4 Schematic view of the application of in-situ gasification [25].

The idea of coal gasification at underground was probably first introduced in England in 1868 by German scientist Sir William Siemens. Siemens claimed that gas production directly from underground coal would be efficient and that the resulting gas pressure would help transport the gas produced to the desired distance with the pipe. In the same years, the famous Russian scientist Dmitriy Ivanovich Mendeleev examined several fires resulting from spontaneous combustion in underground coal mines. As a result of his investigations he decided that by directing these fires under control, energy could be generated from underground coal deposits. Mendelev argued that in the future coal will not be produced from underground but will be fully gasified and that this gas will be transferred to long distances and used at the desired location.A detailed design has been developed and the studies have been published during the 1880s and 1890s. In 1909, American scientist Betts received a patent for the underground coal gasification process developed by Mendelev based on underground coal gasification models. First-time Nobel laureate Sir William Ramsay conducted field tests in 1912 in Durham, England [26]. Due to World War I and Ramser's death, the work had to be terminated. Intensive work on the process was initiated in the Soviet Union in 1930 and commenced commercial production in 1950. During the 40-year underground coal gasification tests in the Soviet Union, high-humidity lignite, large amounts of coal, including bituminous and semi-anthracite containing high volatile matter, were gasified [27,28]. The only remaining commercial underground coal gasification plant after the collapse of the Soviet Union was Angren in Uzbekistan. The discovery of very large natural gas resources in Siberia is considered to be the reason for the decline of the underground coal gasification process in Russia. As a result of the operation of the plant in Uzbekistan, research and development activities for the underground coal gasification process in Russia and Ukraine are continuing [29].

Between 1940 and 1960, various studies were conducted in England, USA, Japan, Poland, Czechoslovakia, Belgium, Monoco and Italy, but these studies were generally unsuccessful [30–34]. Between 1972 and 1989, more than 30 trials were performed in various mines and geological conditions in various parts of the country, such as Wyoming, Texas, Washington, and Virginia. An important consequence of the studies of the underground coal gasification process in the US is the development of the CRIP process by LLNL researchers. This technique was used in the Rocky Mountain 1 trial. This trial in Wyoming between November 1987 and February 1988 is the most successful attempt by the United States. With the CRIP method, another connection technology, the ELW technique, has also been tested. In addition, analytical and numerical models have been developed. In the early 1980s, large scale underground coal gasification processes were not maintained in the USA due to the decrease in oil and gas prices. The increase in energy demand in recent years has stimulated the interest in underground coal gasification technology.

In the UK, underground coal gasification technology has been identified by DTI as one of the potential technologies of the future and following the preliminary feasibility study in June 2000, site selection studies were initiated for the underground coal gasification trial. In Poland and many other European Union member countries, the HUGE project is being carried out for the European Union. The fact that China has many patents in the field of underground coal gasification shows that the country has the largest program in this area. Many attempts have been made since the late 1980s and are still being conducted. The results of the experiments carried out at the Wonuushan coal mine show that large-scale hydrogen production is appropriate with the underground coal gasification process [29].

Today, Cougar Energy, Linc Energy and Carbon Energy Ltd. commercial projects are being developed by Australian companies. In recent days, a trial by Ergo Exergy, a Canadian company, has been completed in Australia, and work is underway in South Africa on an underground coal gasification plant (URL 1–2).

In 1996, in collaboration with Cougar Energy and Ergo Exergy, the Chinchilla underground coal gasification trial was initiated in Southeast Queensland [29]. Source characterization was completed by Cougar Energy in the Kingaroy region of Queensland, and regional characterization was initiated for the 400 MW combined cycle power plant project.

In December 2006, Linc Energy signed an agreement between the Skochinsky Mining Institute in Moscow and the Scientific-Technical Mining Partnership. Linc Energy's plan is to exploit the expertise of Russia and Uzbekistan to expand the underground coal gasification process in Australia (URL 3). Carbon Energy plans a large-scale trial at Bloodwood Creek in Surat Basin, Queensland (URL 4). In addition to commercial development, many research projects are being carried out to model the underground coal gasification process in CSIRO, Queensland and New South Wales Universities. Since 2001, Eskom has been conducting underground coal gasification in the Majuba power plant in South Africa. In India, ONGC plans to build a pilot project using the recommendations of experts from the Skochinsky Mining Institute [29]. In Japan, technical and economic studies for a small-scale underground coal gasification plant are being carried out by Tokyo University and coal companies and a trial is planned recently. A feasibility study was conducted for this plan [35].

In-situ gasification applications can be grouped into two groups, taking into account past successful applications: shaft and shaftless [36]. The sub-headings can be listed as follows [36,37]:

<u>Shaft UCG methods:</u>

- *Chamber or warehouse method* (Fig. 4.5)
- *Borehole producer method*
- *Stream method* (Fig. 4.6)
- *Long and large tunnel (LLT) method*

Shaftless UCG methods:

– Linked *vertical wells (LVW) method*
– *Controlled retracting injecting point (CRIP) method*
– *Steeply dipping beds methods*

Currently, the most commonly used methods are the vertical wells (LVW) method and the oncolled retracting injecting point (CRIP) method.

Shaft methods use coal mine galleries and shafts to transport gasification reagents and products, which sometimes entail the creation of shafts and the drilling of large-diameter openings through underground labor [38]. The shaft method was the first technique utilized within UCG systems. Currently, the shaft method is only employed in closed coal mines due to economic and safety reasons.

The following are examples of common UCG shaft methods: *Chamber or warehouse method* utilizes constructed underground galleries with brick walls separating coal panels. Gasification agents are supplied to a previously ignited coal face on one side of the wall, and the syngas is removed from a gallery on the other side. The chamber method strongly relies on the natural permeability of the coal seam to allow for sufficient oxidant flow through the system. The syngas composition may vary during operation, and the gas production rates are often low [39] (Fig. 4.5).

For b*orehole producer method*, parallel underground galleries are created within a coal seam with sufficient distance between them. The galleries are connected by drilling boreholes from one gallery to the other [38]. Remote electric ignition of the coal in each borehole is used to initiate the gasification process. This method is designed to gasify considerably flat-lying seams [38,39]. *Stream method* is designed for sharply inclined coal beds. Parallel pitched galleries following the contour of the coal seam are constructed and are connected at the bottom of the seam by a horizontal gallery also known as a fire-drift. To initiate gasification, fire is introduced within the horizontal gallery. The hot coal face moves up the seam slope with oxidant fed through one inclined gallery and syngas leaving through the other. The main advantage of this method is that the ash and roof material drop down to fill the void space created during the process, which prevents suffocating the gasification process at the coal front [39] (Fig. 4.6).

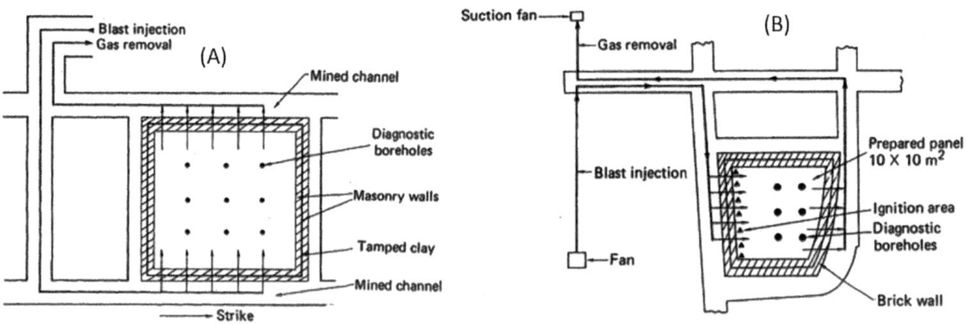

Figure 4.5 First UCG experiments in the former Soviet Union using the chamber method Kirichenko's chamber method (A) and at the Krutova mine (B) [40].

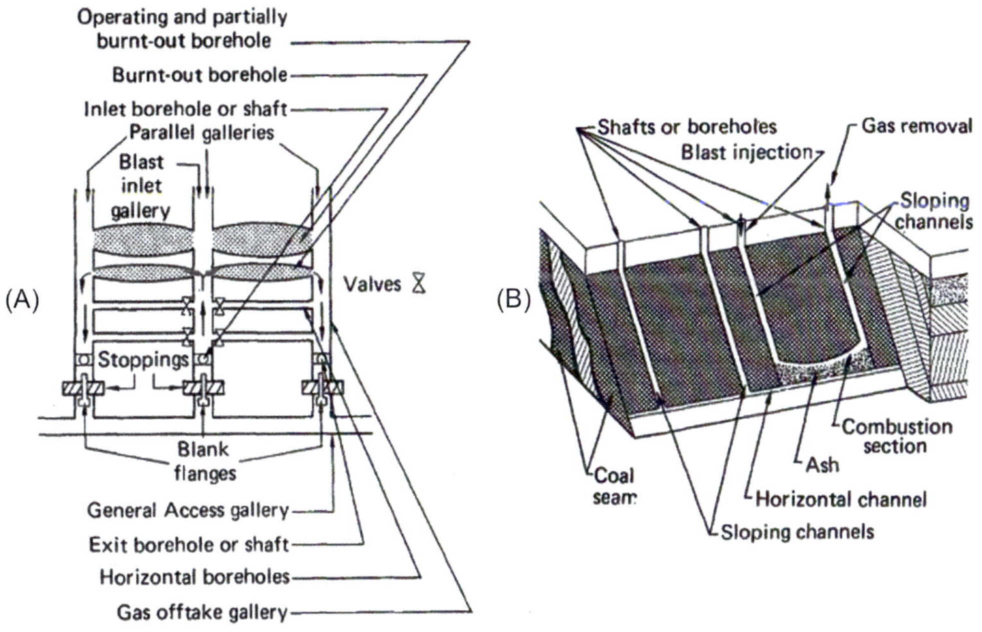

Figure 4.6 Borehole producer method (A) for gasifying coal in flat-lying seams and and the stream method (B) for gasifying coal in steeply dipping beds [40].

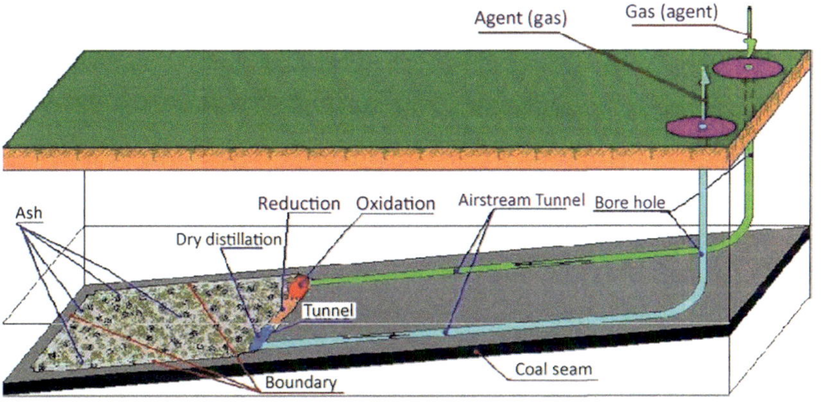

Figure 4.7 Long and large tunnel method (LLT) of gasifying coal seams [44].

LLT gasification method utilizes mined tunnels or constructed roadways to connect the injection well to the production well [41]. Typical long and large tunnel (LLT) systems consist of a gasification channel, two auxiliary holes, and two auxiliary tunnels (Fig. 4.7). The auxiliary holes are arranged between the injection and production wells and are used as malfunction holes for the injection of air and water vapor, or to discharge gas for added gasifier control. LLT also includes an auxiliary tunnel constructed

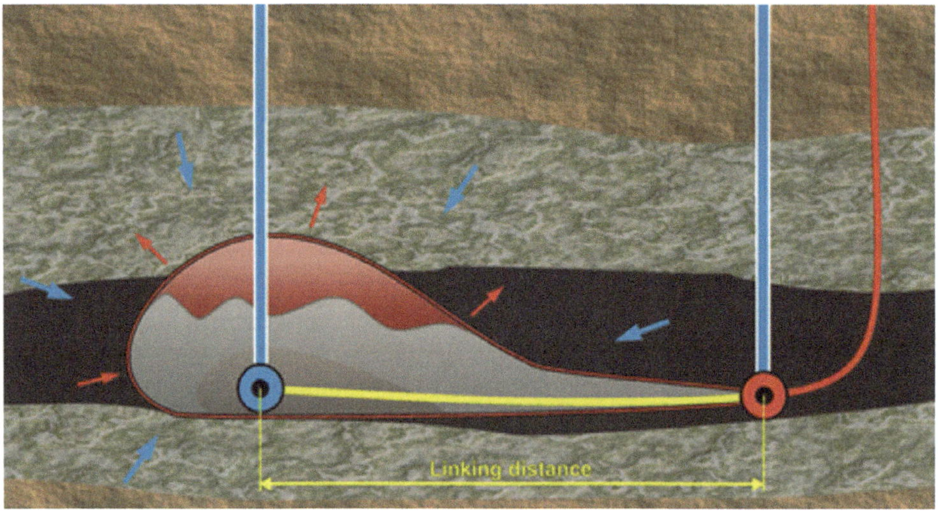

Figure 4.8 Schematic view of linked vertical well (LVW) method [46].

of bricks, which is an auxiliary installation for air injection that prevents blockage in the gasification channel. The mined tunnels are isolated by sealing walls to prevent leakage of combustible gases from the gasifier [42]. The location and height of the oxidant injection points and gas outlet points can be adjusted, allowing for two-dimensional control of oxidant injection and gas production [43].

Recently, most of the focus of research has been on the shaftless methods, which employ directional drilling techniques [45]. With shaftless methods, all preparation and operational processes are carried out through a series of boreholes drilled from the surface into a coal seam and do not require underground labor. Preparation of a shaftless reactor consists of the creation of dedicated in-seam boreholes for oxidant injection and product collection, increasing the coal permeability between the inlet and outlet boreholes, using drilling and completion technology that has been adapted from oil and gas production [38,39]. Currently, there are two main classifications of shaftless UCG methods: linked vertical well (LVW) and controlled retractable injection point (CRIP). Schematic view of LVW method is given in Fig. 4.8, CRIP method is given in Fig. 4.9.

The LVW method is one of the oldest methods for UCG and is derived from technology developed in the former Soviet Union [29]. Vertical wells are drilled into a coal seam, and internal pathways in the coal are utilized to direct the oxidant and product gas flow from the inlet to the outlet borehole. Internal pathways can be naturally occurring or constructed [42]. In its simplest form, the LVW method has inlet and outlet borehole locations that are static for the life of the system. During operation, the coal face migrates and it is found that system control, performance, and

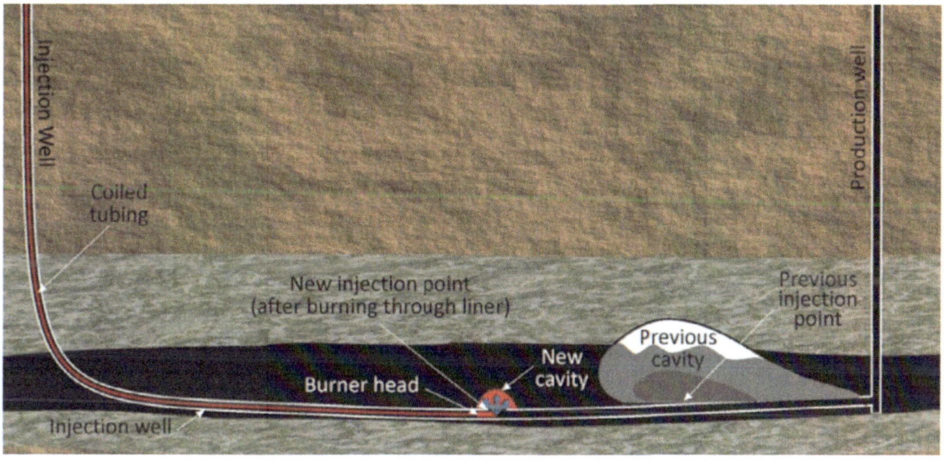

Figure 4.9 Schematic view of linear CRIP method [46].

syngas quality are affected negatively as the distance from the coal face to the oxidant injection point increases [41]; this factor greatly reduces the feasibility of simple LVW systems. A more advanced LVW approach involves a series of dedicated injection boreholes located along the length of a coal seam [39].

Low-rank coals, such as lignites, have considerable natural permeability and can be exploited for UCG without the need for linking technologies. However, high-rank coals, such as anthracites, are far less permeable, making the gas production rate more limited if UCG is employed [42]. For the use of high-rank coals in UCG, a method of linking must be employed to increase the permeability and fracture the coal seam [47]. The boreholes in traditional LVW gasifiers are linked by special methods including forward combustion, reverse combustion, fire linkage, electric linkage, hydrofracturing, and directional drilling to create sizable gasification channels [42,48].

Over the span of a coal seam, the geometry may change, resulting in variable UCG operation and system performance [49]. In the past, this problem was solved by having multiple injection and/or production wells so that static operating conditions could be accomplished through moving the gasifier zones to fresh coal [29]. Controlled retractable injection point (CRIP) offers an alternative approach where the vertical injection well is not moved, but the injection point is moved within the coal seam to fresh coal when necessary [50].

The CRIP method relies on a combination of conventional drilling and directional drilling to access the coal seam and physically form a link between the injection and production wells, without the use of linking technologies utilized in LVW methods [49]. A vertical section of injection well is drilled to a predetermined depth, after which directional drilling is used to expand the hole and drill along the bottom of the coal seam creating a horizontal injection well [51]. At the end of the injection well, a

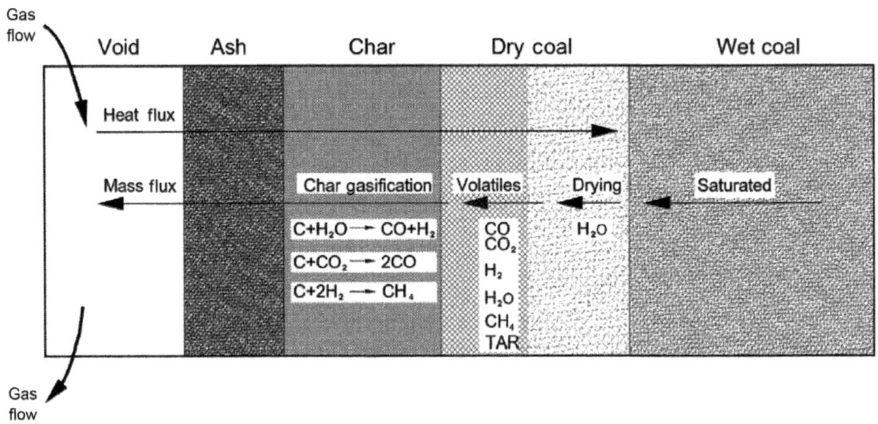

Figure 4.10 General representation of the side wall coal zone of a UCG cavity [41].

gasification cavity is initiated in a horizontal section of the coal seam, creating a localized reactor. The CRIP system utilizes a burner attached to retractable coiled tubing which is used to ignite the coal [50]. The burner burns through the borehole casing to ignite the coal. The ignition point can be moved to any desired location along the horizontal injection well for the creation of a new gasification cavity after a deteriorating reactor has been deserted [49].

Typically, the injection point is retracted using a gas burner, which burns a section of the liner at a desired location [50]. In this manner, accurate control of the gasification process can be obtained. This UCG method has gained popularity in Europe and the USA, but the use of the CRIP method for UCG is fairly new and currently has not become commonly employed [41]. Fig. 4.10 shows the type of reactions that develop into the side wall of the coal.

4.4 Nanocatalysts and sensors use in the process

Catalysis is generally used to accelerate a chemical reaction and the substances used in this process are called catalysts. Inputs and products may degrade at high temperatures. In such cases, the rate of reaction is increased by using catalyst. At the same time, the catalyst is defined as the substance which remains unchanged at the end of the reaction, although it is involved in a chemical reaction and changes the rate of the reaction. The catalyst does not alter the thermodynamics of a reaction. That is, a thermodynamically nonspontaneous reaction does not make the catalyst spontaneous. The catalyst reduces the activation energy of a reaction and makes it faster.

The synthesis and characterization of new and novel functional nanomaterials with well controlled sizes, shapes, porosities, crystalline phases, and structures are of the utmost importance for breakthroughs in several sustainable energy technologies [52]. Nanotechnology is a broad term typically used to describe materials and phenomena at nanoscale, i.e., on the scale of 1 billionth to several tens of billionths of a meter. However, it specifically implies not only the miniaturization but also the precise manipulation of atoms and molecules to design and control the properties of the nanomaterials/nanosystems. Nanomaterials exhibit distinct size-dependent properties in the 1–100 nm range where quantum phenomena are involved. This is one of the major reasons why nanotechnology has a significant impact on energy conversion and storage. These properties are completely different from those possessed by bulk materials, producing custom-made devices with capabilities not found in bulk materials or in nature [53]. It is wellknown that the properties of matter change significantly when the size changes from the macroscale to the microscale and from the microscale to the nanoscale. Nanotechnology is often defined as the science and engineering occurring at dimensions of 100 nm and below. Nanomaterials exhibit distinct size-dependent properties in the 1–100 nm range where quantum phenomena are involved. This is one of the major reasons why nanotechnology has a significant impact on energy conversion and storage.

Important to remember is that the effects of morphology, structure, and composition of nano-materials are equally important. It is also important to consider the synergy of size, morphology, structure, and composition for the design and controlled preparation of nano or nanostructured materials.

Nanoparticles (NPs) are particles sized between 1–100 nano-meters (9–10 meters). Synthesis of stabilized nanoparticles sized between 1–100 nm is the main task of the nanochemistry. Nanoparticles may be synthesized by various means categories in two major heads like (1) top-down technologies and (2) bottom-up technologies (Fig. 4.11) [54–56]. Various sophisticated instruments have been used to characterize the nanomaterials to find out actual size, shape, surface structure, valency, chemical composition, electron band gap, bonding environment, light emission, absorption, scattering and diffraction properties which includes nuclear magnetic resonance spectroscopy (NMR), infrared spectroscopy (IR), ultraviolet and visible spectroscopy (UV-Vis), transmission electron microscopy (TEM), scanning tunneling microscopy (STM), scanning electron microscopy (SEM), energy dispersive X-ray spectrometer (EDS), X-ray diffraction (XRD), X-ray photoelectron spectroscopy (XPS), ultraviolet photoelectron spectroscopy (UPS), extended X-ray adsorption fine structure spectroscopy (EXAFS), X-ray absorption near-edge spectroscopy (XANES), X-ray emission spectroscopy (XES), photoluminescence spectroscopy (PL), small angle X-ray scattering (SAXS), atomic force microscopy (AFM), etc. [57].

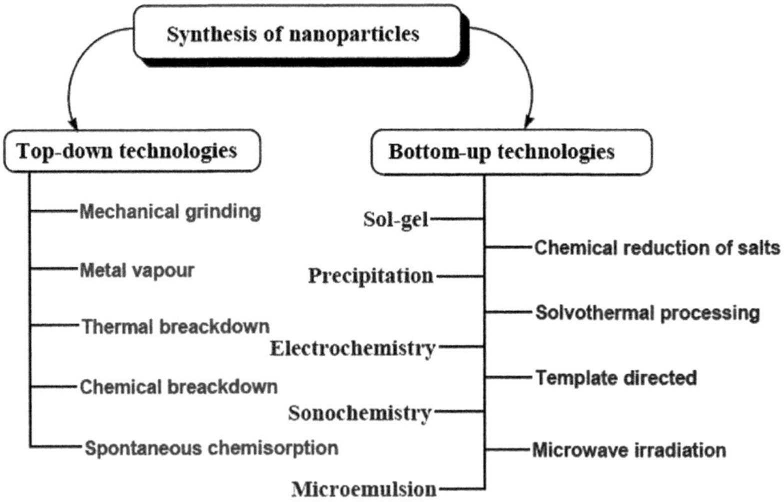

Figure 4.11 Different ways to synthesize nanoparticles [56].

Nano catalytic system allows the rapid, selective chemical transformations with excellent product yield coupled with the ease of catalyst separation and recovery. Because of nano size (high surface area) the contact between reactants and catalyst increases dramatically (this phenomenon is close to homogeneous catalysis). Insolubility in the reaction solvent makes the catalyst heterogeneous and hence can be separated out easily from the reaction mixture (this phenomenon is close to heterogeneous catalysis) [58–62].

Catalysis is one of the pioneer applications of nanoparticles. Various elements and materials like aluminum, iron, titanium dioxide, clays, and silica all have been used as catalysts in nanoscale for many years [56]. These particles are often dispersed on the interior surface of a porous support that stabilizes the particles and allows the reactant molecules to access their surfaces, which is where the catalytic action takes place. But appropriate explanation of its tremendous catalytic behavior showing by NPs still has not been fully understood. Large surface area of nanoparticles has a straight forward positive effect on reaction rate and may also be a reasonable explanation of its catalytic activity. Structure and shape-based properties of any materials at its nanoscale size can also effect the catalytic activity of a material. The fine tuning of nanocatalysts, in terms of composition (bimetallic, core-shell type or use of supports), shape and size has accomplished greater selectivity. Thus the question here is how the physical properties of nanoparticles affect their catalytic properties, and how fabrication parameters can in turn affect those physical properties. By better understanding of these, a scientist can design nanocatalysts which are highly active, highly selective, and highly resilient. All these advantages will enable industrial chemical reactions to become more resource efficient, consume less energy, and produce less waste which help to counter the environmental impact caused by our reliance on chemical process [54,63,64].

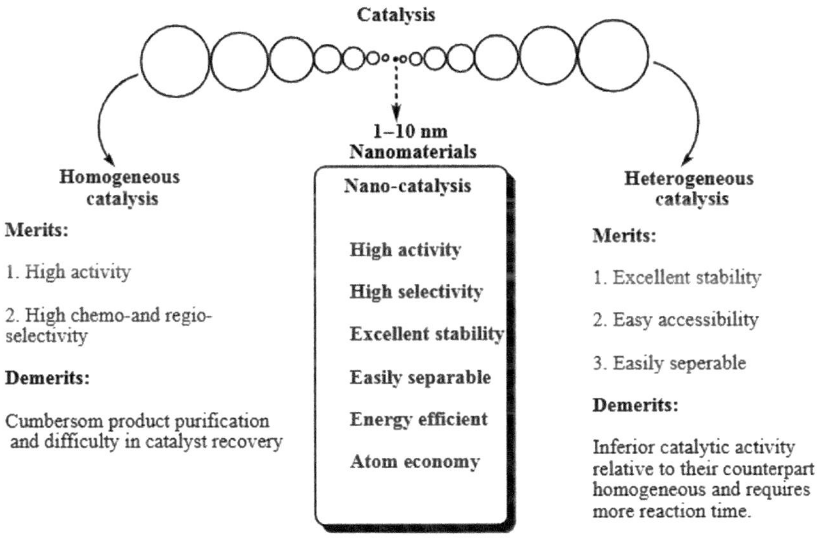

Figure 4.12 Comperative efficiency of homogeneous, heterogeneous and nanocatalysis [56].

Nanoparticles are recognized as the most important industrial catalyst and have wider application ranging from chemical manufacturing to energy conversion and storage. Variable and particle-specific catalytic activity of nanoparticles is due to its heterogeneity and their individual differences in size and shape. Fig. 4.12 represents basic difference in bulk catalysis and catalysis shown by nanoscale materials [56].

If the reactants and the catalyst are in the same phase, such catalysts are called homogeneous catalysts. If they are in different phase the catalysts are called heterogeneous. Heterogeneous catalysts are generally solid or mixture, the reactants and the products are in the liquid or gas phase. Heterogeneous catalytic reactions take place at or near the fluid solid interface [65,66].

In industrial processes; three properties are considered for the suitability of the catalyst [67]:

1. Activity: A measure of how fast one or more reactions in the presence of catalyst is present. It is based on measuring the reaction rates based on temperature and concentration ranges.
2. Selectivity: It is expressed as the ratio of the desired product to the transformed amount of reactant A. Depending on the catalyst used, completely different reactions and products occur.
3. Stability: This is a statement about the life of the catalyst. Catalysts have a certain life span and their lifetime is an important criterion for the economics of a process. During the reaction, the resulting coking, decomposition and some gases released

during the reaction lead to poisoning of the catalyst. This adversely affects the activity of the catalyst.

If the expressed criteria are listed in order of importance in an industrial process:

$$\text{Selectivity} > \text{Stability} > \text{Activity}$$

Nanocatalysts possess superior aspects of "ease separation" and "long useful life" of homogeneous catalysts, and "high efficiency" and "high selectivity" of heterogeneous catalysts. Catalyst preparation is a complex study area. These materials may be used directly or in the form of supported nanoparticles on solids such as oxides, carbon or zeolites.

Solid state reactions, coprecipitation, impregnation, sol-gel process, hydrothermal synthesis, flame spray process (FSP), spray drying and freeze drying are among the major catalyst preparation methods. Precipitation and impregnation methods are simple, cheap, and well-studied but it is difficult to control the size of particles. Chemical vapor deposition is widely used in the electronics industries but it is an expensive method. Electrochemical deposition is an inexpensive method that does not need high temperature and concentration. This method would allow a good control on size and chemical properties of the deposited nanomaterials but usually forms one dimensional nanomaterials [13,68–70].

4.4.1 Catalysts in gasification process and syngas production

The main objectives of the use of catalysts in the gasification process are to increase the carbon conversion rate and to economically reduce the processing time. An important feature of the catalysts is their selectivity to control the process to obtain gas with properties that allow for more use possibilities. The selectivity of the catalysts was investigated by Li et al. (2014) using potassium carbonate to obtain a methane enriched syngas [71]. The same catalyst was used to increase the amount of hydrogen in the gas obtained by Sharma et al. (2008) [72]. The type of catalyst, the properties of the fuel used, the synergy with a catalyst, the concentration of the catalytic components, the degree of distribution, as well as the process parameters play an important role in the gasification process [44,73–83]. In 1867, the first formal results were revealed about the effectiveness of catalysts based on alkali and alkaline earth metals [81], and they were the most effective and at the same time the cheapest catalysts for the gasification process. Otto et al. (1977) investigated the catalytic effect of precious metals such as platinum, ruthenium, rhodium and palladium, and stated that they significantly increased the rate of gasification reaction [84]. However, high cost of catalyst preparation is a disadvantage. The use of transition metals as catalyst in the gasification process brings about the problem of deactivation. The addition of a catalyst to coal is usually carried out in two ways: by physically mixing the coal with the catalyst or by impregnating the coal using a catalytically active substance. The most common method is the physical mixing of the fuel with the catalyst; the resulting mixture is then fed to the gasification reactor. This method is advantageous because it is simple and the application time is short and at the same time allows the correct selection of the amount of catalyst. Thus, physical mixing is a widely used method, especially in the case of catalysts based on alkali and alkaline earth metals [85–88]. Another equally common method is impregnation of the wet impregnation, which provides good contact between the charcoal and the catalyst. It has been found that the addition of a catalyst to the catalyst by physical mixing requires higher temperatures than the wet impregnation process [89]. However, these are not commonly used methods due to their increased workload and time. Song and Kim (1993) investigated the effect of catalyst percentage on char reactivity. In the same operating conditions under the 3% catalyst addition, in the process of gas gasification of bituminous coal, the pure salt catalytic activity was reported as K_2CO_3> Na_2CO_3> $FeSO_4$> K_2SO_4> -Fe(NO_3)3 respectively [77]. They also reported increased reactivity by the addition of up to 6% by weight of catalyst. Hattingh et al. (2011) reported that increased CaO and MgO content, Schobert (1992), Na_2O and K_2O increased the reactivity of coal [79,90]. Therefore, it can be concluded that all catalysts added to the existing or coal may affect the gasification behavior of the coal by altering the reactant reactivity.

Sodium and iron are two good catalysts for coal gasification. However, each of them has its own advantages and disadvantages. To use their advantages and overcome their disadvantages, the composite Na- Fe catalyst was evaluated for its performance on the improvement of coal gasification.

4.4.2 Purification of syngas

Purification of syngas mainly covers mercury removal, desulfurization, and carbon capturing processes. Mercury emissions from coal combustion are an important source of mercury pollution in the air due to the high volatility, toxicity and bioaccumulation of mercury and cause great harm to human health and the environment. H_2S and Hg, which are formed during the gasification process, cause damage not only to pipelines and other equipment, but also to the ecological environment and human health. For these reasons, it is necessary to remove these harmful components from syngas.

Most of the sulfur content of the biomass is converted to H_2S and SO_2 during gasification, which can be then removed by adsorption on limestone, dolomite or CaO (at low temperatures) or calcined at high temperatures [91].

The sulfur compounds (mainly H_2S) can be removed from syngas through chemisorption with the metal oxides [92–94]. The Zn-, Cu-, Fe-, Mn-, and rare earth metal-based oxides are the most commonly used sorbents for syngas desulfurization. However, the utilization of the metal oxides is limited because of the low specific surface area and pore volume, and the sulfide formed during adsorption processes would lead to the blocking of pores, resulting in the decline in H_2S removal. To solve these problems, natural zeolites and rare-earth oxides SiO_2 were added to metal oxides or their mixtures as structure stabilizer [95]. Nano-materials are also used for this purpose because of unique surface effect, electronic effect and quantum size effect. It has been reported that the desulfurization penetration time of nano-ZnO was 40 times longer than the normal analytical ZnO at room temperature. In the desulfurization process, H_2S can be selectively oxidized to elemental S, which will provide favorable conditions to simultaneous removal of H_2S and Hg [96]. The main drawback of employing ZnO sorbent is the difficulty to regenerate the spent ZnO sorbent due to high stability of ZnS. Ni-Zn-based nanocomposite loaded on cordierite mullite ceramic was performed for syngas desulfurization for that drawback. The Ni addition is expected to improve the regenerability of ZnO sorbent, and increase the ZnO sorption capacity. The results confirm that the NiZn-28-HC has promising potential for largescale syngas desulfurization. Fe nanoparticles (NPs) into carbon aerogel desulfurizers were fabricated and their performance for H2S removal in hot coal gas was also studied [97] and promising results were obtained.

CO2 capturing is classified as precombustion in gasification process and focused on advanced solvents, solid sorbents, and membrane systems for the separation of H_2 and CO_2, with specific emphasis on hightemperature/novel materials, process intensification, and nanomaterials. Additionally, novel concepts, such as hybrid technologies that combine attributes from multiple technologies (e.g., sorbents and membranes) are being investigated (URL 5). The current and future CO2 capturing methods are given in Table 4.3 [98].

Although the methods and materials used remain the same, the use of nanoscale materials and filters is increasing to take advantage of the superior aspects of nano-materials.

Most of the nitrogen content of the biomass is converted to ammonia during gasification, which can be further removed by wet scrubbing (at low temperatures) or destroyed at higher temperatures using either dolomites, or Ni-based or Fe-based catalysts [91]. NH_3- selective catalytic reduction (SCR) technology is the most effective

Table 4.3 Current and future precombustion CO_2 capture techniques [98].

Capture Technologies	Current	Emerging
Solvents (Absorption)	Physical solvent Chemical solvents	Improved chemical solvents Novel contacting equipment Improved design of processes
Membranes	Polymeric	Ceramic Palladium Reactors Contactors
Solid sorbents	Zeolites Activated carbon Alumina	Carbonates Hydrotalcites Silicates
Cryogenic	Liquefaction	Hybrid processes

technology for removing NOx from flue gases [99–101]. The commonly adopted commercial catalyst is V2O5–WO3(MoO3)/TiO2 [101–103] and its working temperature must be above 300°C in order to obtain good catalytic activity and to avoid pore plugging caused by the deposition of ammonium-sulfate salts over the catalyst surface.

4.4.3 Advanced product synthesis from syngas

As in the production of the syngas from coal, in later stages, in synthesis of final products such as such as hydrogen, ethanol, methanol, olefins, fuels such as LPG diesel naphtha, waxes, ketones, ammonia, synthetic natural gas (SNG), catalysts are used in large scale.

Ethanol is an important source for the synthesis of various products such as chemicals, fuels and polymers, as well as being considered as a commercial gasoline additive or a potential alternative. Several catalysts have been proposed and tested for the synthesis of ethanol and higher alcohols [104–106]. They can roughly be classified as Rh-based, u-based, Co-based and Mo-based catalysts. Among them, the Rh-based catalysts present the highest selectivity to ethanol. The rest of the catalysts have similar selectivity to ethanol, but higher activity (conversion of syngas per catalyst volume) [91].

Fisher-Tropsch (FT) process, which is one of the most known methods in the production of synthetic fuel, was first used in 1902 by two German scientists, Sabatier and Senderens, when producing methane from hydrogenated CO in the presence of nickel catalysts [107]. Again, in the 1920s, this process was handled in detail by scientists Franz Fischer and Hans Tropsch at the Kaiser Wilheim Institute in Germany, and in 1926, it was patented as a petroleum-derived liquid hydrocarbon production from synthesis gas. The FT process was used to obtain synthetic liquid fuel from coal to meet the oil needs of Germany, which is rich in coal reserves, especially during the Second World War [108].

Fischer-Tropsch synthesis is considered an attractive way to convert natural/shale gas, coal, biomass, and CO_2 into clean liquid fuels and high value-added chemicals via syngas (H_2 + CO). FT synthesis can be catalyzed by various transition metals such as iron, cobalt and ruthenium, but only iron-based and cobalt-based catalysts are known to be commercially viable [109–111]. Iron-based catalysts can be used very effectively in FT synthesis but their preparation creates environmental problems due to discharge of hazardous chemicals. The eco-friendly iron-ore-based catalysts (IO-CAT) with catalytic properties similar to those of conventional precipitated iron-based catalysts (PFe-CAT) were successfully prepared through a combination of a wet-milling process and a wet impregnation method. For the production of hydrocarbons between C5 and C12, which are gasoline-range hydrocarbons (GRHs), further processes are required [112]. An attractive approach is the in-situ upgrading of the primary FT synthesis products by using hybrid catalysts, consisting of the conventional FT synthesis catalyst together with an acidic zeolite [113].

Effective reduction of the active phase plays an important role in optimizing catalyst performance, with the addition of small quantities of promoters during the formulation of the catalyst found to significantly enhance the reducibility of Co and Fe. Furthermore, promoters improve the activity and selectivity of heterogeneous catalysts by influencing the catalyst's structural properties through varying the active phase structure, or modifying the electronic character of the active phase. Generally, metal oxides (MnO and ZnO), alkali metals (Na, K, Rb, Cs) and certain transition metals (Cu, Pd, Pt, Ru) or carbonates are applied as promoters for the Fe and Cobased FT catalysts [112,114]. The third crucial factor that must be considered for catalyst design is the mechanical resistance, morphology and porosity of the support. Accordingly, the selection of support which stabilises the resulting Co or Fe nanoparticles is critical in determining FT-catalyst activity and stability. Altering the support's surface structure and pore size may improve the metal dispersion, reducibility and the diffusion coefficients of reactants and products. Al_2O_3, SiO_2, TiO_2 and ZrO_2 are used as catalyst supports in the FTS due to their high surface area and strong mechanical strength [112]. Bimetallic catalysts also show great promise for application in FTS and their use is reviewed by Calderone et al. (2011) [115].

References

[1] T. Nayır, (2012). Gasification of coal and biomass mixtures and simulation with aspen Hysys® program, Istanbul Technical University Institute of Science and Technology, Master thesis, Istanbul, 103.

[2] P.R. Solomon, D.G. Hamblen, R.M. Carangelo, M.A. Serio, G.V. Deshpande, General model of coal devolatilization, Energy Fuels. 32 (3) (1987) 83–98.

[3] S. Bilge, (2017). Catalytic effect of nanometal oxide catalysts on gaseous product distribution obtained from low-rank coal pyrolysis, Ankara University, graduate school of natural and applied sciences, Department of Chemistry, Master thesis, Ankara, 100.

[4] W.H. Wiser, L.L. Anderson, S.A. Qader, G.R. Hill, Kinetic relationship of coal hydrogenation, pyrolysis and dissolution, J. Appl. Chem. Biotechnol. 21 (3) (2007) 82–86.

[5] A. Tarakcioglu, (2015). Gasification of some Turkish lignites in a fludized bed gasifier, Hacettepe University, Department of Chemical Engineering, Master Thesis, Ankara, 120.

[6] ASTM, (2012). ASTM D388: standard classification of coals by rank name of standards organization: American Society for Testing and Materials.

[7] IEA, (2012). International Energy Agency, IEA Statistics - COAL Information, ISBN, 978-92-64-17470 2, 2012.

[8] WEC, (2010). Energy Report 2010, World Energy Council, Turkish National Committee, 201.

[9] TKI, (2014). Coal Sector Report (Lignite), Turkiye Komur Isletmeleri, 69.

[10] M. Sudiro, A. Bertucco, Production of synthetic gasoline and diesel fuel by alternative processes using natural gas and coal: process simulation and optimization, Energy 34 (12) (2009) 2206–2214.

[11] K. Barbour, (2011). The challenges and benefits of coal gasification, Technical White Paper, Pepperl + Fuchs, 8.

[12] A. Tarakcioglu, (2015). Gasification of some Turkish Lignites in a Fludized Bed Gasifier, Hacettepe University, Department of Chemical Engineering, Master Thesis, Ankara, 120.

[13] K. Liu, C. Song, V. Subramani, Hydrogen and syngas production and purification technologies, A John Wiley & Sons, Inc., Publication, Hoboken, NJ, 2010, p. 533.

[14] L.D. Smoot, P.J. Smith, Coal combustion and gasification, Plenum, New York, 1985, p. 245.

[15] F. Scala, Fluidized bed technologies for near-zero emission combustion and gasification, Woodhead Publishing, 2013, pp. 765–812. Chapter 7: Fluidized bed gasification in (Author: Arena U.).
[16] M. Canel, Coal gasification, Madencilik 25 (2) (1986) 35–41.
[17] S. Fleck, U. Santo, C. Hotz, T. Jakobs, G. Eckel, M. Mancini, et al., Entrained flow gasification Part 1: gasification of glycol in an atmospheric pressure experimental rig, Fuel 217 (2018) 306–319.
[18] J. Brown (2006). Biomass gasification: fast internal circulating fluidised bed gasifier characterisation and comparison, Chemical and Process Engineering, University of Canterbury, Master thesis.
[19] J.G. Speight, Gasification, The chemistry and technology of Coal, 3rd ed., CRC Press, Boca Raton, FL, 2012. Chapter 2.
[20] B.G. Miller, Coal energy systems, Elsevier, USA, 2005.
[21] J. Rezeriyan, P.C. Nicholas, Gasification technologies: a primer for engineers and scientists, Taylor Francis Group, 2005, p. 360.
[22] M.A. Elliott, Chemistry of coal utilization, John Wiley & Sons, New York, 1945, pp. 1602–1752. Second Supplementary Volume (Auth. Herbden D. and Stroud H. J. F.).
[23] L. Shadle, D. Berry, M. Syamlal, (2007). Coal gasification, Kirk-Othmer Encyclopedia of Chemical Technology.
[24] S. Golbasi, (2009). Analysis of the gaseous products forming during thermal decomposition of biomass, Istanbul Technical University Institute of Science and Technology, Master thesis, Istanbul, 111.
[25] M. Laciak, K. Kostúr, M. Durdan, J. Kacur, P. Flegner, The analysis of the underground coal gasification in experimental equipment, Energy 114 (2016) 332–343.
[26] D. Olness, D.W. Gregg, (1977). The historical development of underground coal gasification, Technical Report, UCRL-52283, Berkeley, CA: Lawrence Livermore National Laboratory, University of California, 56.
[27] J.L. Dossey, (1976). Underground coal gasification technology in the USSR, Sandia Laboratories, Albuquerque, SAND 76-0380, 106.
[28] A. Booz, (1977). Underground coal gasification program, ERDA 77-51/4 (Overview), Dist. Category UC-90C, FE · 2343 · 19, 31.
[29] E. Shafirovich, A. Varma, Underground coal gasification: a brief review of current status, Ind. Eng. Chem. Res. 48 (2009) 7865–7875. 2009.
[30] H.H. Lowry, Chemistry of coal utilization, John Wiley and Sons, Inc, New york, 1945.
[31] K. Dziunikowski, Oxidation with water vapor in underground gasification of coal, Prz. Gorniczy (3)(1953) 107–111.
[32] T. Ishikura, F. Ebuchi, Forward burning process and backward burning process in a model experiment on underground gasification of coal, Kyushu Daigaku Seisankogaku Kenkyusho Hokoku 39 (1965) 8–19.
[33] J. Koranda, (1973). Research in underground gasification of coal in Czechoslovakia.
[34] ECFECC, (1976). Economic commission for Europe coal committee, report of the symposium on gasification and liquifaction of coal, Düsseldorf.
[35] S. Shimada, K. Ohga, A. Tamari, E. Ishii, Cost estimation of underground coal gasification in Japan, Miner. Resour. Eng. 5 (03) (1996) 241–252.
[36] S.J. Self, B.V. Reddy, M.A. Rosen, Review of underground coal gasification technologies and carbon capture, Int. J. Energy Environ. Eng. (2012) 3–16.
[37] G. Perkings, Underground coal gasification-Part I: field demonstrations and process performance, Prog. Energy Combust. Sci. (2018) 67.
[38] M. Wiatowski, K. Stańczyk, J. Świądrowski, K. Kapusta, K. Cybulski, E. Krause, et al., Semi-technical underground coal gasification (UCG) using the shaft method inExperimental Mine "Barbara", Fuel 99 (2012) 170–179.
[39] S. Lee, J.G. Speight, S.K. Loyalka, Handbook of alternative fuel technologies, CRC, Boca Raton, FL, 2007.
[40] D. Olness, D.W. Gregg, (1977). The historical development of underground coal gasification, Technical Report, UCRL-52283, Berkeley, CA: Lawrence Livermore National Laboratory, University of California, 56.
[41] D.J. Roddy, P.L. Younger, Underground coal gasification with CCS: a pathway to decarbonising industry, Energy Environ. Sci. 3 (2010) 400–407.

[42] J. Liang, S. Liu, L. Yu, Trial study on underground coal gasification of abandoned coal resource, in: H. Xie, T.S. Golosinki (Eds.), Proceedings of the '99 International Symposium on Mining Science and Technology, Beijing, August, Mining Science and Technology, A.A. Balkema, Rotterdam, 1999, pp. 271–275.
[43] L. Yang, J. Liang, L. Yu, Clean coal technology—study on the pilot project experiment of underground coal gasification, Energy 28 (2003) 1445–1460.
[44] Y. Li, X. Liang, J. Liang, An overview of the Chinese UCG program, Data Sci. 6 (2007) 460–466.
[45] G.P. Hammond, Energy, environment and sustainable development: a UK perspective, Trans. Inst. Chem. Eng. 78 (2000) 304–323. Part B.
[46] CCPC, (2014). In-situ Coal Gasification, A Final Phase IV Report, Canadian Clean Power Coalition (CCPC) Technical Committee, Appendix C, 11.
[47] M.S. Blinderman, A.Y. Klimenko, Theory of reverse combustion linking, Combust. Flame 150 (2007) 232–245.
[48] M.S. Blinderman, D.N. Saulov, A.Y. Klimenko, Forward and reverse combustion linking in underground coal gasification, Energy 33 (2008) 446–454.
[49] H. Nourozieh, M. Kariznovi, Z. Chen, J. Abedi, Simulation study of underground coal gasification in Alberta reservoirs: geological structure and process modeling, Energy Fuel 24 (2010) 3540–3550.
[50] A.Y. Klimenko, Early ideas in underground coal gasification and their evolution, Energies 2 (2009) 456–476.
[51] G.X. Wang, Z.T. Wang, B. Feng, V. Rudolph, J.L. Jiao, Semi-industrial tests on enhanced underground coal gasification at Zhong-Liang-Shan coal mine, Asia-Pac. J. Chem. Eng. 4 (2009) 771–779.
[52] I. Hut, S. Pelemis, D. Mirjanic, Nanomaterials and nanotechnology for sustainable energy, Zastita Mater. 56 (2015) 329–334.
[53] S. Pelemis, I. Hut, (2013). Nanotechnology materials for solar energy conversion, contemporary materials (renewable energy sources), IV – 2, 145-151.
[54] K. Yan, X. Wu, X. An, X. Xie, Facile synthesis and catalytic property of spinel ferrites by a template method, J. Alloy. Compd. 552 (2013) 405–408.
[55] H. Luo, T. Klande, Z. Cao, F. Liang, H. Wang, J. Caro, A CO_2-stable reduction-tolerant Nd-containing dual phase membrane for oxyfuel CO_2 capture, J. Mater. Chem. A 2 (2014) 7780–7787.
[56] S.B. Singh, P.K. Tandon, Catalysis: a brief review on Nano-Catalyst, J. Energy Chem. Eng. 2 (3) (2014) 106–115.
[57] Y. Li, G.A. Somorjai, Nanoscale advances in catalysis and energy applications, Nano Lett. 10 (2010) 2289–2295.
[58] H. Luo, H. Jiang, T. Klande, Z. Cao, F. Liang, H. Wang, et al., Novel cobalt-free, noble metal-free oxygen-permeable 40Pr0.6Sr0.4FeO3-δ – 60Ce0.9Pr0.1O2 – δ dual-phase membrane, Chem. Mater. 24 (2012) 2148–2154.
[59] S.G. Babu, R. Karvembu, Copper based nanoparticles-catalyzed organic transformations, Catal. Surv. Asia 17 (2013) 156–176.
[60] P.K. Tandon, S.B. Singh, Catalytic applications of copper species in organic transformations: a review, J. Catalyst Catal. 1 (2014) 1–14.
[61] K. Yan, G. Wu, C. Jarvis, J. Wen, A. Chen, Facile synthesis of porous microspheres composed of TiO_2 nanorods with high photocatalytic activity for hydrogen production, Appl. Catal. B: Environ. 148 (2014) 281–287.
[62] K. Yan, G. Wu, T. Lafleur, C. Jarvis, Production, properties and catalytic hydrogenation of furfural to fuel additives and value added chemicals, Renew. Sustain. Energy Rev. 38 (2014) 663–676.
[63] Z. Cao, H. Jiang, H. Luo, S. Baumann, W.A. Meulenberg, H. Voss, et al., Simultaneous overcome of the equilibrium limitations in BSCF oxygen-permeable membrane reactors: water splitting and methane coupling, Catal. Today 193 (2012) 2–7.
[64] Y. Qiao, H. Li, L. Hua, L. Orzechowski, K. Yan, B. Feng, et al., Peroxometalates immobilized on magnetically recoverable catalysts for epoxidation, Chem. Chem. 77 (2012) 1128–1138.
[65] M. Campanati, G. Fornasari, A. ve Vaccari, Fundamentals in the preparation of heterogeneous catalysts, Catal. Today 77 (2003) 299–314.
[66] H.S. Fogler, Elements of chemical reaction engineering, 3rd ed., Prentice-Hall, New Delphi, 2004, pp. 581–583.

[67] R.Z. Yarbay, (2010). Catalyst development for multi fuel catalytic reformer for hydrogen production, Master Thesis, Yıldız Technical University, Graduate School of Natural and Applied Sciences, Chemical Engineering Department, Istanbul, 115.
[68] A. Aradi, J. Roos, T.C. Jao, (2010). Nanoparticle catalyst compounds and/or volatile organometallic compounds and method of using the same for biomass gasification. US 20100299990A1.
[69] J.P. Wilcoxon, Nanoparticles preparation, characterization and physical properties, Front. Nanosci. 3 (2012) 43–127. Available from: 10.1016/B978-0-08-096357-0.00005-4.
[70] M. Akia, F. Yazdani, E. Motaee, D. Han, H. Arandiyan, A review on conversion of biomass to biofuel by nanocatalysts, Biofuel Res. J. 1 (2014) 16–25.
[71] W.W. Li, K.Z. Li, X. Qu, R. Zhang, J.C. Bi, Simulation of catalytic coal gasifi cation in a pressurized jetting fl uidized bed: effects of operating conditions, Fuel Proc. Technol. 126 (2014) 504–512.
[72] A. Sharma, T. Takanohashi, K. Morishita, T. Takarada, I. Saito, Low temperature catalytic steam gasification of Hyper Coal to produce H2 and synthesis gas, Fuel 87 (2008) 4–5.
[73] A.K. Agarwal, J.T. Sears, The coal char reaction with CO_2–CO gas mixtures, Ind. Eng. Chem. Process. Des. Dev. 79 (1980) 364 371.
[74] L.R. Radovic, P.L. Walker Jr, R.G. Jenkins, Carbon active sites in coal char gasification fuel, Fuel 62 (1983) 849–856.
[75] T. Takarada, Y. Tamai, A. Tomita, Effectiveness of K_2CO_3 and Ni as catalysts in steam gasification, Fuel 65 (1986) 679–683.
[76] K. Miura, K. Hashimotot, P.L. Silveston, Factors affecting the reactivity of coal chars during gasification and indices representing reactivity, Fuel 68 (1989) 1461–1475.
[77] B.H. Song, S. Kim, Catalytic activity of alkali and iron salt mixtures for steam-char gasification, Fuel 72 (6) (1993) 797–803.
[78] J. Ochoa, M.C. Cassanello, P.R. Bonelli, A.L. Cukierman, CO gasification of Argentinean coal chars a kinetic characterization, Fuel Process. Technol. 74 (2001) 161–176.
[79] B.B. Hattingh, R.C. Everson, H.W.J.P. Neomagus, J.R. Bunt, Assessing the catalytic effect of coal ash constituents on the CO_2 gasification rate of high ash, South African coal, Fuel Process. Technol. 92 (2011) 2048–2054.
[80] H. Namkung, X. Yuan, G. Lee, D. Kim, T.J. Kang, H.T. Kim, Reaction characteristics through catalytic steam gasifi cation with ultra clean coal char and coal, J. Energy Inst. 87 (3) (2014) 253–262.
[81] J. Tang, J. Wang, Catalytic steam gasifi cation of coal char with alkali carbonates: a study on their synergic effects with calcium hydroxide, Fuel Proc. Technol. 142 (2016) 34–41.
[82] X. Qi, X. Guo, L. Xue, C. Zheng, Effect of iron on Shenfu coal char structure and its infl uence on gasifi cation reactivity, J. Anal. Appl. Pyrol. 110 (2014) 401–407. Available from: https://doi.org/10.1016/j.jaap.2014.10.011.
[83] P. Parthasarathy, K.S. Narayanan, Hydrogen production from steam gasification of biomass: influence of process parameters on hydrogen yield – A review, Renew. Energ. 66 (2014) 570–579. Available from: https://doi.org/10.1016/j.renene.2013.12.025.
[84] K. Otto, M. Shelef, Catalytic steam gasification of graphite: effects of intercalated and externally added Ru, Rh, Pd and Pt, Carbon 15 (5) (1977) 317–325.
[85] I. Mochida, Low-temperature catalytic conversion of lignite: 1. Steam gasifi cation using potassium carbonate supported on perovskite oxide, J. Ind. Eng. Chem. 20 (1) (2014) 216–221.
[86] Q. Waheed, C. Wu, P. Williams, Hydrogen production from high temperature steam catalytic gasification of bio-char, J. Energy Inst. 89 (2) (2015) 222–230.
[87] D. Supramono, D. Tristantini, A. Rahayu, R. Suwignjo, D. Chendra, Syngas production from lignite coal using K_2CO_3 catalytic steam gasifi cation with controlled heating rate in pyrolysis step, Procedia Chem. 9 (2014) 202–209.
[88] J. Mazumber, H. Lasa, Fluidizable Ni/La2O3- γAl2O3 catalyst for steam gasifi cation of a cellulosic biomass surrogate, Appl. Catal. B 160–161 (2014) 67–79.
[89] F. Huhn, J. Klein, H. Jüntgen, Investigationson the alkali catalysed steam gasifi cation of coal: kinetics and interactions of alkali catalyst with carbon, Fuel 62 (2) (1983) 196–199.

[90] H.H. Schobert, Catalytic and chemical behaviour of coal mineral matter behaviour in the coal conversion process, in: Y. Yurum (Ed.), Clean utilization of coal: coal structure and reactivity, cleaning and environmental aspects, Kluwer Academic Publishers, Dordrecht, 1992, pp. 65–73.
[91] L.G.L. Nina, (2017). Catalytic conversion of syngas to ethanol and higher alcohols over Rh and Cu based catalysts, Doctoral Thesis in Chemical Engineering, KTH Royal Institute of Technology, School of Chemical Science and Engineering, Department of Chemical Engineering and Technology, Stockholm, Sweden, 98.
[92] B. Dou, C. Wang, H. Chen, Y. Song, B. Xie, Y. Xu, et al., Research progress of hot gas filtration, desulphurization and HCl removal in coal-derived fuel gas: a review, Chem. Eng. Res. Des. 90 (2012) 1901–1917.
[93] V. Girard, D. Chiche, A. Baudot, D. Bazer-Bachi, I. Clémençon, F. Moreau, et al., Innovative low temperature regenerable zinc based mixed oxide sorbents for synthesis gas desulfurization, Fuel 140 (2015) 453–461.
[94] O. Wen-Da, L. Junxi, V. Andrei, G. Apostolos, L. Grzegorz, W.C.C. Victor, et al., Influence of surface morphology on the performance of nanostructured ZnO loaded ceramic honeycomb for syngas desulfurization, Fuel 211 (2018) 591–599.
[95] N.K. Park, D.H. Lee, H.J. Jin, J.D. Lee, O.R. Si, T.J. Lee, et al., Two stage desulfurization process for hot gas ultra cleanup in IGCC, Fuel 85 (2006) 227–234.
[96] J. Zhou, P. Qi, W. Hou, S. You, X. Gao, Z. Luo, Elemental mercury removal from syngas by nano-ZnO sorbent, J. Fuel Chem. Technol. 41 (11) (2013) 1371–1377.
[97] T. Huan, W. Jiang, Z. Wenbo, Y. Siyuan, L. Fangqin, Q. Yongfeng, et al., High performance of Fe nanoparticles/carbon aerogel sorbents for H2S Removal, Chem. Eng. J. 313 (2017) 1051–1060.
[98] B. Metz, O. Davidson, H. Coninck, M. Loos, L. Meyer (Eds.), IPCC special report on carbon dioxide capture and storage, Cambridge University Press, New York, 2005, p. 443.
[99] F. Han, Y. Gao, Q. Huo, L. Han, J. Wang, W. Bao, et al., Characteristics of vanadium-based coal gasification slag and the NH_3-selective catalytic reduction of NO, Catalysts 8 (2018) 327.
[100] H. Bosch, F. Janssen, Formation and control of nitrogen oxides, Catal. Today 2 (1988) 369–379.
[101] S.M. Cho, Properly apply selective catalytic reduction for NOx removal, Chem. Eng. Prog. 90 (1994) 39–45.
[102] J.P. Chen, R.T. Yang, Mechanism of poisoning of the V2O5/TiO2 catalyst for the reduction of NO by NH3, J. Catal. 125 (1990) 411–420.
[103] J.P. Chen, R.T. Yang, Selective catalytic reduction of NO with NH_3 on SO4 2 − /TiO_2 superacid catalyst, J. Catal. 139 (1993) 277–288.
[104] R.S. Paris, L. Lopez, J. Barrientos, F. Pardo, M. Boutonnet, S. Jaras, Chapter 3 Catalytic conversion of biomass-derived synthesis gas to fuels, Catalysis 27 (2015) 62–143. The Royal Society of Chemistry.
[105] J.J. Spivey, A. Egbebi, Heterogeneous catalytic synthesis of ethanol from biomass-derived syngas, Chem. Soc. Rev. 36 (2007) 1514–1528.
[106] V. Subramani, S.K. Gangwal, A review of recent literature to search for an efficient catalytic process for the conversion of syngas to ethanol, Energy Fuels 22 (2008) 814–839.
[107] R.H. Perry, D.W. Green, J.O. Maloney, Perry's chemical engineers' handbook, 7th ed., McGraw Hill, New York, 1997.
[108] B. Gurunlu, (2012). Development of zeolite-supported iron catalysts for Fischer Tropsch synthesis, Istanbul Technical University Institute of Science and Technology, Master thesis, Istanbul, 115.
[109] R.B. Anderson, The Fischer-Tropsch synthesis, Academic Press Inc., New York, 1984.
[110] E. Steen, M. van, Claeys, K.P. Möller, D. Nabaho, Comparing a cobalt-based catalyst with iron-based catalysts for the Fischer-Tropsch XTL-process operating at high conversion, Appl. Catal. A Gen. 549 (2018) 51–59.
[111] A.P. Steynberg, M.E. Dry, Fischer-tropsch technology, Elsevier, Amsterdam, 2004.
[112] M. Feyzi, M. Irandoust, A.A. Mirzaei, Fuel Process. Technol. 92 (2011) 1136–1143.

[113] H. Jahangiri, J. Bennett, P. Mahjoubi, K. Wilson, S. Gu, A review of advanced catalyst development for Fischer–Tropsch synthesis of hydrocarbons from biomass derived syn-gas, Catal. Sci. Technol. 4 (2014) 2210–2229.

[114] Q. Zhang, J. Kang, Y. Wang, Development of novel catalysts for Fischer–Tropsch synthesis: tuning the product selectivity, Chem. Cat. Chem. 2 (2010) 1030–1058.

[115] V.R. Calderone, N.R. Shiju, D.C. Ferre, G. Rothenberg, Bimetallic catalysts for the Fischer–Tropsch reaction, Green. Chem. 13 (2011) 1950–1959.

CHAPTER FIVE

Advanced materials for next-generation fuel cells

Mesut Yurukcu[1], Fatma M. Yurtsever[2], Serkan Demirel[3], Jorge Saldaña[4] and Mufrettin Murat Sari[5]

[1]Department of Physics and Astronomy, University of Arkansas at Little Rock, Little Rock, AR, United States
[2]Department of Chemistry, University of Arkansas at Little Rock, Little Rock, AR, United States
[3]College of Computer and Information Sciences, Regis University, Denver, CO, United States
[4]Department of Political Science, University of Houston, Houston, TX, United States
[5]Texas A&M University, Commerce, TX, United States

5.1 Introduction

To improve the quality of life across the world, we need more energy production and consumption. However, energy produced from fossil fuels has several negative effects on the environment and economy such as air pollution, global warming, and unstable cost of fossil fuels [1]. The question, then, is how to find highly effective ways to produce, deliver, and use energy that improves quality of life but does not threaten the environment and climate [2]. One of the solutions for the next generation of environmentally friendly energy is fuel cell technology. If solar energy, fuel cell, and other types of renewable energy sources can replace fossil fuels, emissions of greenhouse gases can be mostly eliminated. Fuel cell technology is more fuel-efficient than internal combustion engines while also being cleaner and quieter [3–6].

The main focus of this chapter is on enabling a general perspective regarding the types of fuel cells, mechanisms, and applications, along with the nanostructures that are used to produce the catalysts. In Sections 5.2–5.4, we reveal a fundamental knowledge on fuel cells and oxygen reduction reaction that will help the reader to figure out the more elaborate discussions developed in the following sections. Section 5.4 gives information about nanostructures principles and types. Then, simulation methods will be presented in detail in Section 5.5. Finally, applications of fuel cells and future possible works will be given in Sections 5.6 and 5.7, discussing numerous examples from the literature.

5.2 Fuel cells

For the electrochemical devices (i.e., fuel cells), an oxygen from an air and a fuel such as hydrogen supplies are needed to convert the chemical energy into electrical

Sustainable Materials for Transitional and Alternative Energy.
DOI: https://doi.org/10.1016/B978-0-12-824379-4.00003-3

energy [7]. Oxidizing and reducing reagents are performed independently through arbitration by the ions and electrons in the external circuit. A fuel cell has two electrodes, an *anode* (where hydrogen or a related fuel reacts) and a *cathode* (where oxygen from air reacts). These fuel cells are both highly efficient and low environmental impact power generation devices. In addition, a fuel cell has gas-flow structures and separators that supply air and fuel flow to the electrodes [6]. The output is electrical energy in the form of direct current with the by-product of water [5,6,8–10]. Most fuel cell power systems comprise several components such as unit cells, stacks, and a balance of plant (BOP).

5.2.1 Fuel cells versus batteries and heat engines

Despite fuel cell being a kind of battery technology, which does not require recharging, its battery does. The active materials are supported from an external source for the fuel cell, while the battery is stored inside. The fuel cell's electrode materials are inert and are not consumed like the battery [11]. In comparison, heat engines are limited by thermodynamic limitations due to steps of producing heat and mechanical work. Because of these reasons, fuel cells produce less pollution power. As can be seen in Fig. 5.1, not all

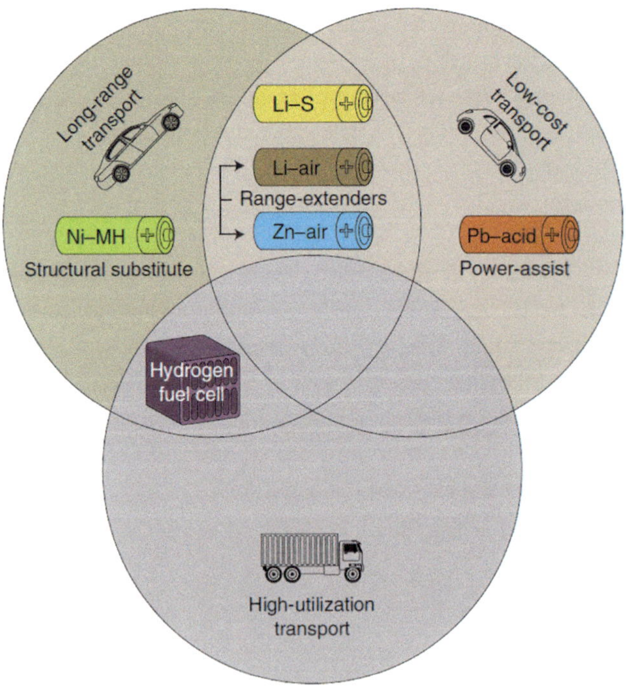

Figure 5.1 Suitability of alternative batteries and fuel cells to emerging electric vehicle markets [12].

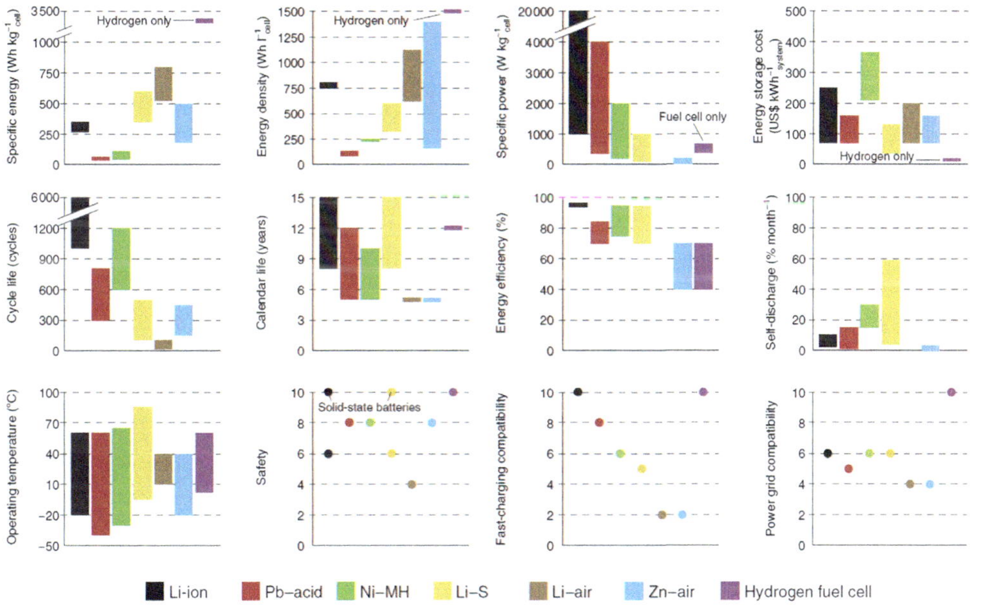

Figure 5.2 Characteristics of rechargeable batteries and hydrogen fuel cells [12].

technologies are suitable for every application of transportation. The enhanced specific energy, cost, and safety of fuel cells and batteries are needed to electrify long-range, low-cost, and high-utilization transportation sectors as seen in Figs. 5.1 and 5.2.

Fig. 5.2 shows that Ni-MH batteries can improve the range of Li-ion–powered electric vehicles (EVs) by changing the structural and energy adsorption components, while Pb-acid batteries could supply a less-power and more specific energy battery with low-cost EVs. Li-ion batteries [13] can be replaced by Li-sulfur batteries to increase the range and reduce the cost in EV markets, while Zn-air and Li-air batteries work as range extenders, further helping the sector. In addition, fuel cells are a natural fit for high utilization transportation due to fast refueling. Thus a combination of fuel cell and battery-operated EVs can increase the chance of clean and low-carbon transportation [12].

Unlike a heat engine (e.g., combustion engine) that converts chemical energy to electrical energy in multiple steps by first converting it to heat, then to mechanical energy, and finally to electrical energy, fuel cells perform this conversion directly in a single step. Fuel cells do not share any characteristics with heat engines, whose maximum efficiency can be calculated by a reversible Carnot Cycle as shown in Fig. 5.3. The Carnot efficiency does not apply to fuel cells because a fuel cell is not a heat engine, but rather an electrochemical energy converter [14].

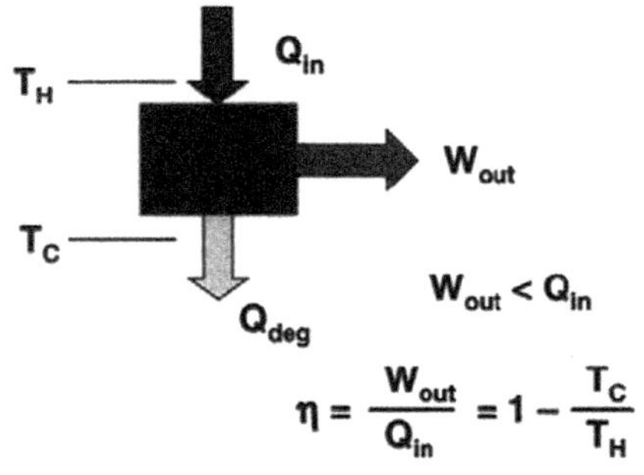

Figure 5.3 Carnot process efficiency [14].

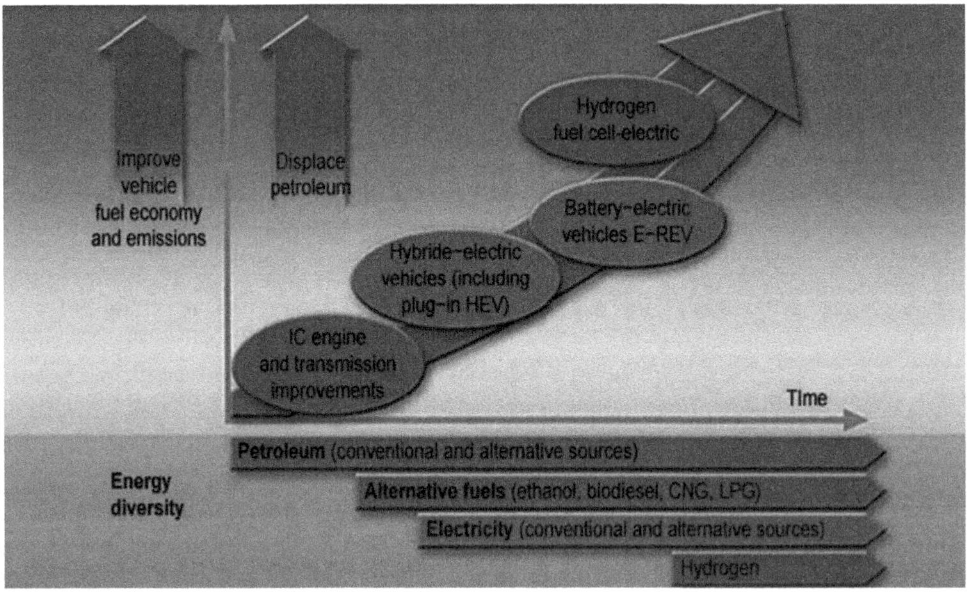

Figure 5.4 GM's advance propulsion strategy [16].

Battery and/or fuel cells-operated zero-emission vehicles [15] have more advantages compared to conventional vehicles regarding the ultimate target of the GM strategy that using performance and ease-of-use scales as can be seen in Fig. 5.4 [16].

Last but not least, an example of an electrical car is shown in Fig. 5.5. This EV energy storage is heavier when using an advanced Li-ion battery system (usable system energy density: 120 Wh/kg; current technology is closer to 85 Wh/kg) [16].

5.2.2 Hydrogen

Hydrogen has a high reactivity feature when high-active catalysts are used, and once produced, it is environmentally friendly, thus having no emissions. It uses fuel cells to

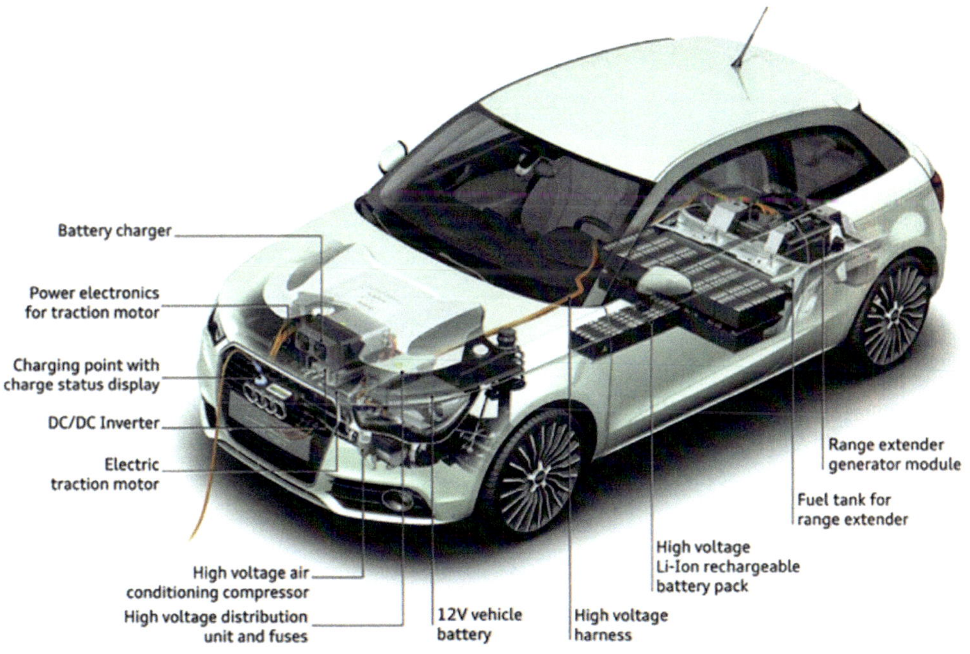

Figure 5.5 Battery charged electric cars [16].

operate effectively at all scales, like automobile, neighborhood, and utility applications such as personal electronics and emergency power [17].

5.2.3 Platinum (Pt) and electrochemistry

Platinum is a unique, noble, and precious metal which is required to interact the electrode processes. Its atomic number and weights are 78 and 195.08 amu, respectively. Pt bulk metal forms a face-centered-cubic (FCC) crystal structure, resulting in surfaces generally based on the 111, 100, and 110 crystal planes. Pt bulk crystal of different crystal orientations for Oxygen Reduction Reaction (ORR) activity in liquid electrolytes has been studied by Markovic et al [18–21]. The ORR in 1 M $HClO_4$ was unresponsive to the Pt surface structure, however was sensitive in bisulphate or Cl^- solutions [22,23]. Fig. 5.6 shows the cyclic voltammetry (CV) profile for a Pt electrode in an acidic medium. The ORR is the major contributor to the loss of effectiveness in an operating fuel cell due to the slow kinetics, which results in using more Pt compared to the anodic reaction (HOR) [24–28].

Platinum is also capable of effectively releasing the intermediate to form the final product as seen in Fig. 5.7. One of the most important obstacles that fuel cell commercialization is facing is the high cost of platinum and lowering Pt catalyst levels is an on-going effort.

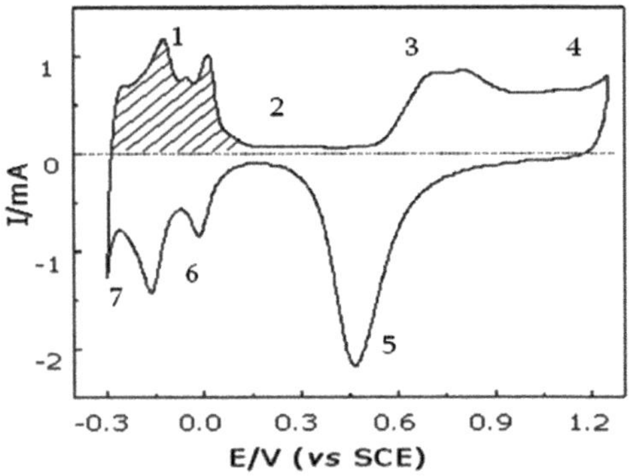

Figure 5.6 Current-potential curve (cyclic voltammetry) for Pt electrode in an acidic media. Hydrogen desorption (1), double layer charge/discharge (2), water activation (3), oxygen evolution (4), Pt-oxide reduction (5), hydrogen adsorption (6), hydrogen evolution (7) [24].

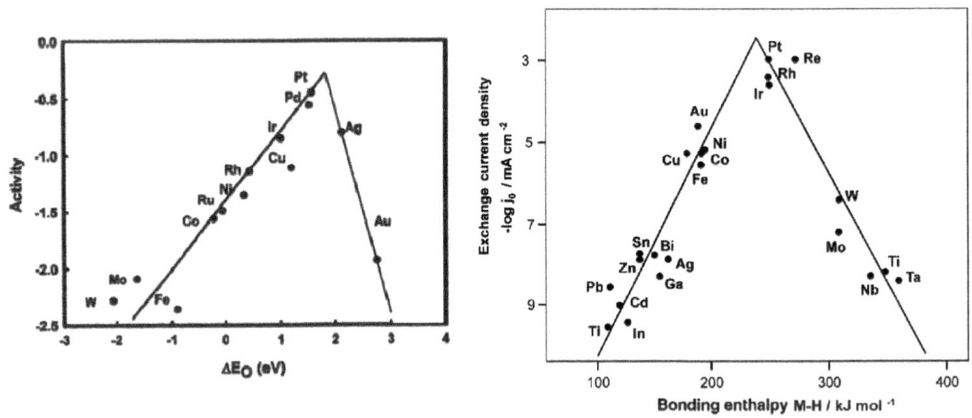

Figure 5.7 Trends in oxygen reduction activity plotted as a function of the oxygen binding energy (left) and hydrogen volcano curve: logarithm of exchange current density of H_2 reaction vs. enthalpy change of hydrogen adsorption at different metals (right) [24].

5.2.4 Carbon as a catalyst support

Carbon supports are very prone to oxidation, which causes a reduction in the chemical and electrical activity in the electrode. For a practical PEMFC cathode potentials, Carbon has very low oxygen reduction activity. However, even at a lower potential, Carbon has some activity when the reduction of oxygen incompletes to peroxide. Because of the thermodynamically nonstable property of the Carbon, it can be a problem for the long-time durability of the electrodes. The carbon corrosion rate increases with the level of humidification and influences the rate with the surface chemistry of the carbon support since graphitized carbons generally show greater stability. The synthesis of new

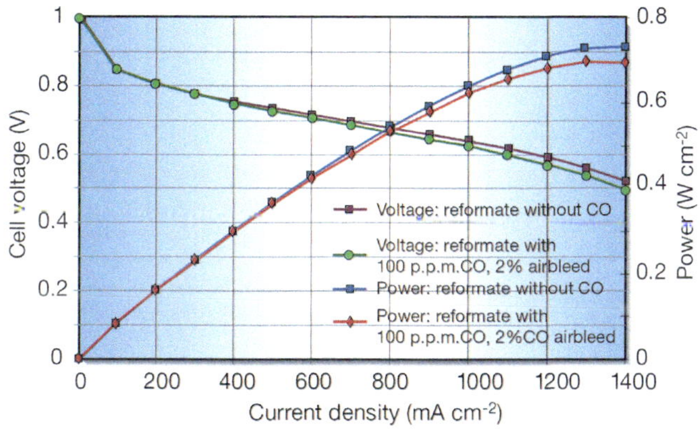

Figure 5.8 CO tolerance on Pt/Ru anode electrodes. Active cell area, 50 cm^2; precious metal loadings, <0.4 $mgcm^{-2}$; temperature, 80°C; pressure, 3 bar; anode fuel reformate, 60% H_2, 25% CO_2, 15% N_2; cathode, air [38].

nanostructures in the CO-tolerant catalysts field with Pt submonolayers on carbon-supported Ru nanoparticles [29–35], Ni nanowires [36] and Ni powder [37] is also studied.

CO-tolerant catalyst is required for fuels which is containing residues of CO or methanol in the DMFC [33,38–42]. To develop of the commercial PEMFC systems, a requirement of the CO-tolerant catalyst still remains as a most challenging task [40]. Pt/Ru and other composed of the two and three alloys with these noble metals have been investigated Intensively [41], and performance values have increased significantly. The loss in performance is usually stated in mV for a certain CO content of the fuel, and recent publications show promising results, as indicated, for example, in Fig. 5.8. There are few works that published about the CO resistance and electrocatalytic activity of WC which is highly sensitive to the synthesis technique [43–46].

5.2.5 Types of fuel cells

Many types of fuel cells (seen in Fig. 5.9) are being used to provide energy solutions for different technologies. As noted, the fuel cell's long-term feature expressively touches the current economics and technology challenges [6,10,48–51].

Every type of fuel cell has its own advantages and disadvantages. The solid oxide fuel cells (SOFC) and molten carbonate fuel cells (MCFC) have very high temperature which cause sluggish the start-up times during the operation. Contrary of this limitation, using nonprecious metal catalysts in these fuel cells catalysts are becoming an advantage to reduce the cost. The produced heat can be captured and reused to increase the theoretical overall efficiency up to 85% [52–55]. Besides the high-temperature fuel cells, the low-temperature fuel cells' work temperatures are ~80°C, ~120°C, and ~100°C for proton exchange membrane fuel cells (PEMFC) [56,57],

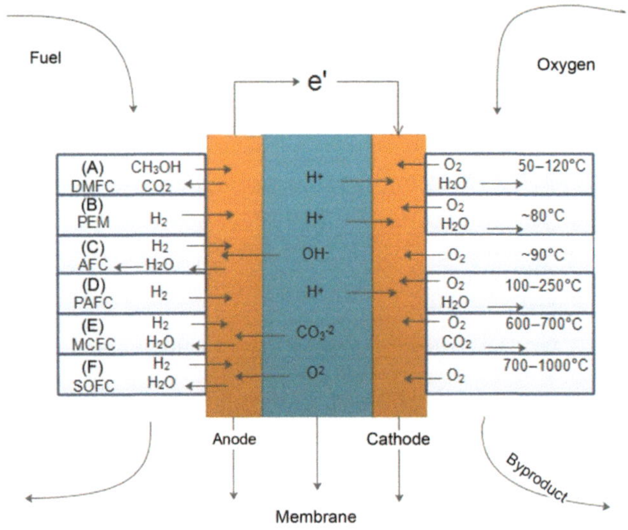

Figure 5.9 Various fuel cell types [47].

direct methanol fuel cells (DMFC) [39] and alkaline fuel cells (AFC), respectively. Moreover, phosphoric acid fuel cells (PAFC) working temperatures are between 100°C and 250°C [47,51].

5.2.5.1 Polymer electrolyte membrane fuel cells

The polymer electrolyte membrane fuel cell (PEMFC), which have aforementioned advantages, is going to be the prevailing technology in the near future [2,8,9,58–64].

Proton exchange membrane fuel cells (PEM FCs) have advantageous features when compared to other varieties. These advantages include: low-operating temperature (~80°C), sustained operation at high-current density, compactness, low weight, low cost, and volume, long stack life, fast start-ups and suitability for automotive applications where quick start-up is required [65]. Hydrogen gas is a crucial component of the PEM fuel cells. The membrane electrode assembly (MEA) is made with a solid organic polymer and the formed-composites which consists of a perfluorinated polymer backbone along with sulfonic acid. The very thin porous catalysts are combined around the proton exchange membrane as an anode and a cathode [66]. When fully humidified, the proton exchange membrane is also an excellent proton conductor.

As can be seen in Fig. 5.10 [47], the MEA includes proton exchange membrane, anode, and cathode side catalyst layers and gas diffusion layers (GDL) for both sides of the membrane [66–68]. The catalysts serve as a barrier to gas diffusion layers (GDL) and only the hydrogen ions allowed to transfer through the proton exchange

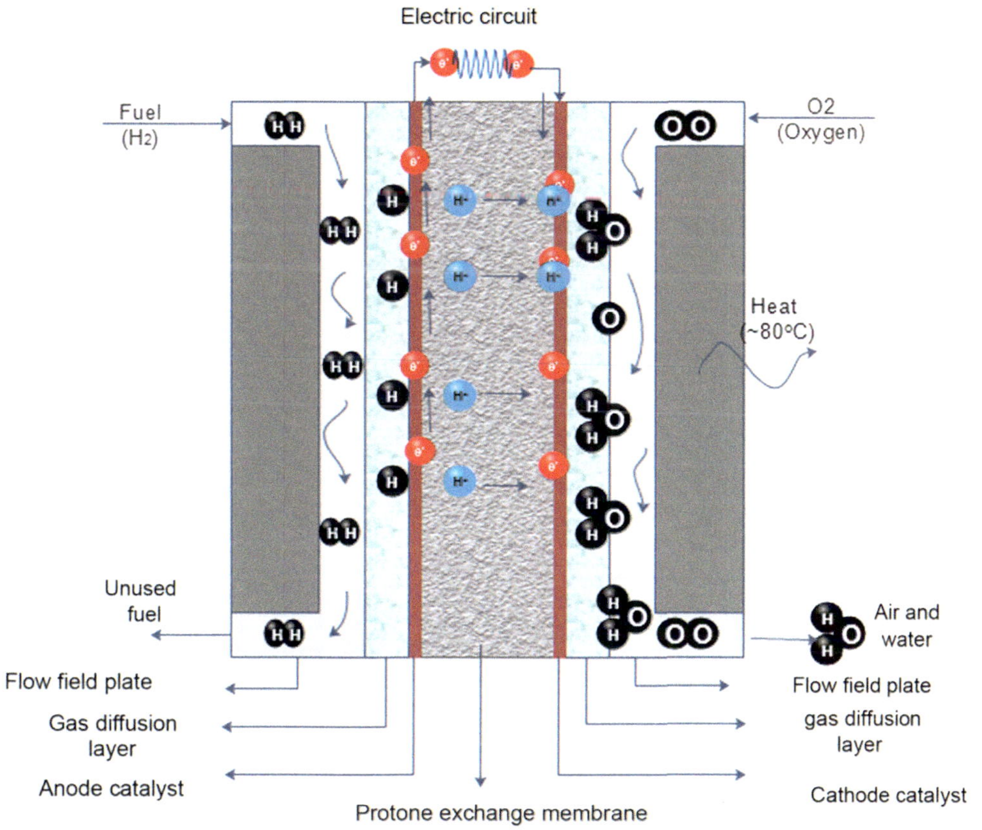

Figure 5.10 Schematic of a PEM fuel cell [47].

membrane [68]. While the hydrogen ions transfer through the membrane, the MEA needs moisture for a successful ion transfer. However, if there is too much water in the MEA, it will stay in the negative side of the electrode and cause the delivery issue of the oxygen (Fig. 5.11) [62,69].

The overall reaction is exothermic and can be written as Eq. (5.3) [17,66,68,70,71].

As a summary, the PEM fuel cell reactions are:

$$\text{Anode: } 2H_2 \rightarrow 4H^+ + 4e^- \quad E0 = 0V \tag{5.1}$$

$$\text{Cathode: } O_2 + 4H^+ + 4e^- \rightarrow 2H_2O \quad E0 = 1.23V \tag{5.2}$$

$$\text{Overall: } 2H_2 + O_2 \rightarrow 2H_2O \tag{5.3}$$

The PEM fuel cell technology has limitations that need to solve such as durability, cost, and performance while it is started to produce in a commercial industry [72]. Due

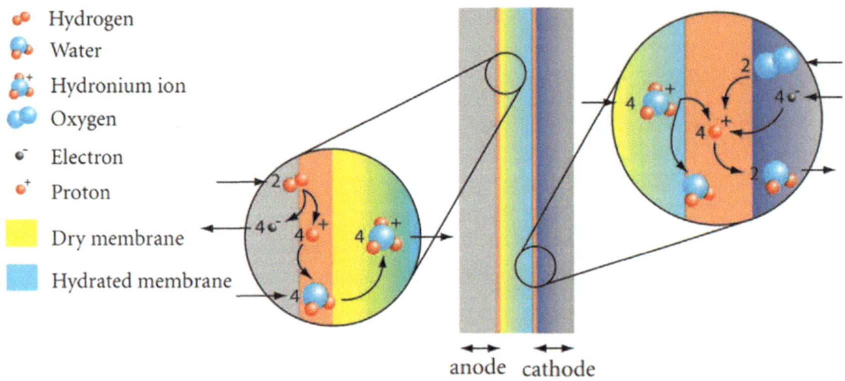

Figure 5.11 Proton exchange membrane structures with nano-scale dynamic interface [62].

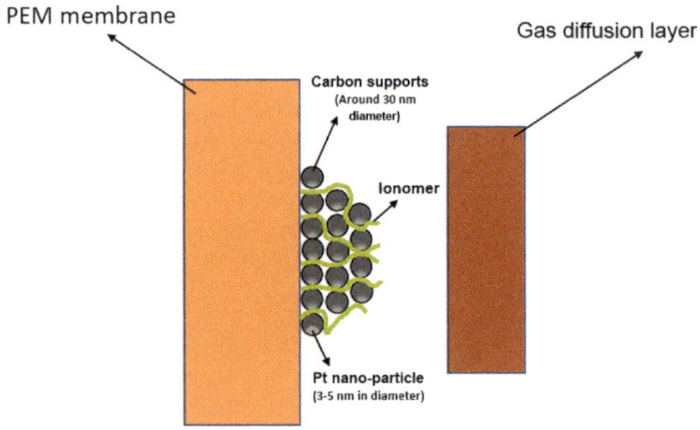

Figure 5.12 Conventional electrode design for PEM fuel cells [47].

to these limitations, researchers, and developers are working on novel catalysts that show superior improved performance compared to previous developed catalysts [73,74].

5.2.5.1.1 Limitations of traditional PEM fuel cell electrodes

The ionomer—solution of the membrane- were filled around the platinum nanoparticles-3 to 5 nm in size- lay on the carbon nanoparticles to diffuse through this complex of nanoparticles (Fig. 5.12).

The conventional type of fuel cell has more serious limitations [2,59,60]; the agglomeration of the Pt nanoparticles [33,70,75]; the problem of the oxidation of carbon [76,77]; the effectiveness problem due to carbon disconnects from the ionomer [78]. Cathode catalyst layers acquire a major part of the efficiency losses during operating of PEMFC. The consumption of the catalyst material is an important issue in the

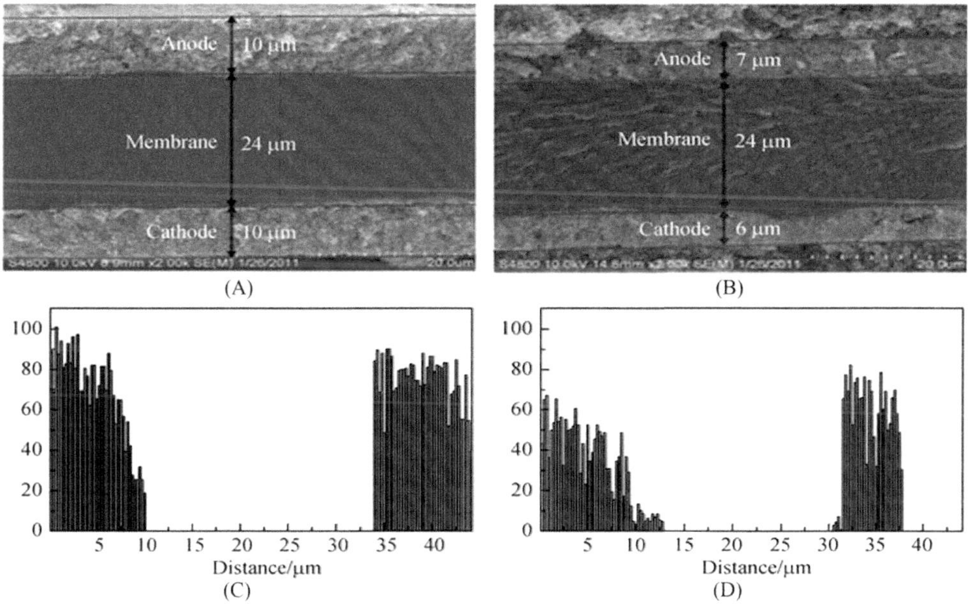

Figure 5.13 Cross-sectional SEM picture of the MEA before (A) and after (B) vehicle accelerated test and distribution of the elements in MEA before (C) and after (D) vehicle accelerated test [81].

catalyst layer, which acquires a major portion of the overall costs of the fuel cells system [79,80].

The Tian et al. study also shows that the loss of efficiency in catalysts is seen clearly before and after the test (Fig. 5.13), which was a new enhanced test technique to analyze the durability of MEA during actual running of the fuel cell in vehicle [81]. Aside from the cost of the precious metal, another major problem is the core-shell structure's durability which makes the inclusion of Pt quite necessary. Due to these issues, massive efforts are on track to create high-performance, durable, carbon-free [24,82–89], and low-cost electrocatalysts with an alternative support material [13,82] such as columnar titanium (Ti), Tungsten Carbide (WC) [90], and chromium nitride (CrN) supports [91–93], or an organic whisker support layer [72].

Also, operating PEM fuel cells at high temperature (HT-PEM) is a promising technology, because of the enhanced electrochemical kinetics, a simplified water management and cooling, and enhanced carbon monoxide tolerance [62,94–96]. Also, the Ni leaching reduced and the stability along with higher catalytic ORR activity improved by the smaller size, annealing, and less-oxidative acid treatment [97].

5.2.5.1.2 Polarization curve of PEM fuel cells

Fig. 5.14 illustrates a typical polarization curve of PEM fuel cells. This curve results from both the anodic HOR and the cathodic ORR reactions. The actual cell voltage

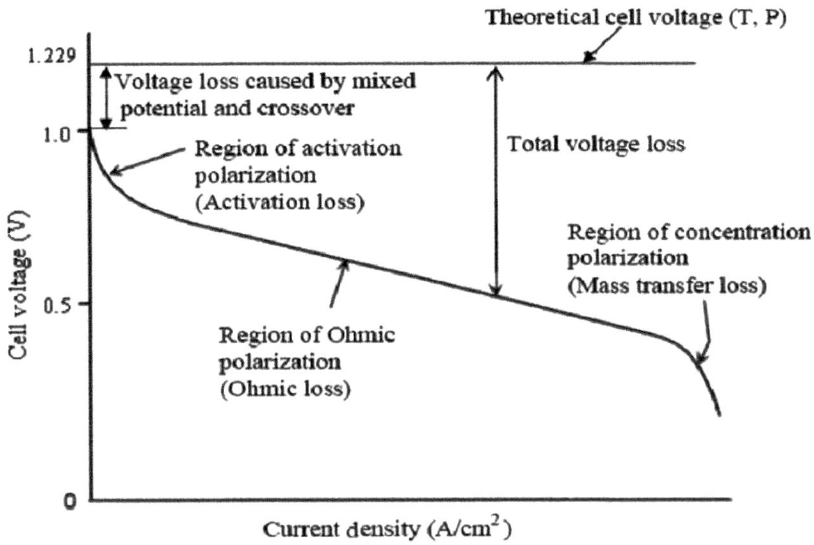

Figure 5.14 Typical polarization curve of PEM fuel cells [98].

is much lower than the theoretical cell voltage. When the current is extracted from a fuel cell, the actual cell voltage will drop from its theoretical value due to several types of irreversible losses. The drop is mainly caused by mixed potential and fuel crossover, activation overpotential, ohmic overpotential, as well as mass transfer (concentration) overpotential [82,98,99].

Obermaier, M., et al., worked on the uncomplicated model for the catalyst layers of the PEM fuel cells' cathode side with the nonappearance of the oxygen [100].

5.2.5.2 Microbial fuel cells

The researchers realized that the electricity could be produced from the organic waste by using microbially catalyzed anodic and cathodic electrochemical reactions in the last couple decade. Microbial fuel cells (MFCs), as seen in Fig. 5.15, and bio-electrochemical systems (BES) were interested and studied by the researchers. MFC and BES are a combination of the biological catalytic redox activity with classic abiotic electrochemical reactions and physics [101,102].

The major differences between MFC and conventional low-temperature fuel cells such as direct methanol fuel cells (DMFC) and PEMFC are: a biotic anode catalyst [103]; the changing temperature scales between 15 Co to 45 Co [104]; neutral pH operational environments [105]; the application of complicated biomass as an anodic fuel [106]; and a promising moderate environmental impact assessed through life cycle analysis [107].

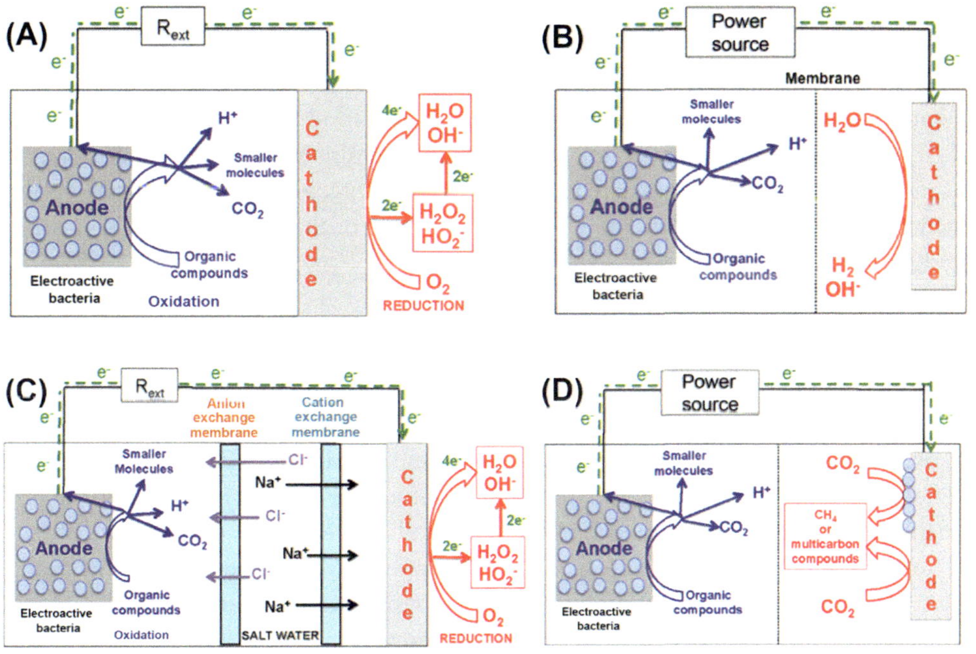

Figure 5.15 Schematic of a microbial fuel cell (A), microbial electrolysis cell (B), microbial desalination cell (C) and general microbial electro synthesis cell (D) [101].

5.2.5.3 Alkaline fuel cells

An alkaline fuel cell (AFC) has two different concentrated KOH in its electrode. The one has 85 wt. percent KOH in electrode operated at ~250°C high temperature and the other one which has ~50 wt. percent KOH operated at less than 120°C. A broad scale of electro-catalysts can be used, and the electrodes maintained in a matrix. During the operation, CO_2 is reacting with KOH and forming K_2CO_3. Only nonreactive suppliers can be used as a fuel for this type of fuel cell. However, there is an exception for the hydrogen.

$$\text{Anode: } H_2 + 2(OH)- = 2H_2O + 2e- \tag{5.4}$$

$$\text{Cathode: } \tfrac{1}{2}O_2 + H_2O + 2e- = 2(OH)\_ \tag{5.5}$$

The advantages of these fuel cells are the tremendous performance on H_2 and O_2 electrodes, flexibility to apply a wide-range using area of catalysts. On the other side, to eliminate the CO_2 issue, the system needs very high pure H_2 to use during operation. Therefore, the use of a reformer would require a highly effective CO and CO_2 removal system. In addition, if ambient air is used as the oxidant, the CO_2 in the air must be removed [26].

5.2.5.4 Phosphoric acid fuel cell

Phosphoric acid fuel cell (PAFC) uses 100 percent concentrated electrolyte with the ~150 to 220°C operation temperature. Phosphoric acid is becoming a poor ionic conductor at low temperature, CO poisoning in the anode catalyst is becoming severe. PAFC can be able to run at the ~220°C temperature range, which minimizes the water vapor pressure in the system. Thus, water management in the cell can be easily handled. Pt is used for both anode and cathode side of the electrodes, and silicon carbide is used to retain the acid.

$$\text{Anode: } H_2 = 2H^+ + 2e^- \quad (5.6)$$

$$\text{Cathode: } \tfrac{1}{2}\, O_2 + 2H^+ + 2e^- = H_2O \quad (5.7)$$

CO poising is less sensitive in this fuel cells compared to PEFC and AFC. Common construction materials can be used due to still low operating temperatures. The operating temperature also provides considerable design flexibility for thermal management. PAFC has high efficiency—around 40 percent—compared to PEMFC. The waste heat can be used in other areas. In contrast to these advantages, the fuel cell system is very corrosive, and mobile electrodes require a high amount of Nickel and a high quality of stainless steel as the cell hardware. Due to higher temperatures, mechanical stability and stack life are other issues [108].

5.2.5.5 Solid oxide fuel cell

The solid oxide fuel cell has nonporous metal oxide-Y_2O_3-stabilized ZrO_2—electrodes which can operates between 600°C and 1000°C. The cathode side's electrode is made by Sr-doped $LaMnO_3$, and the anode side's electrode is made up with X-ZrO_2 (X may be Co or Ni elements). SOFC operating temperature were dropped from

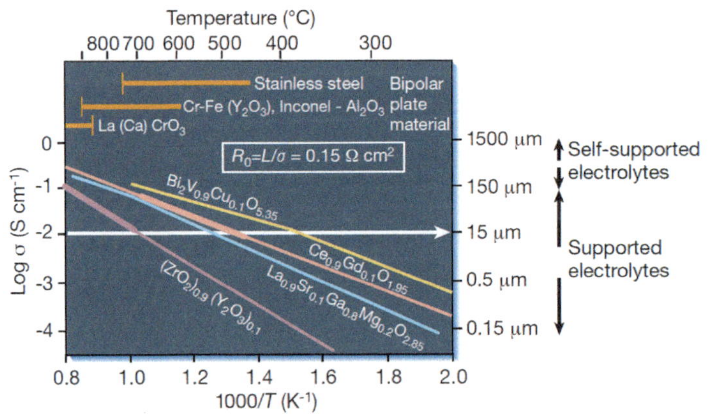

Figure 5.16 Specific conductivity versus reciprocal temperature for selected solid-oxide electrolytes [38].

1000°C to around 600°C by using improved cathodes, and scientists are trying to drop the working temperature even lower compared to 600°C.

As can be seen from Fig. 5.16, the reciprocal of temperature relies on the specific ionic conductivity of solid electrolytes [38]. Some of the advantages of this type fuel cells are: the cell can be cast into various shapes due to solid electrolyte; the corrosion issue happens less with solid ceramic construction; the kinetic of the cell is fast, CO can be reuse as a fuel in the MCFC. The efficiency range is around 40 to 60 percent. On the other hand, the major limitation is the high temperature of the fuel cell. Due to this, the sealing between cells is very hard in flat plate configurations. The high operating temperature puts serious limitations on resources choice and outcomes in hard fabrication process [109,110].

5.2.5.6 Protonic ceramic fuel cells

Protonic ceramic fuel cell (PCFC) allows through electrochemical transition of chemical fuels to electricity at superior productivity and with zero emission which incorporate a proton-conducting oxide as the catalyst. The working temperature of this type fuel cell is around 400°C–600°C which is key to reduce the cost. Few works showed high power densities of PCFC exceeding 200 mW cm^{-2} at 500°C. The poor rate of oxygen electroreduction at the cathode of PCFCs has been recognized as one of the key factors limiting power densities in such fuel cells [111].

The reaction is written globally as;

$$\frac{1}{2} O_2(gas) + 2e'(cathode) + 2H + (electrolyte) \text{-----------------} H_2O(gas) \quad (5.8)$$

where it is implied, from the formalism of the expression, that the cathode, is electronically conductive but ionically insulating. The reaction is limited to the triple-phase boundary lines at which cathode, electrolyte, and gas are in mutual contact. Experience with traditional SOFC indicates that the best cathode electrocatalysts are those with mixed ionic and electronic conductivity [111].

5.3 Mechanism and kinetics of oxygen reduction reaction

The high overpotential for oxygen reduction, however, is a long-standing problem and, so far, research on the fundamental processes of oxygen reduction and catalysis has not yielded a breakthrough. To tackle this shortcoming, new alternatives for finding cheaper materials with lower overpotential are summarized below.

As depicted in Fig. 5.17, for the case of $Pt_3Ni(111)$, the Pt(111)-skin structure is 90-fold more active than the state-of-the-art Pt/C catalysts for PEMFC. This huge enhancement in activity arises from a substantial reduction in the surface coverage by OHad relative to Pt(111). This was first observed experimentally and then confirmed

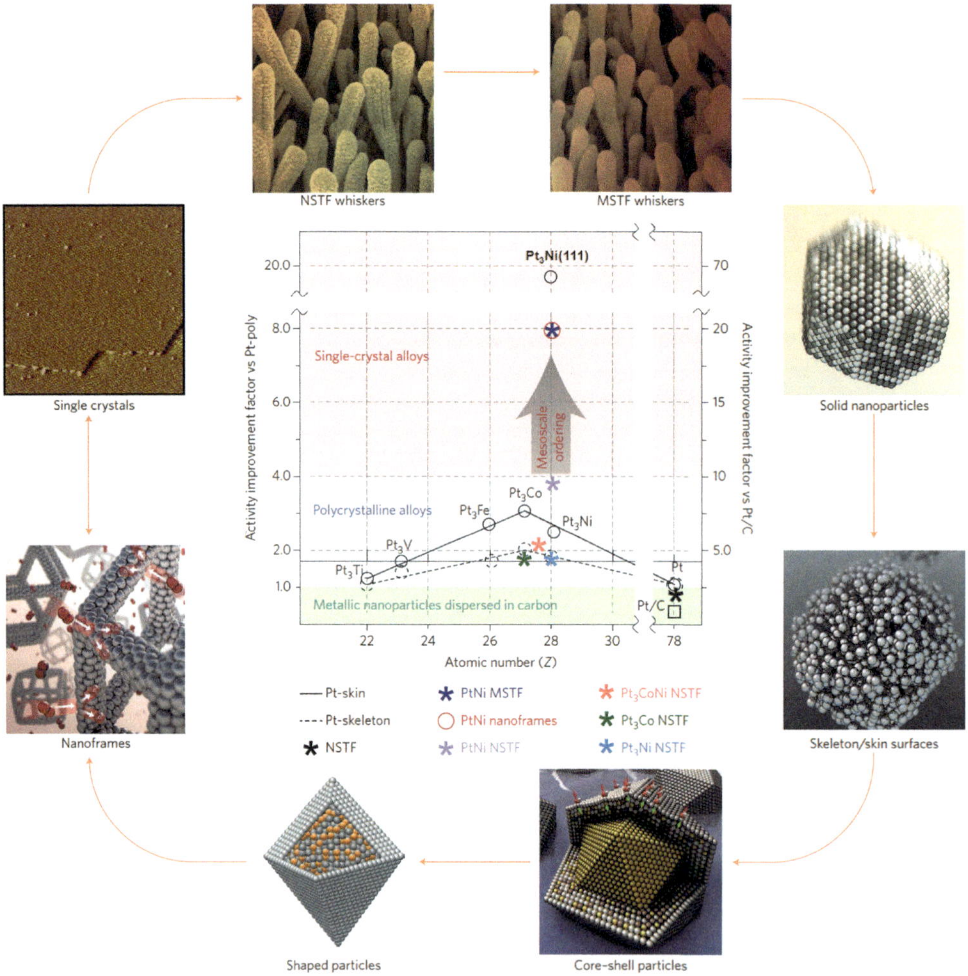

Figure 5.17 Materials-by-design strategy for development of the most active materials for ORR in PEMFC. Extended single-crystalline well-defined surfaces, thin-film-based catalysts with nano-and mesostructured surface morphology, solid nanoparticles with tailored skin and/or skeleton surfaces, core-shell nanoparticles, shape-controlled nanoparticles, and hollow nanoscale systems, such as nanoframes, represent a full circle in development of efficient ORR catalysts in acidic media. NSTF, nanostructured thin-film; MSTF, mesostructured thin-film. The central graph shows an ORR activity map for Pt-alloys depending on alloying element and class of materials [112].

by DFT calculations, the latter showing the optimal range of binding energy between OHad* and Pt-skin surface atoms. This insight has since been extended to a number of real-world catalyst systems, including nano- and meso-structured thin films [113,114], and nanomaterials with tailored particle sizes [115], compositional profiles [116,117], surface morphology [118], shape and architecture [119]. Taken together, fundamental and practical insights from these studies have resulted in the recent successful integration of PtCo alloy catalysts into the PEMFC that powers the Toyota Mirai fuel-cell electric vehicle [120]. Although the catalytic activity of the Pt(111)-skin has still not been fully implemented in real-world catalysts, it serves as a permanent inspiration for designing real-world nanoparticles. The most recent success is the design of PtNi nanoframe catalysts with a 3D network of Pt-skin surfaces, which exhibit the best activity of all real-world materials developed thus far: more than 20-fold improvement over the Pt/C catalyst (Fig. 5.17) [119].

Aligned perylene red substrate (Fig. 5.18, center of the schematic) were grown and coated by a metallic thin film with an changeable thickness and composition of (Pt, M, and/or NDNi, Co, Fe, Ti, V) [114].

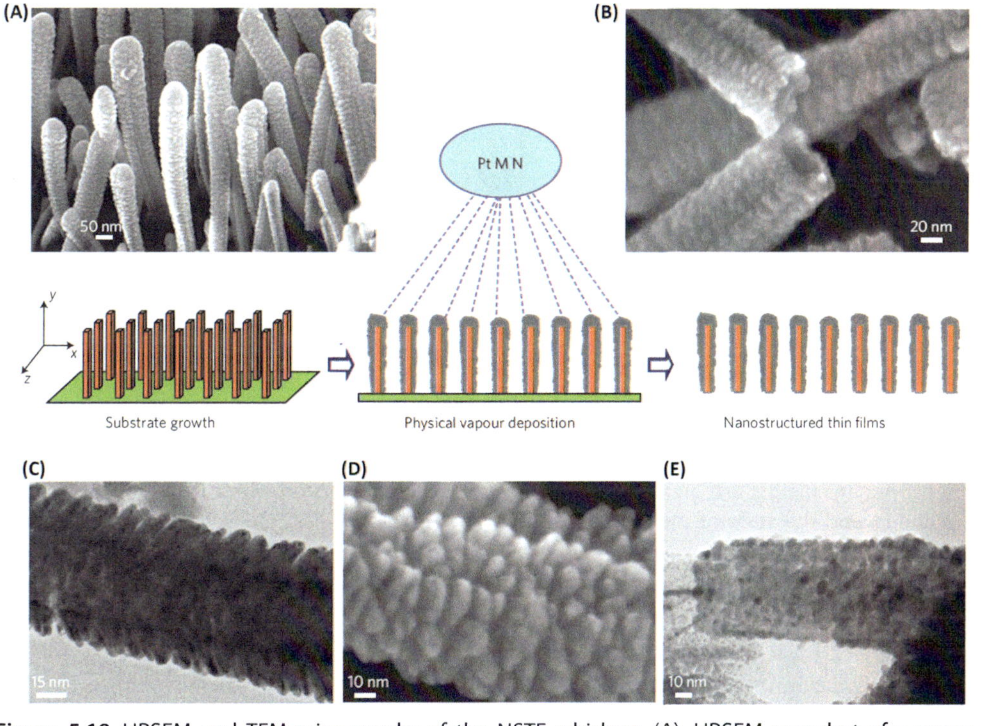

Figure 5.18 HRSEM and TEM micrographs of the NSTF whiskers. (A), HRSEM snapshot of a group of whiskers that indicates their length, shape, and alignment after thin-film deposition. (B), HRSEM close-up of an intentionally broken single whisker that demonstrates the thickness of the metallic film over perylene red substrate. (C), HRTEM close-up of a single whisker side, which reveals growth of whiskerettes along the whisker. (D), HRSEM insight into the whisker's surface showing a close-packed formation of whiskerette tips of 5 nm in diameter that facilitates a highly corrugated morphology. (E), TEM micrograph of a whisker side that confirms the grained texture of the sputtered thin film and the average diameter of the whiskerettes [114].

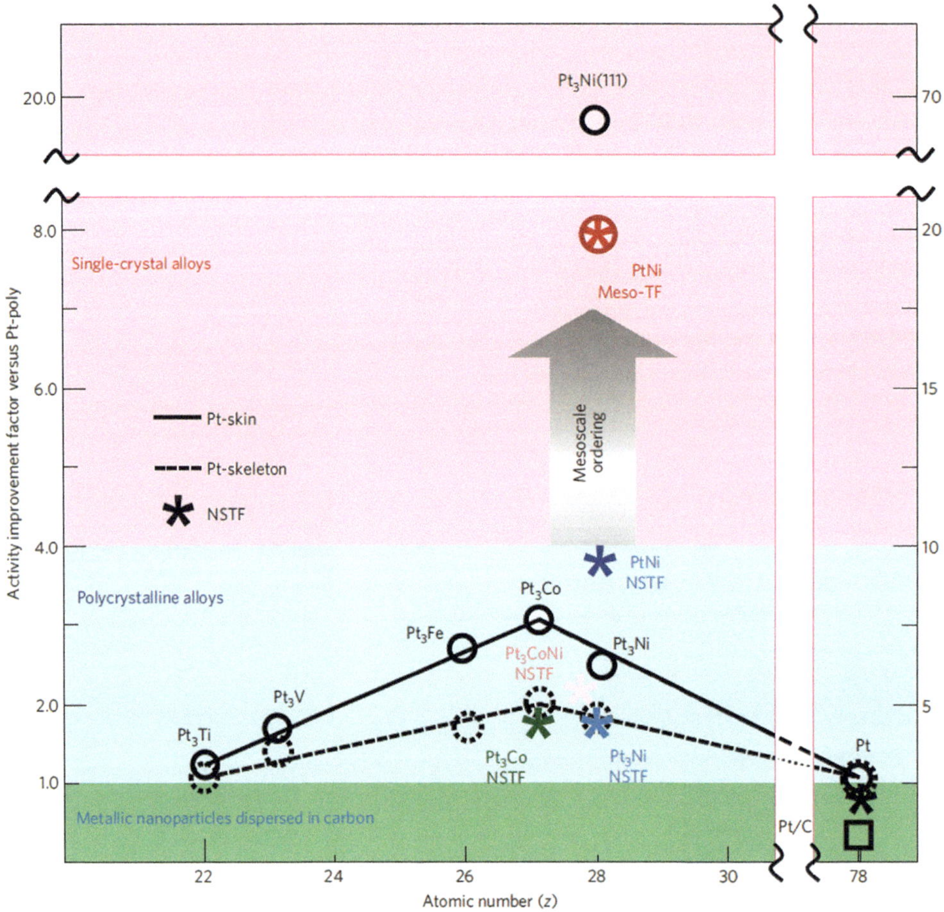

Figure 5.19 Activity map for the ORR obtained for different classes of Pt-based materials. Improvement factors are given on the basis of activities compared with the values for polycrystalline Pt and the state-of-the-art Pt/C catalyst established by RDE measurements in 0:1MHClO4 at 0.95 V [114].

Based on the principles shown in Fig. 5.19, the NSTF catalysts can effectively imitate the catalytic performance of polycrystalline bulk materials, while Pt-alloy mesostructured thin films surpass the range specified for polycrystalline structures. This is the first useful catalyst to deal with the levels of activity heretofore retained only for bulk single-crystalline surfaces, and near-surface form like that of the ideal Pt3Ni (11_1)-skin. These bimetallic Meso-TF materials maintain adequately high specific surface areas, which allow improved consumption of precious metals [114].

ORR is a significantly more complex reaction compared to HOR. It includes the following possible steps in an acidic medium: [24,98]. Direct four-electron reduction,

$$O_2 + 4H^+ + 4e^- \rightarrow 2H_2O \quad E^o = 1.229V \tag{5.9}$$

or, indirect reductions

$$O_2 + 2H^+ + 2e^- \rightarrow H_2O_2 \quad E^o = 0.682V \tag{5.10}$$

and

$$H_2O_2 + 2H^+ + 2e^- \rightarrow 2H_2O \quad E^o = 1.76V \tag{5.11}$$

The ORR is somewhat complex and contains several intermediates, mainly depending on the spirit of the electrode material, electrolyte, and catalyst. It is desirable to have the O_2 reduction reaction occurring at potentials as close as possible to the reversible electrode potential (thermodynamic electrode potential) with a satisfactory reaction rate. The current-overpotential relationship is represented by Butler-Volmer equation: [14,24,121,122]

$$j = j_o\left[\exp\left(\frac{\alpha_a nF}{RT}\eta\right) - \exp\left(\frac{-\alpha_c nF}{RT}\eta\right)\right] \tag{5.12}$$

where j is the oxygen reduction current density, j_o is the exchange current density, n is the number of electron transferred in the rate determining step, α_a and α_c are the anodic and cathodic charge transfer coefficient, respectively, η is the overpotential of ORR ($E-E_e$), F is the Faraday constant, R is the gas constant, T is the temperature in Kelvin. The Butler-Volmer equation applies to all single-step electrochemical reactions. This equation basically states that the current produced by any electrochemical reaction increases exponentially with activation overvoltage. For the cathodic reaction (ORR) with negative η, the cathodic branch will exponentially increase, while the anodic branch will be a diminishing function:

$$j = j_o\exp\left(\frac{-\alpha_c nF\eta}{RT}\right) \tag{5.13}$$

Solving this equation for η_c yields:

$$\eta = \frac{2.303RT}{\alpha_c nF}\log j_o - \frac{2.303RT}{\alpha_c nF}\log j \tag{5.14}$$

If this equation can be generalized in the following form:

$$\eta = a + b\log j \tag{5.15}$$

This is known as the *Tafel equation*, and b is called the *Tafel slope*. Experimentally, the exchange current density and charge transfer coefficient are found with a Tafel plot, which is a plot of the log of current density versus overpotential for a given reaction. From the slope of a semilog plot of voltage versus current, the charge transfer coefficient can be determined, and from the intercept the exchange current density cab be found. Thus, for an electrochemical reaction to obtain a high rate current at low overpotential, the reaction should exhibit a low Tafel slope or a large α_o n. For ORR, usually two Tafel slopes are obtained, 60 and 120 mV/decade at low and high current densities, respectively, depending on the electrode materials used and on the potential range [14,24]. Wang et al. [123] developed a class of Pt highly concave cubic nanocrystals and durability of these nanocrystals improved by using graphene. Besides using Pt self to prepare catalysts, Pt-Co-Mn-Ni-Cu-Mo-W targets used to prepare catalysts to enhance the ORR activity [75,124,125]. N. Karthikeyan et al., combined the effect of oxygen and nitrogen functional groups on highly crystalline carbon supports like multiwalled carbon nanotubes and few layer graphene hybrid structures to investigate the RR performance in PEM Fuel Cells [126].

5.4 Nanostructures materials for fuel cells

The timeline in Fig. 5.20 is highly subjective and is meant to give only a general ranking of technology readiness for each catalyst family. No specific dates are provided as this would be rather presumptuous, depending largely on where current and future research efforts are focused.

As can be seen in Fig. 5.21, several developments and improvements concerning the increase in hydrogen production [131], improvements in cell design [132], removal of membrane [133], utilization of microbial catalysts [134] or Pt-free catalysts [102] have been successfully demonstrated.

Zhengetal [134] prepared nano-catalysts and used a high-resolution transmission electron microscopy, a synchrotron radiation X-ray absorption spectroscopy, and a high-sensitivity, low-energy ion-scattering spectroscopy to study the Pt-FeNi(OH)x interface synergistic mechanism for the oxidation of CO. Fig. 5.22 shows that artificial technique utilized for Pt-based catalyst which extends the catalyst reactivity interface from one-dimension to three-dimensions [135]. Gustavsson, M., et al., work on the thermally evaporated thin Pt/TiO2 films showed that TiO2 can both increase or decrease the performance of a PEMFC electrode, depending on geometrical placement [136].

While PEM fuel cell development is more thrilling, the traditional catalysts -Pt nanoparticles on Carbon (Pt/C)- are still being used in current fuel cells; however, for the commercial production, it is needed more than four times cost and supply constraints in catalytic activity per mass of precious metal [82]. Some of the works showed that the nonprecious metal catalysts could be used in catalysts [137,138] which focus on nitrogen-coordinated iron in a carbon matrix such as the combination of these

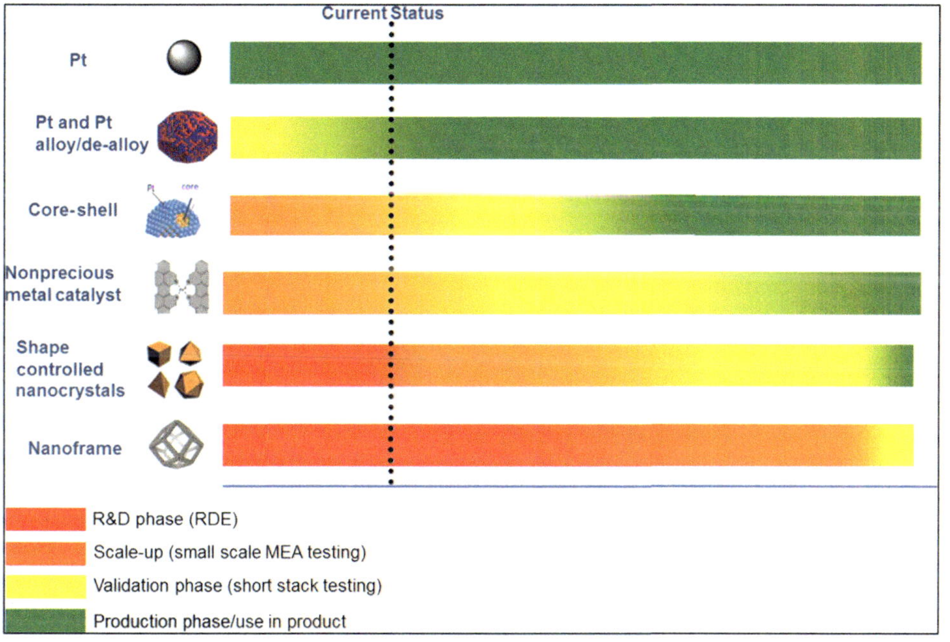

Figure 5.20 Development timelines for Pt, Pt alloy/dealloy, core – shell [127], nonprecious metal [128], shape-controlled [129], and nanoframe [119] ORR electrocatalysts [130].

metals' Fe/N/C (FeN_4 or FeN_2) [139–141]. However, greatly divergent syntheses generated practically identical catalyst activities [142].

In parallel with the nonprecious metal approach, Pt-based oxygen reduction electrocatalysis has also seen important advances. Pt-based catalysts must have very high mass-activities due to cost and supply limitations. As a traditional method, PtM alloy nanoparticles turnover frequencies are only ~60 s^{-1} [82], on the other side, the "de-alloying" of PtM nanoparticles turnover frequencies can up to ~160 s-1 [116,117] which are achieved for ~10 monolayer-thick Pt films on support structures [118]. These works offer large Pt dispersions while approaching turnover frequencies of bulk Pt surfaces resulted four times more mass activities for de-alloyed PtM compared to current Pt/C catalysts [117]. (Fig. 5.23).

Turnover frequencies for the electrochemical conversion of oxygen to energy and water for different oxygen reduction catalysts. Lefèvre et al. reported that previous inventions of Fe/N/C catalysts had much lower turnover frequencies than the novel Fe/N/C catalyst [143]. Solution-based colloidal nanoparticles were prepared to control particle size, shape, composition and structure for the catalytic applications [144,145].

Bostjan Genorio, et al., used chemical preparation methods to prepare the catalysts for the HOR and ORR reactions by pattering of platinum with calix[4]arene molecules which were demonstrated a chemically modified Pt electrode with a SAM of calix[4]arene molecules can selectively block the ORR, but in such a way that the

Catalyst Type	Benefit	Remaining Challenges
Pt 	(1) Mature technology	(1) Unable to meet long term automotive Pt loading and catalyst layer durability targets
Pt alloy/de-alloy 	(1) Mature technology (2) Improved performance over Pt/C (3) Enhanced membrane/MEA durability	(1) Difficult to meet long term automotive Pt loading target
Core-shell[16] 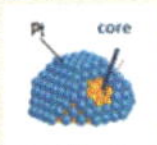	(1) Improved mass activity over Pt alloy (2) Improved durability over Pt/C (3) Highest reported ECSA	(1) Difficult to maintain quality of 'shell' (2) Dissolution of 'core' still a concern
Shape controlled nanocrystal[18] 	(1) Significantly higher mass activity (~ 15 x) over Pt (2) Chemical synthesis (vs. electrochemical) may allow for easier scale up vs. core-shell	(1) Scale up is at an early stage (2) Conflicting data on stability (3) MEA performance has not been demonstrated yet
Nanoframe/nanocage[19] 	(1) Significantly higher mass activity (> 20 x) over Pt (2) Highly stable (improved durability over Pt/C)	(1) Scale up is at an early stage (2) Ionomer penetration into nanocage will likely be difficult (3) MEA performance at high current density may be challenging
Non-precious metal catalyst[17] 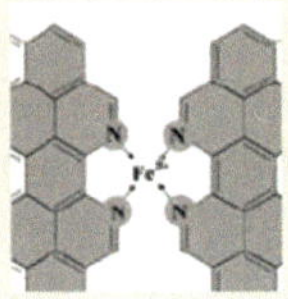	(1) Potentially offer the largest benefit (significant cost reduction) (2) Tolerant to common contaminants	(1) Close to meeting targets for non-automotive applications (e.g. backup power) but far from meeting automotive targets. (2) At current volumes, PGM loading is not a major concern for non-automotive applications.

Figure 5.21 Benefits and remaining challenges for each of the primary categories of electrocatalysts [130].

HOR proceeds with Pt-like activity [134]. For the anode side of the PEM fuel cell, the catalysts which made without Pt have a strong interaction between PdNi alloys and WC compared to conventional Pt/C catalyst [80].

5.4.1 Nanoparticle materials

Gasda, M.D., et al. have built PEM fuel cell cathode electrodes (Fig. 5.24) comprising sputtered Pt layers that are supported by vertically oriented CrN particles grown by GLAD and normal angle deposition [93].

5.4.2 Nanoframes materials

Recently, Chen et al. presented a novel class of electrocatalysts that converted in solution by center loss into Pt_3Ni Nanoframes (Fig. 5.25A) with shells that offers three-dimensional

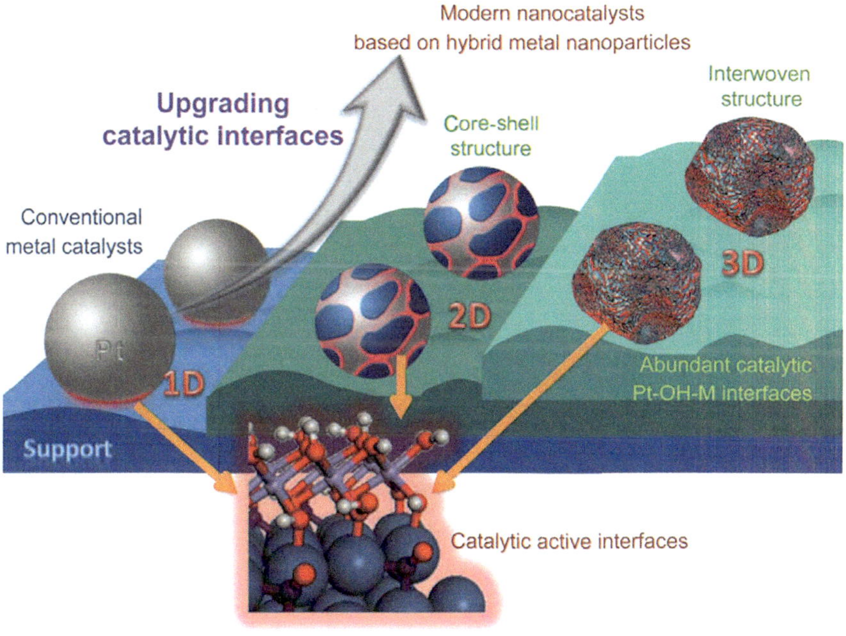

Figure 5.22 Schematic of development from one-dimensional to three-dimensional catalyst [135].

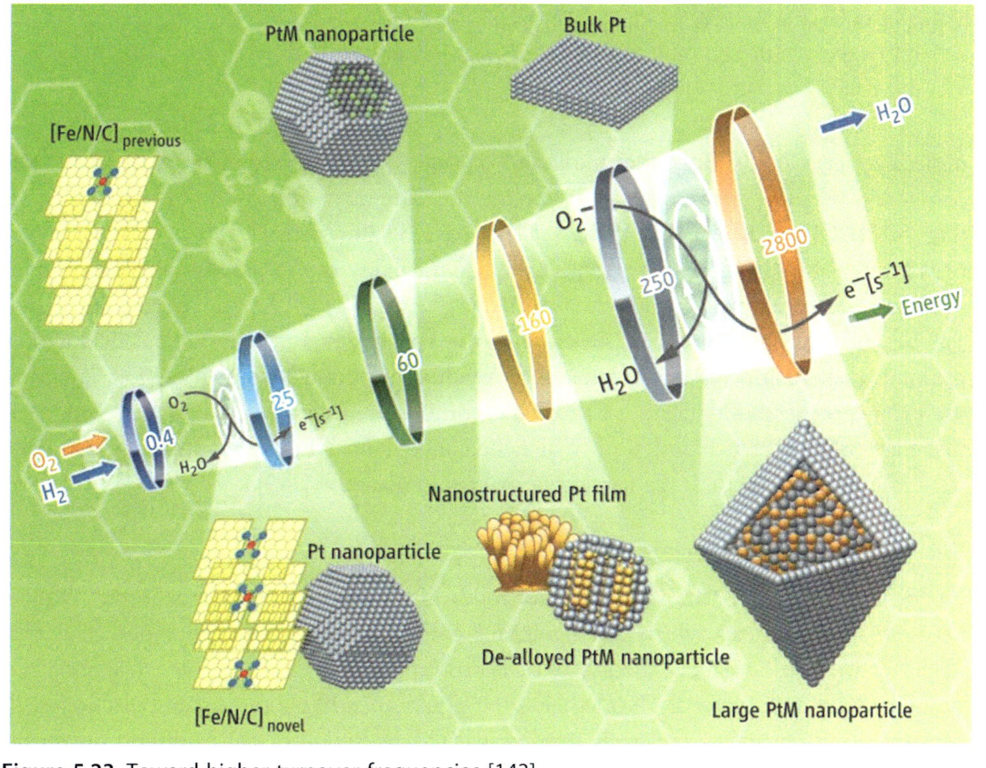

Figure 5.23 Toward higher turnover frequencies [143].

Figure 5.24 Plan-view SEM micrographs showing GDL substrates that were successively coated with 0.29 mg/cm2 CrN and (A) 0.05 and (B) 0.25 mg/cm2 Pt [93].

molecular accessibility. Both the inside and outside shell of this open-structure shape were composed of the Nano segregated Pt-skin structure and enhanced the ORR activity (Fig. 5.25B). The Pt_3Ni Nano frame catalysts have 36 times and 22 times enhancement in mass activity and specific activity, respectively. The stability of this pt3Ni structures were also maintained after potential cycling durability test, (Fig. 5.25C) [119]. The stability and activity can be enhanced by control of structure at the atomic level. The structural evolution of platinum-nickel nanocrystals is exploited by synthesizing the electrocatalysts. The interior of the $PtNi_3$ polyhedral-crystalline structure transformed by decomposition in the solution and the edges of the Pt-rich structures maintained in the final Pt3Ni Nanoframes. As a final stage, the Pt-rich structures are segregated to Pt-skin structure [119].

5.4.3 Nanorod materials

Reducing the size of the catalyst's electrodes affects the space-charge at the interface between small particles which can significant enhancements of the properties. This challenges us to develop a new theory, or at least adapt and develop theories that have been established for bulk materials. Nonetheless the significance of nanostructured electrodes, the performance of practical fuel cells remains limited by scale-up, tack housing design, gas

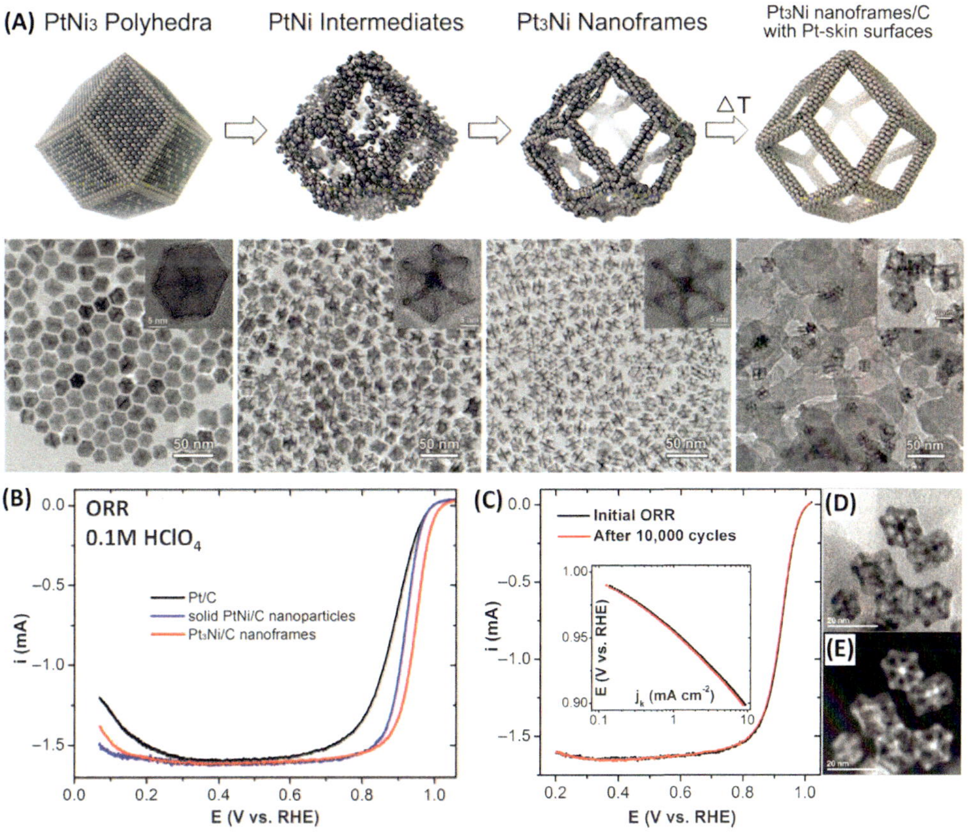

Figure 5.25 (A) Schematic illustrations and corresponding TEM images of the samples obtained at four representative stages during the evolution process from polyhedral to Nanoframes. (B) ORR polarization curves. The ORR activity (C) and TEM images of Pt_3Ni Nanoframes before (D) and after (E) the stability test [119].

manifold, and sealing. Even though some of the fuel cell's processes run by using the electrochemical oxidation of hydrocarbon and alcohol [146–148], still hydrogen is the main supplier for this process. There are some works focused on nanostructured hydrides such as carbon nanotubes [2,149–153], nanomagnesioum-based hydrides and metalhydride/carbon nanocomposites [144,154,155]. Even using these types of structures in hydrogen storage may not enough to store the DOE's targeted amount of the hydrogen [156]. Recent advancements in this area include change of Mg hydrides with transition metals [157,158], and the investigation of boron-nitride nanostructures [31]. Magnesium hydride, MgH2, is often modified by high-energy ball-milling with alloying elements including [157] Ni, Cu, Ti, Nb, and Al to obtain, after 20 h of refining, nanoparticles providing hydrogen-storage capacities of about ~9 weight percent by the size of the of ~25 nm [157,158].

Durable multilevel MEAs are being developed (3 M Corporation) using high-speed precision coating technologies and an automated assembly process. The platinum-coated nano-whiskers, part of the MEA, are highly oriented, have high-aspect-ratio single-crystalline whiskers of organic pigment materials which are measured around $4x10^9$ whiskers per cm^2 by the researchers as seen on the Fig. 5.26 [159].

The NSTF catalyst particles consist of high-aspect-ratio stretched particles created by vacuum-coating catalyst thin films onto a monolayer of whiskers. The organic whiskers are practically inert and predominantly encapsulated by the polycrystalline catalyst film, so that the issues with oxidatively unstable support is eliminated. The thin film catalyst coatings consist of relatively large crystallite domains or nanoscopic particles, which give to the NSTF catalysts both, enhanced specific activity [160,161] and resistance to loss of surface area by Pt dissolution and agglomeration [162]. The catalyst coated NSTF support whiskers are thermally very stable. This is illustrated in Fig. 5.27. The bottom row of SEM micrographs show the as-made Pt coated PR149 whiskers at original magnifications of 15 K, 50 K, and 150 K. The whisker samples were then baked in air at 250°C for 3.5 hours, and re-examined. As shown in the top row of micrographs, there is no apparent change in the NSTF catalyst particles under the same magnifications [72].

Even though the 3 M Company achieved high durability and specific-area activity of nanostructured thin film Pt electrocatalyst layers by using the polycrystalline Pt thin films deposited on the organic pigment's whiskers [72], the polycrystalline nature of Pt thin films has some limitations of being disposed to oxidation at the grain boundaries [163]. Markovic et al., showed that the ORR activity in perchloric acid electrolyte for different crystal planes of platinum follows the trend Pt(110) > Pt(111) > Pt(100), with relative ORR activities of approximately 1.5,1.2,

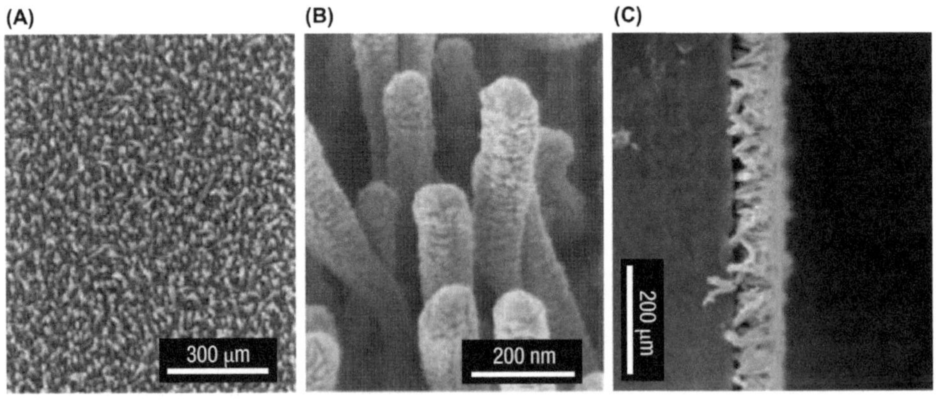

Figure 5.26 Platinum-coated nanostructured whisker supports (0.25 mg cm^{-2}). (A) Plane view; (B) 45° view (higher magnification). The nanostructured film of the MEA (C) shows the Pt-coated nanowhiskers sandwiched between the PEM and the gas-diffusion layer [159].

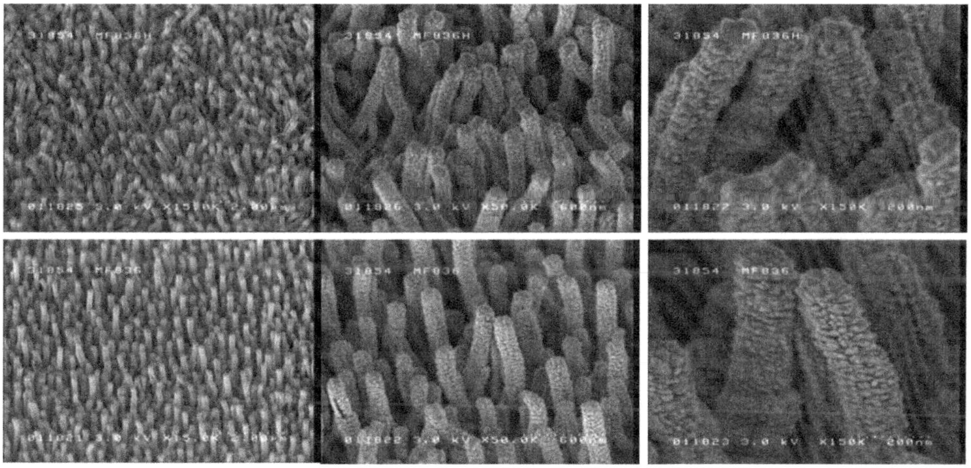

Figure 5.27 SEM micrographs of the NSTF Pt catalyst whiskers before (bottom row) and after (top row) baking in air at 250°C for 3.5 hours [72].

and 0.6 as compared to polycrystalline Pt [164,165]. In contrast to perchloric acid environment, in sulfuric acid environment, the activity of the Pt showed that Pt(110) > Pt(100) > Pt(111) [163].

Kadiri et al., [22] employing the same method, found that the ORR in perchloric acid (a nonadsorbing anion) is relatively insensitive to the Pt surface structure, but is structure-sensitive in solutions containing adsorbing anions like bisulfate, phosphate, or chloride, which impede the reaction. Kadiri et al., reported that the ORR in perchloric acid is nonsensitive to Pt surface structure but sensitive to solutions which has adsorbing anions like chloride, phosphate, etc. [22]. On the other hand, Subbaraman et al, focused on the perfluorosulfonic acid fuel cells, and showed that sulfonate ion in this fuel cell didn't consider as a nonadsorbing but this sulfonate ion adsorbed for deactivation of ORR is same as the bisulfate anion [166,167]. Chen et al., used the Pt nanotubes in 0.5 M H_2SO_4 and received the enhanced ORR electrocatalytic activity. This could be qualified to superior revelation of current crystal facets of nanotubes. Thus, controlling the crystal orientation endorsing from Pt(110) to Pt(111), and the electrode roughness factor along with the changing the single-crystal property should enhance the electrochemical activity of PEM fuel cell cathodes.

In order to further enhance the Pt utilization, GLAD chromium (Cr) were used as a low-cost catalyst supports to be coated conformally with Pt thin film deposited at a small angle deposition (SAD). In addition to that, one potential disadvantage of 3 M approach [72] support material is the decomposition of the organic whiskers at elevated temperature (≥350°C), which makes the thermal treatment, such as catalyst annealing, of the catalyst is limited to temperatures lower than 350°C [91]. Based on that, Bonakdarpour et al. [91] fabricated titanium (Ti) nanocolumns on glassy carbon (GC) disks, using the glancing angle deposition (GLAD) physical vapor deposition technique. The GLAD Ti nanocolumns were used as supports for normal deposited ($\theta = 0°$) Pt oxygen reduction electrocatalysts which were tested in cyclic voltammetry (CV) and rotating-ring disk electrode (RRDE) in 0.1 M $HClO_4$ at room temperature. The area-specific activity of this catalyst was reported to be 0.5 mA/cm^2 at 0.9 V, which is higher than that of the conventional Pt/C electrocatalyst, while it is lower than that of 3 M electrocatalyst [72].

Gasda et al. [168] compared the fuel cell performance of conventional polycrystalline Pt thin films and Pt particles formed by GLAD and verified that the particle-electrodes display a higher mass activity than continuous-layer film and conventional Teflon-bonded Pt-black (TPBE) electrodes at high current densities. Their results proposed that GLAD technique combines the advantages of thin film electrodes with the ability to fabricate porous electrodes allowing efficient reactant flow especially for high-current density operations. This approach has been extended by the same group [93,169,170] who has performed fuel cell measurements on normal-angle deposited Pt on CrN nanoparticles and electrochemically etchable carbon nanorod array supports fabricated by the GLAD technique. Compared to 3 M electrocatalyst (NSTF Pt), one of the potential drawbacks of the above approaches [91–93,169,170] is that Pt tends to accumulate mainly on the tips of relatively dense nanoparticles/nanorods/nanocolumns supports, which results in lower catalyst utilization.

To be able to lower the resistance to oxygen mass transport inside the cathode electrode, electrode porosity needs to be controlled; however this can be end up with high power densities with high efficiency at the high current density region in the polarization curve of operating PEM fuel cells [24]. It has been observed that enhanced Pt utilization with low oxygen mass transport resistance on very thin catalyst layered cathode electrodes by sputter deposited [86,171–173]. The electrodes listed above [91,93,168–170] typically suffer from poor porosity, which may have a negative effect on oxygen mass transport in an working PEM fuel cell.

Moreover, depositing the Pt thin film with the normal angle on the GLAD nanostructures [91,93,169,170] has disadvantages. Pt atoms will gather mostly on the tips of the nanorod layer and this will cause a low utilization of the catalyst. Although the solid GLAD Pt nanorod electrocatalysts exhibited good electrocatalytic activity toward ORR, there is still an urgent need to enhance and control the electrode porosity in order to improve oxygen transport and further enhance the ORR electrocatalytic activity of nanostructured GLAD fuel cell electrodes especially at the low potential region of a polarization curve. This can be achieved by utilizing substrate surface pattering approaches, which can lead to the formation of periodic and well isolated GLAD nanostructures [174]. This will also provide insight on which roughness values of GDL and PEM substrates would result in nanorod coatings with the best electrochemical properties. Such and similar surface engineering approaches may open the way for utilizing GLAD nanorods as model electrodes with controlled porosity to evaluate the relationship between the oxygen mass transport limitations and electrode porosity.

In addition, some researchers also studied that an increase in polytetrafluoroethylene (PTFE), commonly known as Teflon, film surface roughness increases the contact angle of water and therefore hydrophobicity without altering the surface chemistry [175,176].

5.4.4 Core-shell materials

The basic synthesis approaches for the producing of core–shell nanoparticle catalysts is illustrated in Fig. 5.28. Electrochemical dealloying results in Pt bimetallic core–shell nanoparticles (Fig. 5.28A), and leaching end up in Pt-skeleton core–shell nanoparticles (Fig. 5.28B).

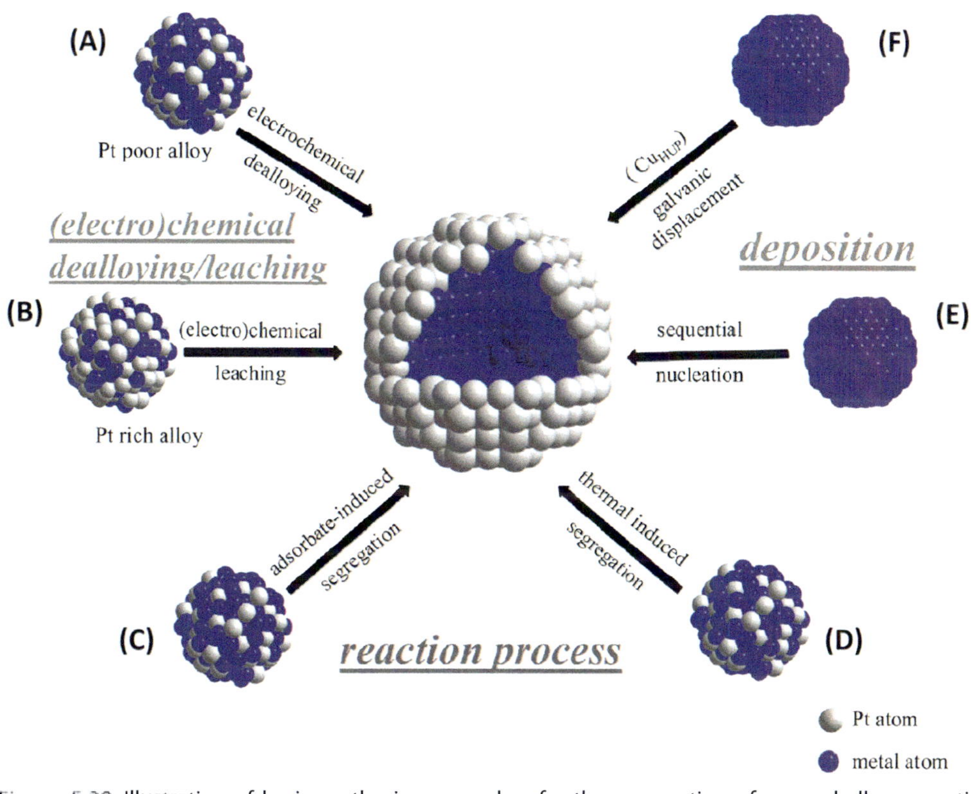

Figure 5.28 Illustration of basic synthesis approaches for the preparation of core–shell nanoparticle catalysts. (A) dealloyed Pt bimetallic core–shell nanoparticles, (B) Pt-skeleton core–shell nanoparticles, (C) strong binding to adsorbates (D) thermal annealing, (E) heterogeneous colloidal core–shell nanoparticles, and (F) Pt monolayer core–shell nanoparticles [177].

Reaction process routes generate segregated Pt skin core–shell nanoparticles induced either by strong binding to adsorbates (Fig. 5.28C) or thermal annealing (Fig. 5.28D). The core–shell nanoparticles (Fig. 5.28E) and Pt monolayer core–shell nanoparticles (Fig. 5.28F) are prepared by the heterogeneous nucleation and galvanic displacement, respectively [177].

Core/shell nanomaterials consist of two separate matters closely interacting. However, aggregation of nanoparticles eventually leads to loss of catalytic activity in nanoscale practical applications. This nanostructured design will significantly impact fuel cell technology by improving catalysts preparations while simultaneously being more cost-effective [178,179]. Core/shell nanostructures [47,125,136,180–186] will allow for further control of size, shape, and composition at such a precise scale that it will perhaps lead to improvement of current or discovery of new catalytic properties.

The core-shell structured catalysts can be prepared by three methods which are Pt-based nanorods, only Pt shell coated structures, and Pt with other types of metal shell coated on the nonmetal core structures. Glancing Angle Deposition (GLAD) can optimize the cost-effective on preparation of the core-shell structures. Pt shell thin film can be deposited on the GLD growth Ni, WC, Cr core structures [28,44,187–192].

The Khudhayer, WJ et al., showed that the stability and electrochemical surface area of the catalysts enhanced by using the Pt core-base nanorods compared to traditional way of the Pt/C nanoparticles' catalysts. However, using only Pt core nanorods resulted the lower mass specific activity due to high amount of the Pt loading value compared to the only Pt-shell coating on the nonmetal core structures. The Pt(110) plane is primarily positioned at around 80 from the surface normal, which corresponds to the nanorod sidewalls (Fig. 5.29) [192].

Seongyul, K., et al., results showed that increasing the nanostructured-electrode's active area can lower the working current density of the electrolyzer, but this may instigate the overpotential to demonstrate a consequent drop [193]. The focus of the working on the core-shell structures is the producing thin-films with high-active and more stable catalysts which are coated on vertically aligned rod arrays. For instance, some researchers used Platinum-Nickel shell thin films with the several Pt:Ni ratios on the Nickel core structures [184], and tungsten carbide (WC) core structures [194]. Takimoto, D., et al, showed that Ru@Pt core/shell nanostructures with thickness of 1.5–4.5 Pt monolayers' electrochemical active surface area are larger more than 1.4 compared to typical Pt/C catalysts [195]. Jiang R. et al., explained the core/shell electrocatalysts, which have advanced nanostructures, have shown notable activity and stability in PEM Fuel Cells [196].

When the core structures coated with monolayer Pt thin film can increase the mass activity at least 4 to 8 times compared to traditional methods [197]. However, if carbon coated Titania used as a core structures and coated with Pt thin film, it can enhance the of the electrochemical surface area nearly by 50% compared to the conventional Pt-C structures [198]. Some of the researchers were focusing on Carbon

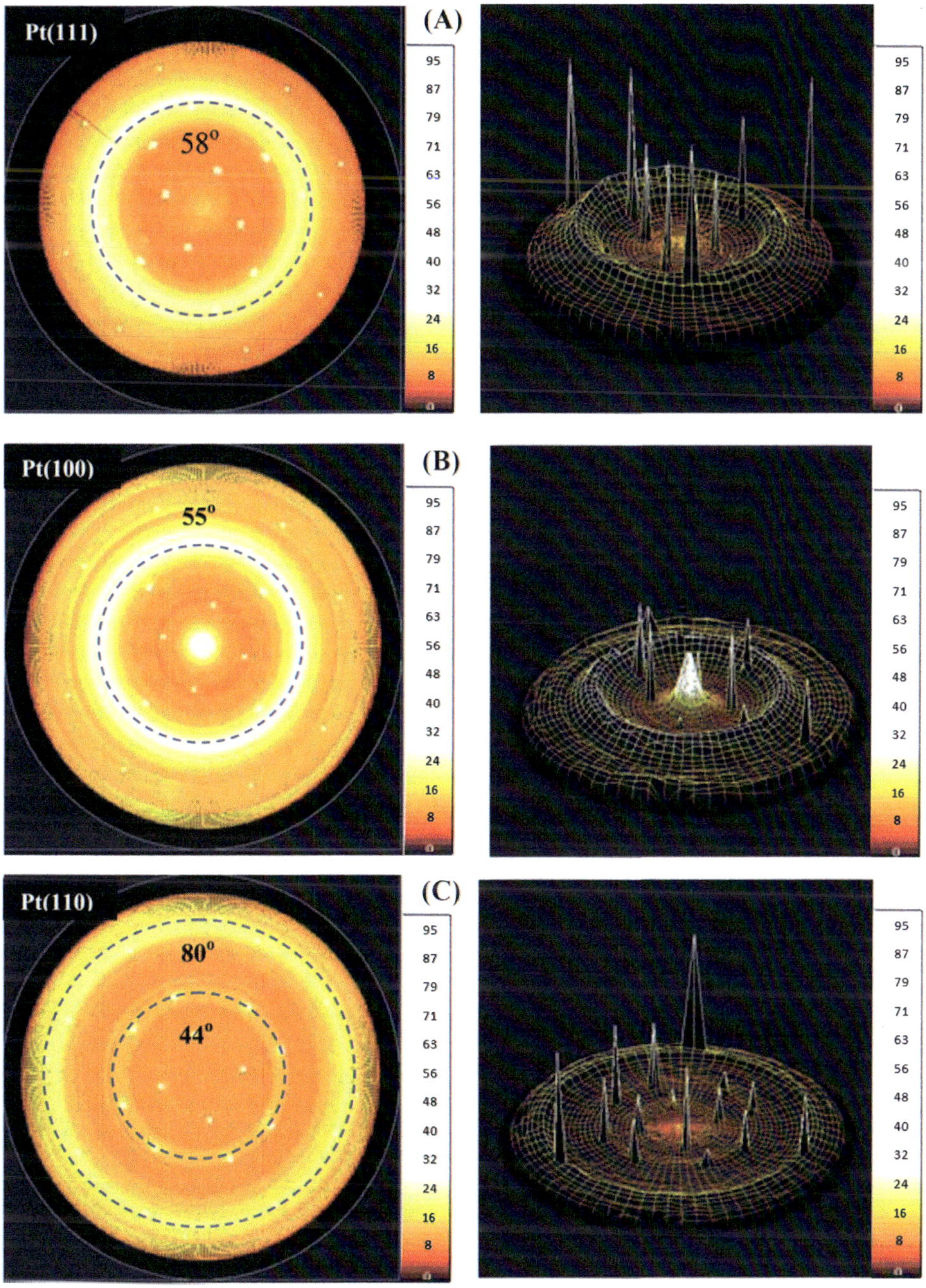

Figure 5.29 XRD pole-figure results for the Pt(111), Pt(100), and Pt(110) planes for glancing angle deposited (GLAD) Pt nanorods [192].

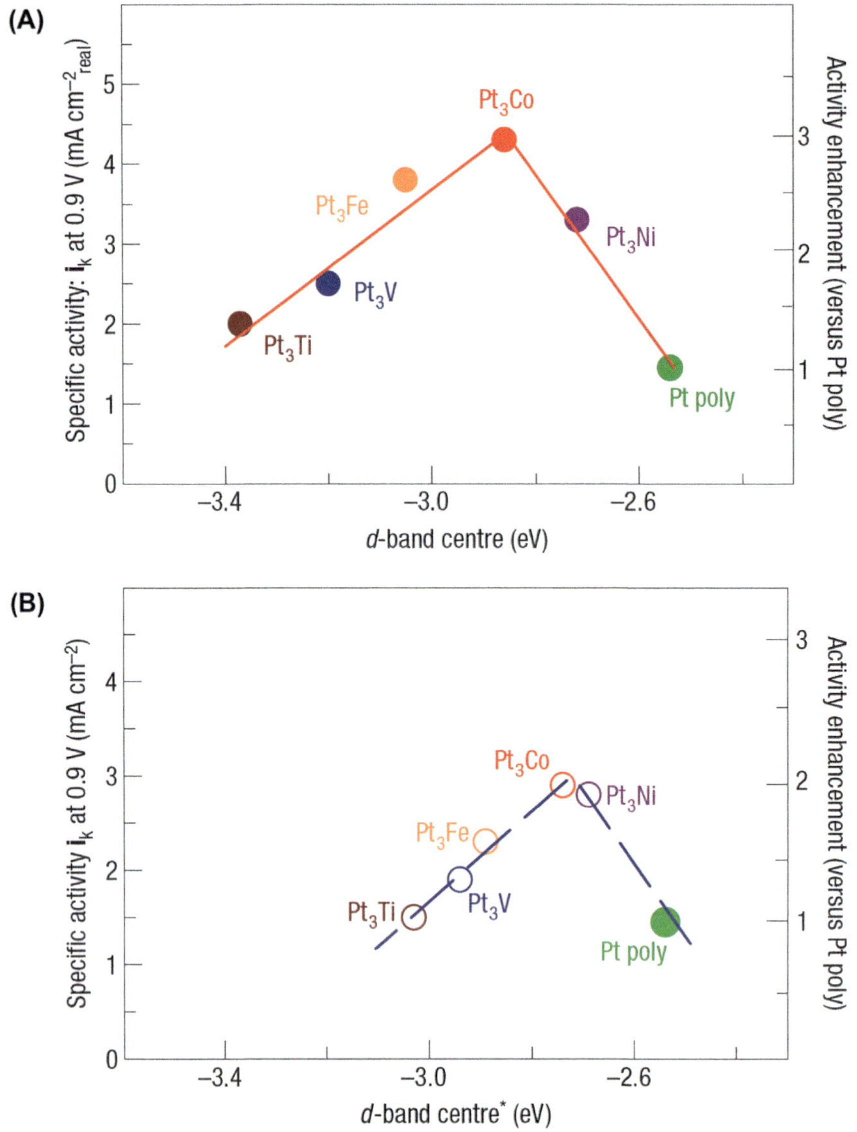

Figure 5.30 Relationships between the catalytic properties and electronic structure of Pt3M alloys. (A,B), Relationships between experimentally measured specific activity for the ORR on Pt3M surfaces in 0.1 M HClO4 at 333 K versus the d-band center position for the Pt-skin (A) and Pt-skeleton (B) surfaces. (B) shows the d-band center values* established in UHV, which may deviate in the electrochemical environment due to dissolution of nonPt atoms [203].

NanoTubes as core structures and coated by PtCu alloy [199], and Palladium-Platinum alloy core-shell tree shape structures [200]. In last decade, many studies showed that electrocatalysts with new and unexpected properties can be achieved by alloying two or more metals [201,202].

As can be seen from the volcano curves established by Stamenkovic et al., in Fig. 5.30, in order to obtain the maximum activity, the catalyst should strike the optimum balance between the kinetics of the O−O bond breaking and the electroreduction of the oxygenated intermediates or O−H formation [203]. However, unfortunately for Pd, due to the high-lying d -bandcenter, Pd binds to oxygen strongly and thus the O−O bonds can be easily broken. Though, this generates a high coverage of oxygenated intermediates that are slowly desorbed on Pd surface [204]. On the other hand, due to its low-lying d -bandcenter, Ag can bind to the oxygen less strongly, which leads to less efficient splitting of the O−O bond and a lower coverage degree of oxygenated intermediates [205,206].

Considering the bimetallic effects on the catalytic activity, we chose PdAg alloy NPs to tune and optimize Pd catalysis for the ORR and investigate the effects of surface composition and structure on their ORR catalytic performance [207,208]. Due to the large standard reduction potential difference between Pd and Ag, a galvanic displacement reaction and/or a seed-mediated method is usually adopted to synthesize PdAg alloy nanoparticles, which, however, constrains the potential application of PdAg catalysts in the field of catalysis [209].

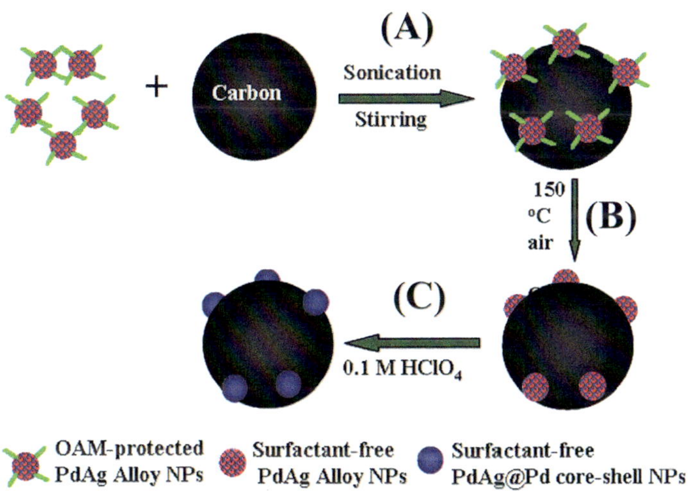

Figure 5.31 Synthetic strategy of carbon supported PdAg@Pd core−shell nanoparticles (PdAg@Pd/C), (A) Loading of PdAg NPs on Vulcan XC-72R carbon support (PdAg/C), (B) Thermal treatment to obtain carbon-supported surfactant-free PdAg alloy nanoparticles (PdAg/C), (C) Electrochemical treatment to obtain PdAg@Pd/C [210].

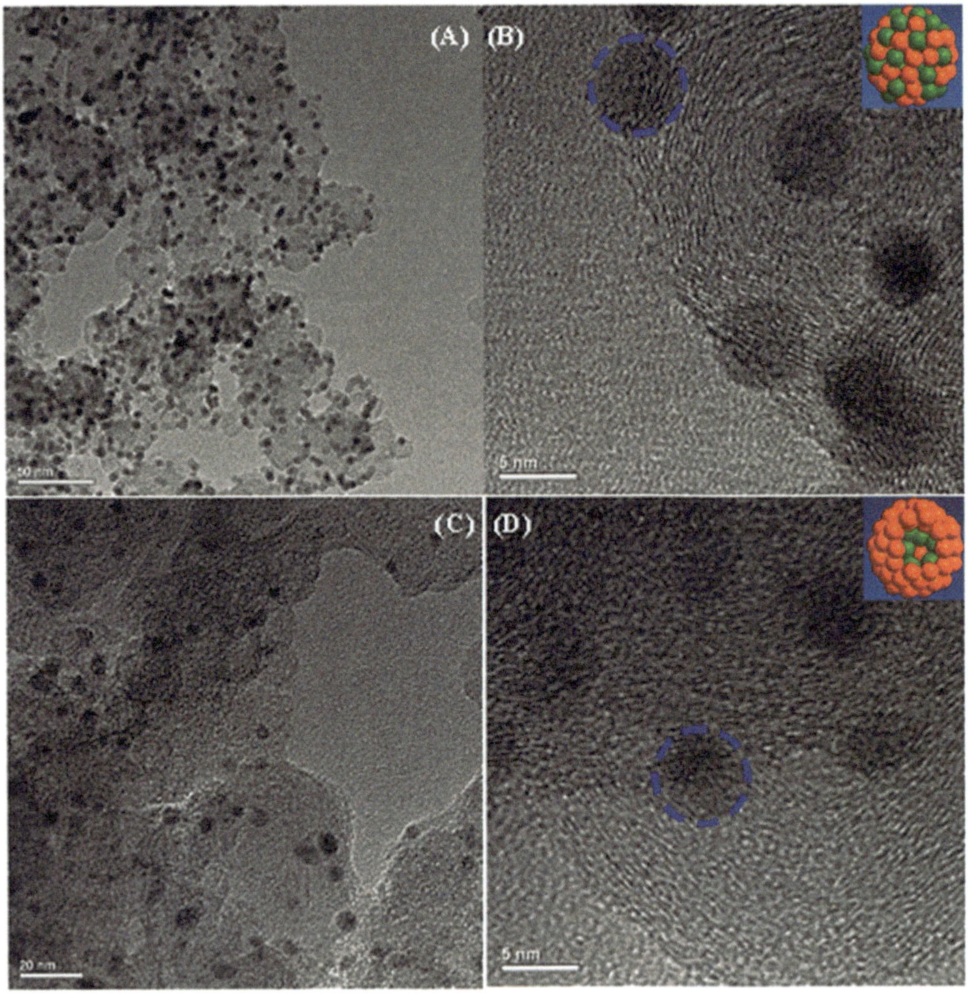

Figure 5.32 (A,B) TEM (A) and HRTEM (B) images of carbon-supported PdAg NPs (PdAg/C) after thermal treatment at 150°C overnight. The inset at the top right in (B) shows the corresponding surface structure model of single PdAg alloy NP supported on carbon. (C,D) TEM (C) and HRTEM (D) images of carbon-supported PdAg@Pd core–shell NPs (PdAg@Pd/C) after electrochemical treatment in 0.1 M HClO 4 (100 cycles). The inset at the top right in (D) shows the corresponding surface structure model of single PdAg@Pd core–shell NP supported on carbon [210].

As can be seen in Figs. 5.31 and 5.32, the prepared PdAg@Pd/C nanostructured catalyst enhanced the electrocatalytic activity and durability for the ORR in alkaline media [210]. The stability of Pt/C and Pt/CNTs under simulated PEMFC conditions are also investigated by Yuyan Shao, et al. [211].

The core-shell structure was confirmed by a number of techniques, including scanning transmission electron microscopy, energy-dispersive X-ray spectroscopy mapping,

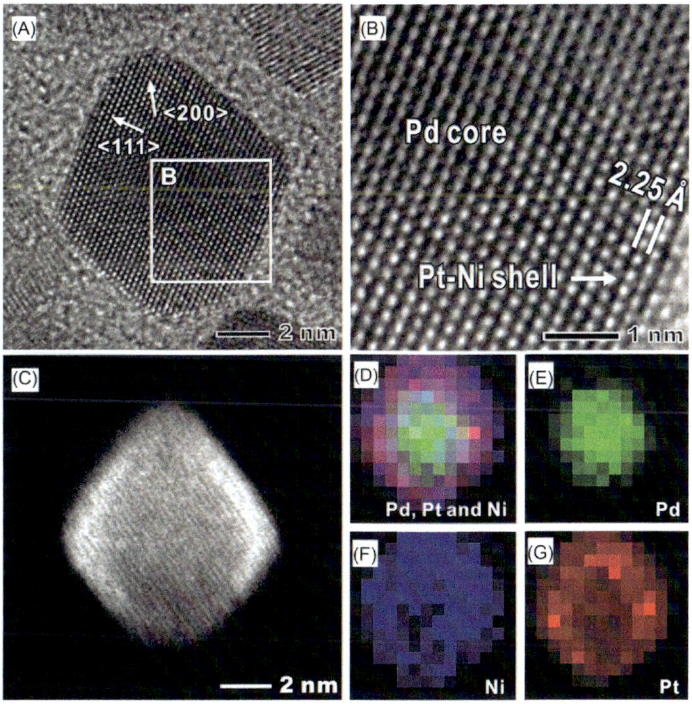

Figure 5.33 (A) High-resolution TEM image taken from a single Pd@Pt_Ni octahedron with an edge length of 8 nm. (B) Highresolution TEM image of the region marked by a box in (A). The lattice spacing of 2.25 Å for both the Pd core and the Pt_Ni shell confirms the epitaxial overgrowth of Pt_Ni on Pd. (C) HAADF-STEM image of a single Pd@Pt_Ni octahedron. The brighter shell on a relatively darker core indicates the difference in atomic number between Pt and Pd. (D_G) EDS mapping of elemental distributions for (D) Pd, Pt, and Ni, (E) Pt and Ni, (F) Pd and Ni, and (G) Pd and Pt [212].

in situ X-ray diffraction under H_2 and He, and electrochemical measurements. Relative to the state-of-the-art Pt/C catalyst, the Pd@Pt_Ni/C catalyst showed mass and specific ORR activities enhanced by 12.5- and 14-fold, respectively. The formation of a core-shell structure helped increase the electroactive surface area in terms of Pt and thus the mass activity. During an accelerated durability test, the mass activity of the Pd@Pt_Ni/C catalyst only dropped by 1.7% after 10 000 cycles (Fig. 5.33) [212].

5.5 Simulations/computational works

A combined experimental and theoretical studies such as Density functional studies (DFT) [213–217], Monte Carlo Simulations (MC) [13,216–220], numerical analysis

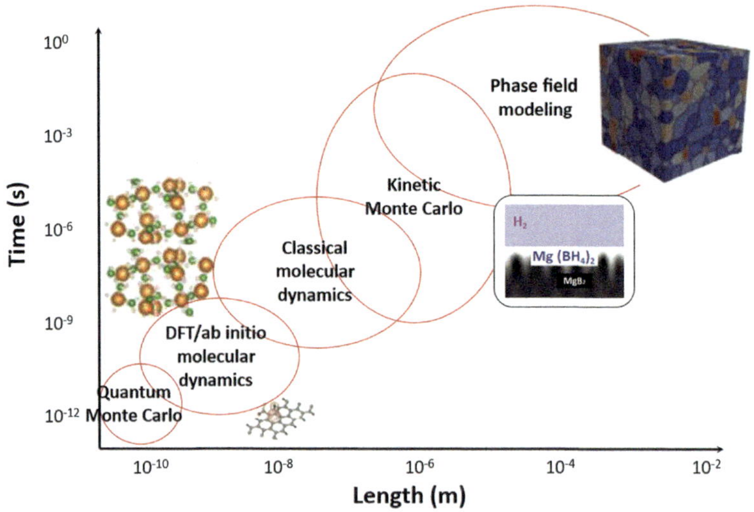

Figure 5.34 Techniques for multiscale simulation to understand properties of hydrogen storage materials [226].

[221], COMSOL [222], mathematical models [223], Volume of Fluid (VOF) [224], and a finite Markov chain analysis [225], would be appropriate to provide a new platform for the design of more selective and impurity-tolerant catalysts for the fuel cell technology.

Leveraging multiscale modeling approaches for the properties of hydrogen storage materials, as shown in Fig. 5.34, will play a major role in reaching this understanding.

As can be seen in Fig. 5.35, after leaching away the Ni atoms from the surface, 3 atomic-layer thickness of the Pt-skeleton formed by relaxed-low-coordinated surface atoms. This method is supposed to generate favoured structure of highly active (111) surface, due to the higher atomic coordination [227]. Wang, C., et al, investigated the parallel-serpentine flow fields with different channel geometries, including the channel/rib width and total channel cross-sectional area using computational Fluid Dynamics (CFD) Modeling [228]. Ghanbarian, A., et al, investigated the flow channel indentation of PEM fuel cell by CFD simulation [229].

5.6 Fuel cell applications

Fig. 5.36 illustrates the FCTO mission and framework along with specific technoeconomic targets for fuel cells, on board hydrogen storage, and hydrogen production and delivery which are foundational to the mission thrusts.

Important progress in research and development, demonstration, and deployment is shown in Fig. 5.37. For example, FCTO research and development activities have

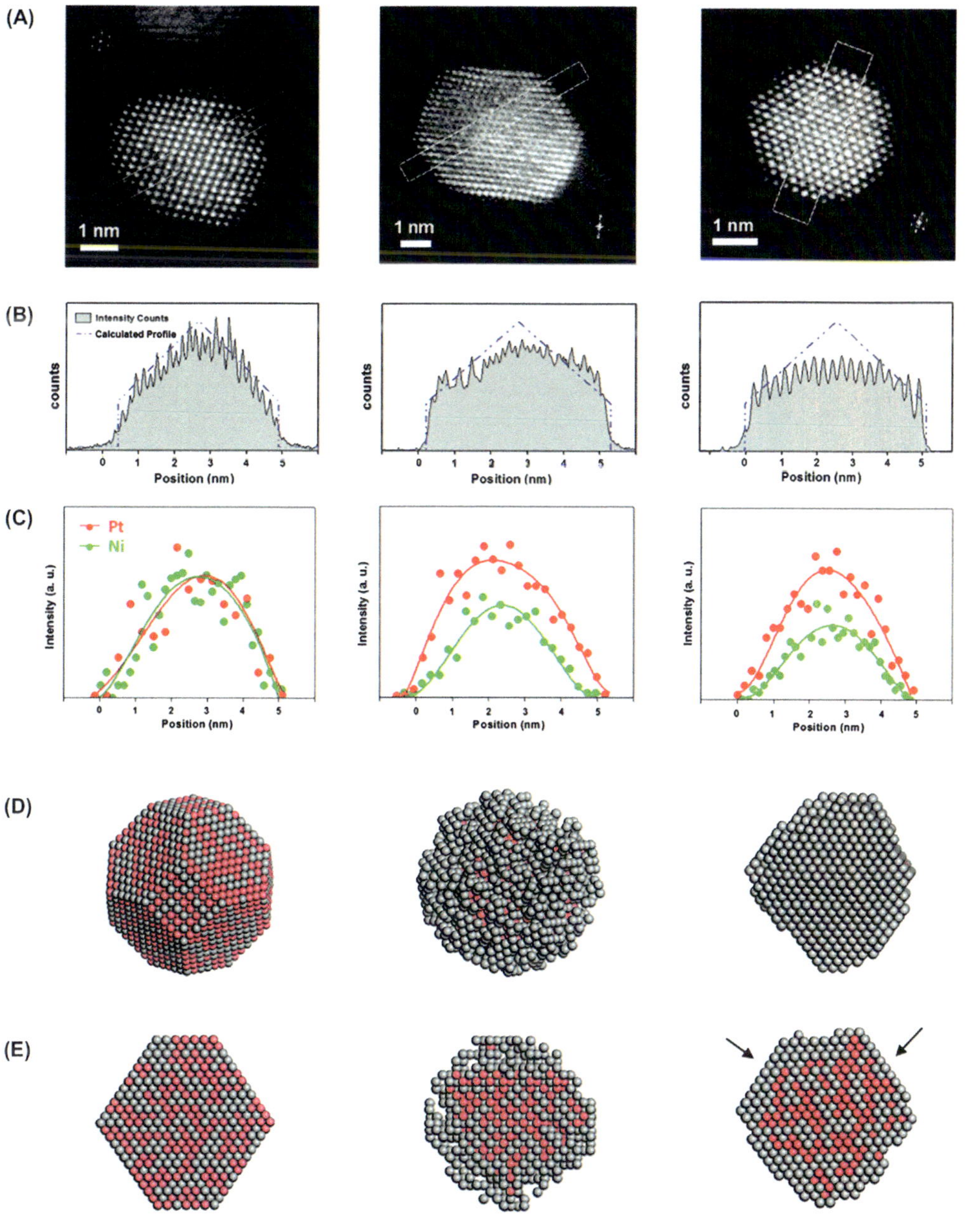

Figure 5.35 Microscopic characterization and theoretical simulation of nanostructure evolution in the PtNi/C catalysts: (A) Representative high-angle annular dark-field scanning transmission electron microscopy (HAADF-STEM) images taken along the zone axis Æ110æ, as confirmed by the fast Fourier transfer (FFT) patterns of the STEM images (shown as insets); (B) background subtracted, normalized intensity line profiles extracted for the regions marked in (A); (C) composition line profiles (normalized for PtL peaks) obtained by energy-dispersive X-ray spectroscopy (EDX) with an electron beam (~2 Å in spot size) scanning across individual catalyst particles; (D) overview; and (E) cross-section views of the nanostructures depicted by atomistic particle simulation. The figure is also organized in columns for the as-prepared (left), acid treated (middle), and acid treated/annealed (right) PtNi/C catalysts, respectively [227].

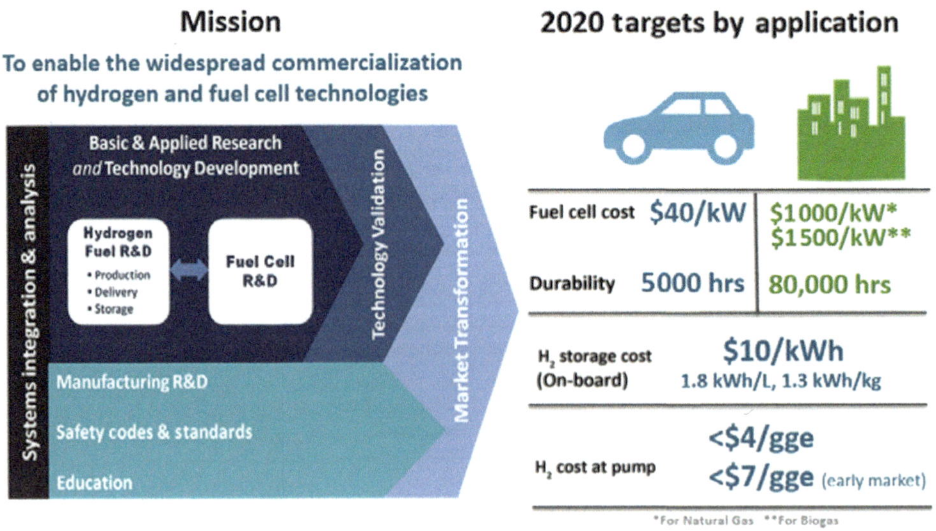

Figure 5.36 Integrated target-driven approach to widespread commercialization of fuel cell technologies [226].

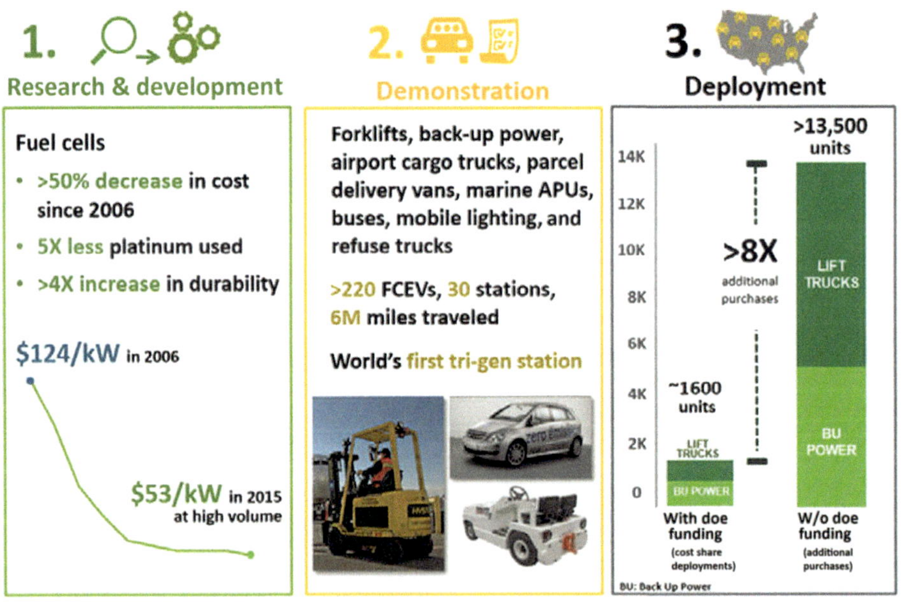

Figure 5.37 DOE is engaged in R&D, technology demonstration, and supporting commercial deployment of fuel cell technologies [226].

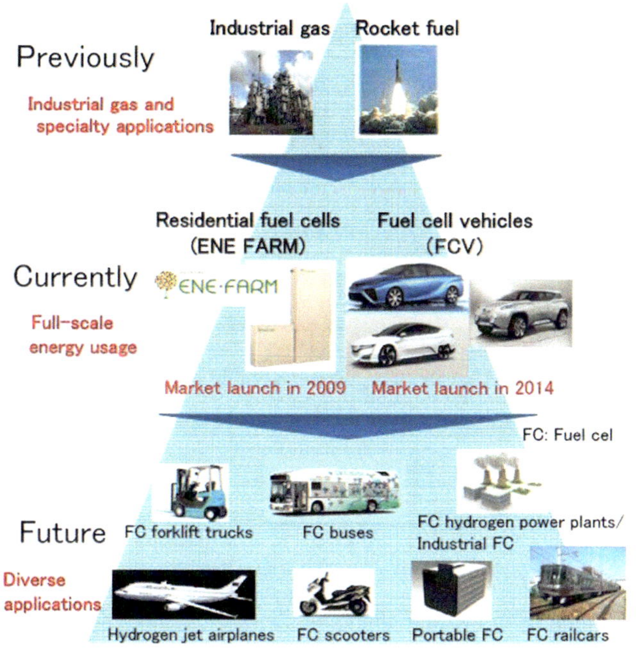

Figure 5.38 Cartoon for potential fuel cell applications [230].

resulted in a 50% decrease in fuel cell cost since 2006, ultimately leading to a modeled high-volume cost of $53 per kW cost in 2015. Fuel cell technology has been successfully demonstrated in forklifts, airport cargo trucks, marine auxiliary power units, and several other applications. Significantly, deployment of commercially viable fuel cell lift trucks and backup power units reached over 13,000 units without the need for further DOE funding [226].

Diverse potential application includes hydrogen air jets, portable FC, rail cars, scooters as depicted in the cartoon (Fig. 5.38).

Recent advances in the application of nanostructured carbon-based materials have suggested the possibility of using carbon nanotubes as novel electrocatalysts supports. Studies shown that Pt nanoparticles supported on carbon nanotubes display remarkably higher electrocatalytic activity toward the direduction of oxygen than Pt nanoparticles supported on carbon black, which would contribute to substantial cost reduction in PEM fuel cells. The finite size of nanoscale materials such as carbon nanotubes positively influences the thermodynamics and kinetics of oxygen reduction due to their length scale and specific properties. Overall, their unique characteristic encourages the use of nanosized catalyst materials instead of their bulk counter parts to enhance the oxygen electroreduction performance [231–233].

Table 5.1 Summary of future R&D requirements for fuel-cell materials [3].

Application	Size (kW)	Fuel cell	Fuel	Critical materials issues
Power systems for portable electronic devices	0.001–0.05	PEMFC DMFC SOFC	H_2CH_3OH CH_3OH	Membranes exhibiting less permeability to CH3OH, H2O Novel PEN structures
Micro-CHP	1–10	PEMFC SOFC	CH_4 LPG CH_4 LPG	CO-tolerant anodes, novel membranes, bipolar plates More robust thick-film PENs operating at 500°C–700°C
APU, UPS, remote locations, scooters	1–10	SOFC	LPG Petrol	More robust thick-film PENs operating at 500°C–700°C; rapid start-up
Distributed CHP	50–250	PEMFC MCFC SOFC	$CH_4CH_4CH_4$	CO-tolerant anodes, novel membranes, bipolar plates better thermal cycling characteristics cheaper fabrication processes; redox-resistant anodes
City buses	200	PEMFC	H_2	Cheaper components

Probable applications of fuel cells in the next decade together with a selection of critical materials issues are summarized in Table 5.1.

5.6.1 PEM fuel cells

5.6.1.1 Transportation applications

The focus of PEMFC applications today is on prime power for cars and light trucks. PEMFC is the only type of fuel cell considered for prime motive power in on-road vehicles. Early prototypes of fuel cell vehicles have been released to controlled customer groups in Japan and USA. Hence, rear body modifications are necessary to integrate the hydrogen storage vessels. Fig. 5.39 shows the design of the Chevrolet Sequel by GM providing enough space for three large 70 MPa CGH2 vessels (Total fuel capacity: 8 kg of hydrogen). Through this approach, the FCEV operating range could increase tremendously by more than 300 miles as demonstrated on public roads between suburban Rochester and New York City in May 2007. The fuel cell system of the Sequel has been packaged into the

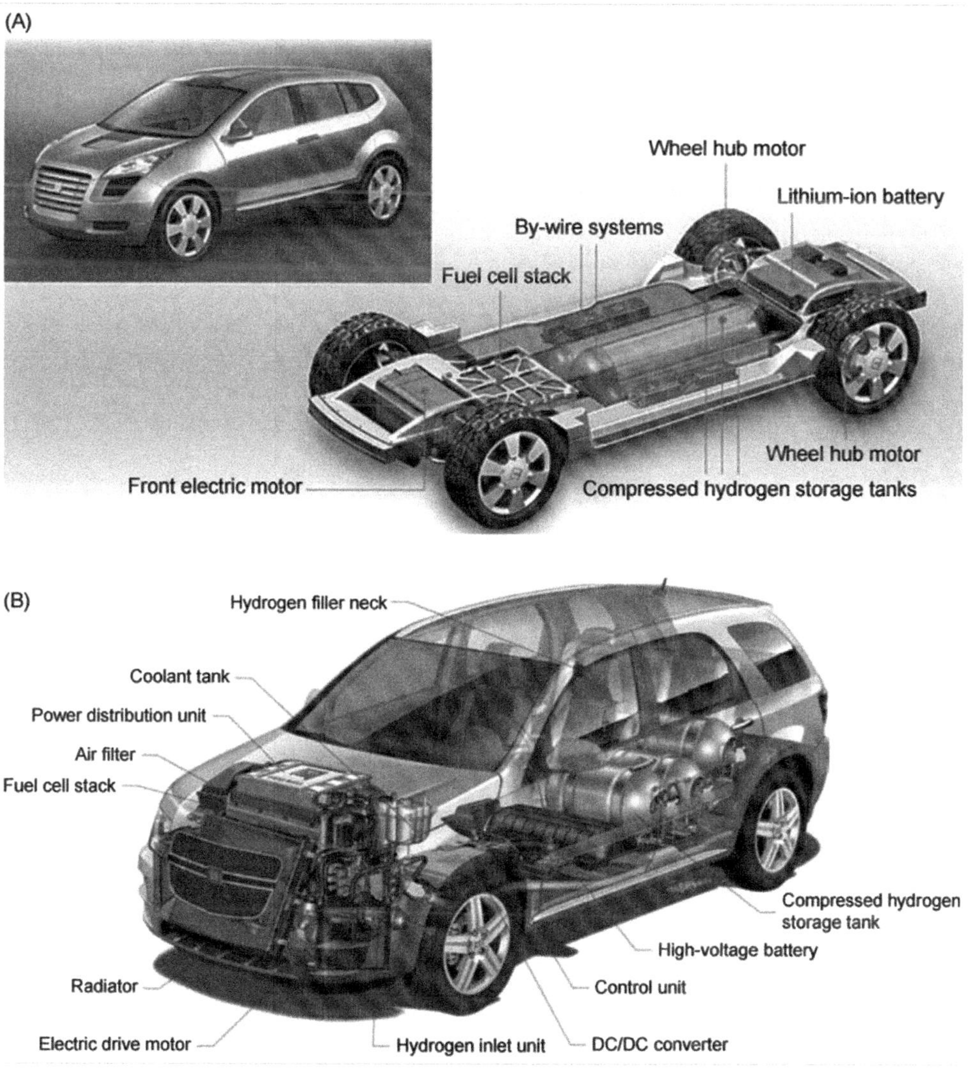

Figure 5.39 (A) GM sequel and the skateboard chassis (B) GM HydroGen4 vehicle [234].

vehicle underbody as well, offering flexibility for the interior design. Although the Sequel was only a concept vehicle with no production intent, it could be developed and optimized for the specific characteristics and opportunities that fuel cells and H_2 can offer. Since autumn 2007, within the framework of Project Driveway, more than 100 cars of the generation HydroGen4, shown in Fig. 5.39 (B), were developed and demonstrated in countries like Germany and the United states [234].

5.6.1.2 Stationary applications

Several developers are also advancing PEMFC systems for stationary applications. These efforts are aimed at very small scale distributed generation (1 to 10 KW AC). The vast majority of systems are designed for operation on natural gas or propane. System efficiency typically ranges from 25% to 30%. In addition, system operating life has been extended to about 8000 hours for a single system with a single stack, with degradation of about 5% per 1000 hours [68].

5.6.2 AFC

AFC are being considered for automobile applications because of their sensitivity to poisoning, which requires use pure hydrogen and oxygen. It also has been used since mid-1960 by NASA in the Apollo and space shuttle programs. The fuel cells on board these spacecrafts provide electrical power for an-bard systems, as well as drinking water [68].

5.6.3 PAFC

More than 265 of these "first generation" power units were placed in operation in stationary power applications in USA and overseas. Most are the 200 KW PC-25 fuel cell power plants manufactured by united technologies corporation (UTC). UTC has installed over 75 MW of PAFC systems, operating for over 8 million hours in 85 cities and 19 countries [68].

5.6.4 MCFC

MCFC are being developed for natural gas and coal-based for industrial, electrical utility and military applications. The potential market for MCFC system is thought to be in cogeneration applications. The electrical efficiency of MCFC unit is 50%. Currently, the MCFC technology is entering the 0.1–2 MW demonstration phase to confirm the initial indications of performance and efficiency [68].

5.6.5 SOFC

The opportunity for application of high and intermediate temperature SOFC range from large scale distributed power generation to small scale domestic heat and power. Applications of high temperature SOFC will focus on large scale static system where downstream energy extraction systems such as steam and gas turbines are feasible. While the applications for intermediate SOFC will be the smaller scale more cost sensitive areas such as distributed SOFC will be the smaller scale more cost sensitive areas such as distributed domestic power generation where low grade heat emitted by the system can be used for hot water generation or space heating [68].

5.7 Future works

5.7.1 Future work water management improvement

Continue to combine the improvement factors now identified that significantly enhance cool operation and load transient start-up; determine best synergistic combinations towards the 2010 best of class MEA for final stack testing [235].

5.7.2 Cathode catalyst mass activity gain

Continue to fabricate and optimize catalyst compositions, structures and fabrication processes to exceed target mass activity of 0.44 A/mgPt while meeting all other performance requirements. Down-select the final catalyst and configuration to be scaled-up for the final stack testing [235].

5.7.3 MEA integration

Down-select the final 2010 Best of Class MEA (catalysts for each electrode, membrane, GDLs for each electrode, processes, etc.) under Task 5, for final stack testing [235].

5.7.4 Stack testing

Continue Task 3 short stack evaluations with 2009 Best of Class MEA, upgrading to 2010 best of class MEA. Identify OEM stack for final stack testing under Subtask 5.3. Start-up conditioning and reversible stability. Continue to develop simplified break-in conditioning protocols and catalyst/membrane components to reduce MEA break-in conditioning time to $<$ 3 hours for full performance [235].

5.8 Conclusion

The condition of our environment is a subject of increasing concern. Modern society and its way of life has had effects on the environment, some of which are evident. Because of these events, topics such as ecology, pollution, population, and the "energy problems" are featured in the public focus constantly. Although the magnitude of today's energy problems has frequently been stressed, the term "energy problems" has different connotations to different people. Some think primarily of blackouts and the need for guaranteed continuity of electric power, without visualizing how difficult and, increasingly, how unacceptable is a continued growth that follows the patterns of the past in meeting growing demands. On the other hand, others think primarily of environmental problems and are impatient with the slow change in these growth patterns. The return-to-renewables is an excellent way to tackle this problem, helping to mitigate energy problems. But in order to ensure a sustainable future for generations to meet their energy needs, this approach needs to be implemented taking all the precautions needed. The term "energy problems" relates to the recent works of the fuel cells processes, present and projected. Our objective is to assess the energy problems by comparing renewable energy sources such as fuel cell technology with fossil fuels energy sources.

Development of future power generation technology allows the implementation of advanced materials into the renewable energy devices. We summarized the application of advanced nanomaterials into one of the most promising environmentally friendly renewable energy devices, fuel cells. Advanced nanomaterials can be used to improve efficiency and durability, while reducing the cost of the fuel cell. Each of the techniques that used to produce the fuel cell electrocatalysts has a series of advantages and disadvantages compared to the other methods. However, core-shell structures showed enhanced results in the cathodes and anodes in fuel cells electrodes. Finally, commercialization of the fuel cell can be expanded.

References

[1] M.J. Harris, B. Roach, A.M. Codur, The economics of global climate change, Tufts Univ. (2017). Available from: http://www.ase.tufts.edu/gdae/education_materials/modules.html.

[2] G.W. Crabtree, M.S. Dresselhaus, The hydrogen fuel alternative, MRS Bull. 33 (2008) 421–427.

[3] T. Bak, J. Nowotny, M. Rekas, C.C. Sorrell, Photoelectrochemical hydrogen generation from water using solar energy, Int. J. Hydrog. Energy 27 (10) (2002) 991–1022.

[4] J. Kibsgaard, Y. Gorlin, Z. Chen, T.F. Jaramillo, Meso-structured platinum thin films: active and stable electrocatalysts for the oxygen reduction reaction, J. Am. Chem. Soc. 134 (18) (2012) 7758–7765.

[5] R.W. Lashway, Fuel cells: the next evolution, MRS Bull. 30 (8) (2005) 581–583.

[6] S. Harris, The Fuel Cell Review, 2004, (June) 31.

[7] R. Allan, MEMS: a new power source for portables, Electron. Des. 53 (6) (2005) 43–47.

[8] http://www.fuelcells.org.

[9] R.L. Bechtold. Alternative Fuels Guidebook, Society of Automotive Engineers, 1997.

[10] Y. Wang, et al., A review of polymer electrolyte membrane fuel cells: technology, applications, and needs on fundamentals research, Appl. Energy 88 (4) (2011) 981–1007.

[11] T.M. Demirkan, M. Yurukcu, B. Dursun, T. Karabacak, Evaluation of double-layer density modulated Si thin films as Li-ion battery anodes, Mater. Res. Express 4 (10) (2017) 106405.

[12] Z.P. Cano, D. Banham, S. Ye, A. Hintennach, J. Lu, M. Fowler, et al., Batteries and fuel cells for emerging electric vehicle markets, Nat. Energy 3 (2018) 279–289.

[13] H. Cansizoglu, M.F. Cansizoglu, M. Yurukcu, W.J. Khudhayer, N. Kariuki, T. Karabacak, SAD-GLAD core-shell nanorod arrays for fuel cell, photodetector, and solar cell electrode applications, Nanoepitaxy Mater. Devices VI 9174 (2014) 917411.

[14] F. Barbir, PEM Fuel Cells: Theory and Applications, Elsevier Academic Press, 2005.

[15] K.A. Starz, K. Ruth, M. Vogt, R. Zuber, Proc. Int. Symp. Fuel Cells for Vehicles (2000), 210–215.

[16] U. Eberle, R. Helmolt, Fuel Cell Electric Vehicles, Battery Electric Vehicles, and Their Impact On Energy Storage Technologies, (2010) ISBN 978-0-444-53565-8. An Overview.
[17] C.A. Ward, J.A. Garcia, Analytical method for determining the internal resistance and electrocatalyst utilization of fuel cells, J. Power Sources 66 (1–2) (1997) 83–88.
[18] N.M. Markovic, H.A. Casteiger, P.N. Ross, Oxygen reduction on platinum low-index single-crystal surfaces in sulfuric acid solution: rotating Ring-Pt(hkl) disk studies, J. Phys. Chem. 99 (11) (1995) 3411–3415.
[19] N.M. Markovic, H.A. Gasteiger, P.N. Ross, Kinetics of oxygen reduction on Pt (hkl) electrodes: implications of the crystalline size effect with supported Pt electrocatalysts, J. Electrochem. Soc. 144 (1997) 1591–1597.
[20] N.M. Markovic, R.R. Adzic, B.D. Cahan, E.B. Yeager, ISE Proceedings, Montreaux (1991), 138.
[21] N.M. Markovic, R.R. Adzic, B.D. Cahan, E.B. Yeager, Structural effects in electrocatalysis: oxygen reduction on platinum low index single-crystal surfaces in perchloric acid solutions, Electroanal. Chem. 377 (1994) 249–259.
[22] F. El Kadiri, R. Faure, R. Durand, Electrochemical reduction of molecular oxygen on platinum single crystals. *Journal of electroanalytical chemistry and interfacial electrochemistry*, J. Electroanal. Chem. 301 (1991) 177–187.
[23] Y.H. Shih, G.V. Sagar, S.D. Lin, Effect of electrode Pt loading on the oxygen reduction reaction evaluated by rotating disk electrode and its implication on the reaction kinetics, J. Phys. Chem. 112 (2008) 123–130.
[24] J. Zhang, PEM Fuel Cell Electrocatalysts and Catalyst Layers, Fundamentals and Applications, Springer-Verlag, 2008.
[25] F.Y. Yurtsever, M. Yurukcu, M. Begum, F. Watanabe, T. Karabacak, Stacked and core–shell Pt: Ni/WC nanorod array electrocatalyst for enhanced oxygen reduction reaction in polymer electrolyte membrane fuel cells, ACS Appl. Energy Mater. 1 (11) (2018) 6115–6122.
[26] R. Jervis, Development of Novel Alloy Electrocatalysts for the Hydrogen Oxidation Reaction in Alkaline Media and their Application to Low Temperature Fuel Cells, UCL Chemical Engineering, 2015.
[27] P. Zoltowski, The mechanism of the activation process of the tungsten carbide electrode, Electrochem. Acta 31 (1) (1986) 103–111.
[28] L.G.R.A. Santos, K.S. Freitas, E.A. Ticianelli, Electrocatalysis of oxygen reduction and hydrogen oxidation in platinum dispersed on tungsten carbide in acid medium, J. Solid. State Electrochem. 11 (2007) 1541–1548.
[29] W. Chrzanowski, A. Wieckowski, Enhancement in methanol oxidation by spontaneously deposited ruthenium on low-index platinum-electrodes, Catal. Lett. 50 (1998) 69–75.
[30] S.R. Brankovic, J.X. Wang, R.R. Adzic, Pt submonolayers on Ru nanoparticles—a novel low Pt loading, high CO tolerance fuel-cell electrocatalyst, Electrochem. Solid-State Lett. 4 (2001) A217–A220.
[31] A.S. Arico, P. Bruce, B. Scrosati, J.M. Tarascon, W.V. Schalkwijk, Nanostructured materials for advanced energy conversion and storage devices, Nature 4 (2005) 366–377.
[32] M.J.S. Farias, W. Cheuquepán, A.A. Tanaka, J.M. Feliu, Nonuniform synergistic effect of Sn and Ru in site-specific catalytic activity of Pt at bimetallic surfaces toward CO electro-oxidation, ACS Catal. 7 (5) (2017) 3434.
[33] A.S. Aricò, S. Srinivasan, V. Antonucci, DMFCs: from fundamental aspects to technology development, Fuel Cell 1 (2001) 133–161.
[34] T.J. Schmidt, et al., Electrocatalytic activity of Pt–Ru alloy colloids for CO and CO/H_2 electrooxidation: stripping voltammetry and rotating-disk measurements, Langmuir 14 (1997) 2591–2595.
[35] O.A. Petrii, Pt-Ru electrocatalysts for fuel cells: a representative review, J. Solid. State Electrochem. 12 (2008) 609–642.
[36] S. Kumar, S.K. Chakarvarti, Large-scale synthesis of uniform nickel nanowires and their characterisation, J. Exp. Nanosci. 5 (2) (2010) 126–133.
[37] K.S. Chou, K.C. Huang, Studies on the chemical synthesis of nanosized nickel powder and its stability, J. Nanopart. Res. 3 (2001) 127–132.
[38] B.C. Steele, A. Heinzel. Materials for fuel-cell technologies, in: Materials For Sustainable Energy: A Collection of Peer-Reviewed Research and Review Articles from Nature Publishing Group, 2011, pp. 224–231.

[39] Z. Chen, M. Waje, W. Li, Y. Yan, Anode catalysts and cathode catalysts of direct methanol fuel cells, Angew. Chem. Int. Ed. 46 (2007) 4060−4063.
[40] S. Gottesfeld, et al., Fuel Cell Seminar 799−802, Courtesy Associates, Washington DC, 2000.
[41] B.D. McNicol, D.A.J. Rand, K.R. Williams, Direct methanol-air fuel cells for road transport, J. Power Sources 83 (1999) 15−31.
[42] A. Hassan, V.A. Paganin, E.A. Ticianelli, Pt modified tungsten carbide as anode electrocatalyst for hydrogen oxidation in proton exchange membrane fuel cell: CO tolerance and stability, Appl. Catal. B Environ. 165 (2015) 611−619.
[43] D.R. Mcintyre, G.T. Burstein, A. Vossen, Effect of carbon monoxide on the electrooxidation of hydrogen by tungsten carbide, J. Power Sources 107 (2002) 67−73.
[44] M. Nie, P.K. Shen, M. Wu, Z. Wei, H. Meng, A study of oxygen reduction on improved Pt-WC/C electrocatalysts, J. Power Sources 162 (2006) 173−176.
[45] S. Kumar, XRD investigation of tungsten carbide nanoparticles prepared by displacement reaction technique, Int. J. Eng. Manag. Res. 4 (4) (2014) 28−30.
[46] D.V. Esposito, J.G. Chen, Monolayer platinum supported on tungsten carbides as low-cost electrocatalysts: opportunities and limitations, Energy Environ. Sci. 4 (2011) 3900.
[47] M. Yurukcu, E.O. Badradeen, S. Bilnoski, F.M. Yurtsever, M. Begum, The effect of the core/shell nanostructure arrays on PEM fuel cells: a short review, Mater. Sci. Eng. 2 (2) (2018) 55−61.
[48] G. McVay. U.S. Dept. of Energy Report on Solid State Energy Conversion Alliance (SECA), U.S. Department of Energy, Washington, D.C, 1999.
[49] M. Doyle, G. Rajendran, Handbook of Fuel Cells; Fundamentals, Technology, and Applications 3 (3) (2003) 351.
[50] K. Hilbert, Hand Book of Fuel Cells: Fundamentals, Technology, and Applications, John Wiley & Sons, 2003, p. 1037.
[51] V.S. Bagotsky Fundamentals of Electrochemistry, second ed., John Wiley & Sons, pp. 361−363.
[52] M. Li, M. Zhao, F. Li, W. Zhou, V.K. Peterson, X. Xu, et al., A niobium and tantalum co-doped perovskite cathode for solid oxide fuel cells operating below 500°C, Nat. Commun. 8 (2016) 13990.
[53] J.A. Cebollero, R. Lahoz, M.A. Laguna-Bercero, J.A. Pena, A. Larrea, V.M. Orera, Characterization of laser-processed thin ceramic membranes for electrolyte-supported solid oxide fuel cells, Int. J. Hydrog. Energy 42 (2017) 13939−13948.
[54] C. Duan, D. Hook, Y. Chen, J. Tong, R. O'Hayre, Zr and Y co-doped perovskite as a stable, high performance cathode for solid oxide fuel cells operating below 500°C, Energy Environ. Sci. 10 (2017) 176−187.
[55] L. Lei, Z. Tao, T. Hong, X. Wang, F. Chen, A highly active hybrid catalyst modified (La0.60Sr0.40)0.95Co0.20Fe0.80O3-δ cathode for proton conducting solid oxide fuel cells, J. Power Sources 389 (2018) 1−7.
[56] Y. Zhang, A. Smirnova, A. Verma, R. Pitchumani, Design of a proton exchange membrane (PEM) fuel cell with variable catalyst loading, J. Power Sources 291 (2015) 46−57.
[57] Y. Li, J. Yang, J. Song, Structure models and nano energy system design for proton exchange membrane fuel cells in electric energy vehicles, Front. Energy 11 (3) (2017) 326−333.
[58] Committee on Review of the Freedom Car and Fuel Research Program, National Research Council. Review of the Research Program of the Freedom CAR and Fuel Cell Partnership: First Report, National Academic Press, Washington, DC, 2005.
[59] J.M. Bowden, J.M. Norbeck. Hydrogen Fuel for Surface Transportation Society of Automotive Engineers, 1996.
[60] International Energy Agency, Hydrogen & Fuel Cells: Review of National R&D Program, Organization for Economic Co-Operation and Development, France, 2004.
[61] D.J. Ludlow, C.M. Calebrese, S.H. Yu, C.S. Dannehy, D.L. JAcbson, D.S. Hussey, et al., PEM fuel cell membrane hydration measurement by neutron imaging, J. Power Sources 162 (2006) 271.
[62] U.R. Salomov, E. Chiavazzo, P. Asinari, Pore-Scale modeling of fluid flow through gas diffusion and catalyst layers for high temperature proton exchange membrane (HT-PEM) fuel cells, Comput. Math. Appl. 67 (2014) 393−411.

[63] https://www.firstenergycorp.com/content/dam/environmental/images/gallery/fuel-cell/Fuel-Cell-Technology-Illustration.jpg.
[64] V. Mehta, J.S. Cooper, Review and analysis of PEM fuel cell design and manufacturing, J. Power Sources 114 (2003) 32–53.
[65] W. Jung-Ho, L. Kwan-Young, K. Sung-Hyun, Fabrication methods for low-Pt loading electrocatalysts in proton exchange membrane fuel cell systems, J. Power Sources 165 (2007) 667–677.
[66] T.D. Berning, M. Lu, N. Djilali, Three-dimensional computational analysis of transport phenomena in a PEM fuel cell, J. Power Sources 16 (1–2) (2002) 284–294.
[67] S. Litster, G. McLean, PEM fuel cell electrodes, J. Power Sources 130 (1–2) (2004) 61–76.
[68] S.M. Haile, Fuel cell materials and components, Acta Materialia 51 (19) (2003) 5981–6000.
[69] A.A. Franco, Polymer Electrolyte Fuel Cells Science, Applications, and Challenges, CRC Press, Boca Raton, 2013.
[70] Z. Ogumi, Polymer Electrolyte Fuel Cells (PEFCs) - State of the Art-, Technical Report, Kyoto University, 2004, 30-35.
[71] Report of the DOE Basic Energy Sciences Workshop on Hydrogen Production, Storage, and Use, Basic Research Needs for the Hydrogen Economy, 2003, at http://www.sc.doe.gov/bes/hydrogen.pdf.
[72] M.K. Debe, A.K. Schmoeckel, S.M. Hendricks, G.D. Vernstrom, G.M. Haugen, R.T. Atanasoski, Durability aspects of nanostructured thinfilm catalyst for PEM fuel cells, ECS Trans. 1 (8) (2006) 51–66.
[73] Report of the DOE Basic Energy Sciences Workshop on the Hydrogen Production, Storage, and Use, http://www.sc.doe.gov/bes/hydrogen.pdf, May 2003.
[74] Z. Ogumi, Technical Report, Kyoto University, 2004, pp. 30–35.
[75] A. Bonakdarpour, K. Stevens, G.D. Vernstrom, R. Atanasoski, A.K. Schmoeckel, M.K. Debe, et al., Oxygen reduction activity of Pt and Pt-Mn-Co electrocatalysts sputtered on nano-structured thin film support, Electrochim. Acta 53 (2007) 688–694.
[76] S. Satyapal, Hydrogen Program Overwide, *DOE Hydrogen Program and Vehicle Technologies Program*, Annual Merit Review Proceedings, 2009.
[77] H. Tang, Z.G. Qi, M. Ramani, J.F. Elter, PEM fuel cell cathode carbon corrosion due to the formation of air/fuel boundary at the anode, J. Power Sources 158 (2006) 1306.
[78] Goodwin, J.G., Zhang, J., Hongsirikarn, K., Martinez, M., Colon-Mercado, H., Greenway, S., and et al. Effects of Impurities on Fuel Cell Performance and Durability, *DOE Hydrogen Program and Vehicle Technologies Program*, Annual Merit Review Proceedings, 2009.
[79] Q. Wang, M. Eikerling, D. Song, Z. Liu, Structure and performance of different types of agglomerates in cathode catalyst layers of PEM fuel cells, J. Electroanalytical Chem. 573 (2004) 61–69.
[80] D.J. Ham, C. Pak, G.H. Bae, S. Hn, K. Kwon, S.A. Jin, et al., Palladium-nickel alloys loaded on tungsten carbide as platinum-free anode electrocatalysts for polymer electrolyte membrane fuel cells, Chem. Commun. 47 (2011) 5792–5794.
[81] T. Tian, J. Tang, W. Guo, M. Pan, Accelerated life-time test of MEA durability under vehicle operating conditions in PEM fuel cell, Front. Energy 11 (3) (2017) 326–333.
[82] H.A. Gasteiger, S.S. Kocha, B. Sompalli, F.T. Wagner, Activity benchmarks and requirements for Pt, Pt-alloy, and non-Pt oxygen reduction catalysts for PEMFCs, Appl. Catal. B Environ. 56 (1–2) (2005) 9–35.
[83] U.A. Paulus, T.J. Schmidt, H.A. Gasteiger, R.J. Behm, Oxygen reduction on a high-surface area Pt/Vulcan carbon catalyst: a thin-film rotating ring-disk electrode study, J. Electroanal. Chem. 495 (2001) 134–145.
[84] D. Gruber, N. Ponath, J. Muller, F. Lindstaedt, Sputter-deposited ultra-low catalyst loadings for PEM fuel cells, J. Power Sources 150 (2005) 67–72.
[85] M.K. Debe, A.K. Schomoeckel, G.D. Vernstrom, R. Atanasoski, High voltage stability of nanostructured thin film catalysts for PEM fuel cells, J. Power Sources 161 (2006) 1002–1011.
[86] D. Gruber, J. Müller, Enhancing PEM fuel cell performance by introducing additional thin layers to sputter-deposited Pt catalysts, J. Power Sources 171 (2) (2007) 294–301.
[87] A. Caillard, C. Coutanceau, P. Braulta, J. Mathias, J.-M. Léger, Structure of Pt/C and PtRu/C catalytic layers prepared by plasma sputtering and electric performance in direct methanol fuel cells (DMFC), J. Power Sources 162 (2006) 66–73.

[88] M. Gustavsson, H. Ekstrom, P. Hanarpa, L. Eureniusa, G. Lindbergh, E. Olsson, et al., Thin film Pt/TiO2 catalysts for the polymer electrolyte fuel cell, J. Power Sources 163 (2007) 671–678.
[89] C. Chien, K. Jeng, Noble metal fuel cell catalysts with nano-network structures, Mater. Chem. Phys. 103 (2007) 400–406.
[90] Zheng, W., Wang, L., Demg, F., Giles, S.A., Prasad, A.K., Advani, S.G., et al. Durable and self-hydrating tungsten carbide-based composite polymer electrolyte membrane fuel cells, Nature Commun. 8, 418.
[91] A. Bonakdarpour, M.D. Fleischauer, M.J. Brett, J.R. Dhan, Columnar support structures for oxygen reduction electrocatalysts prepared by glancing angle deposition, Appl. Catal. A Gen. 349 (2008) 110–115. Available from: https://doi.org/10.1016/j.apcata.2008.07.015.
[92] J.L. Qiao, M. Saito, K. Hayamizu, T. Okada, Degradation of perfluorinated ionomer membranes for PEM fuel cells during processing with H_2O_2, J. Electrochem. Soc. 153 (2006) A967.
[93] M.D. Gasda, G.A. Eisman, D. Gall, Sputter-deposited Pt/CrN nanoparticle PEM fuel cell cathodes: limited proton conductivity through electrode dewetting, J. Electrocem Soc. 157 (1) (2010) B71–B76.
[94] S.S. Araya, F. Zhou, V. Liso, S.L. Sahlin, J.R. Vang, S. Thomas, et al., A comprehensive review of PBI-based high temperature PEM fuel cells, Int. J. Hydrog. Energy 41 (2016) 21310–21344.
[95] R.E. Rosli, A.B. Sulong, W.R.W. Daud, M.A. Zulkifley, T. Husaini, M.I. Rosli, et al., A review of high-temperature proton exchange membrane fuel cell (HT-PEMFC) system, Int. J. Hydrog. Energy 42 (2017) 9293–9314.
[96] T. Ossiander, M. Perchthaler, C. Heinzl, F. Schönberger, P. Völk, M. Welsch, et al., Influence of membrane type and molecular weight distribution on the degradation of PBI-based HTPEM fuel cells, J. Membr. Sci. 509 (2016) 27–35.
[97] B. Han, C.E. Carlton, A. Kongkanand, R.S. Kukreja, B.R. Theobakd, L. Gan, et al., Record activity and stability of dealloyed bimetallic catalysis for proton exchange membrane fuel cells, Energy Environ. Sci. 8 (2015) 258–266.
[98] Fuel Cell Handbook, EG&G Technical Services Inc, seventh ed., U.S. Department of Energy, Office of Fossil Energy, National Energy Technology Laboratory, 2004.
[99] B.G. Levi, The competition is gaining on platinum as a catalyst for hydrogen fuel cells, Phys. Today 62 (5) (2009) 16. Available from: https://doi.org/10.1063/1.3141928.
[100] M. Obermaier, A.S. Bandarenka, C. Lohri-Tymozhynsky, A comprehensive physical impedance model of polymer electrolyte fuel cell cathodes in oxygen-free atmosphere, Sci. Rep. 8 (2016) 4933.
[101] C. Santoro, C. Arbizaani, B. Erable, Microbial fuel cells: from fundamentals to applications: a review, J. Power Sources 356 (2017) 225–244.
[102] R.A. Rozendal, A.W. Jeremiasse, H.V. Hamelers, C.J. Buisman, Hydrogen production with a microbial biocathode, Environ. Sci. Technol. 42 (2) (2008) 629–634.
[103] A.P. Borole, G. Reguera, B. Ringeisen, Z.-W. Wang, Y. Feng, B. Hong Kim, Electroactive biofilms: current status and future research needs, Energy Environ. Sci. 4 (2011) 4813–4834. Available from: https://doi.org/10.1039/C1EE02511B.
[104] A. Larrosa-Guerrero, K. Scott, I.M. Head, F. Mateo, A. Ginesta, C. Godinez, Effect of temperature on the performance of microbial fuel cells, Fuel 89 (2010) 3985–3994. Available from: https://doi.org/10.1016/j.fuel.2010.06.025.
[105] L. Qiang, L.J. Yuan, Q. Ding, Influence of buffer solutions on the performance of microbial fuel cell electricity generation, Huanjing Kexue Xuebao/Acta Scientiae Circumstantiae 32 (2011) 1524–1528.
[106] D. Pant, G. Van Bogaert, L. Diels, K. Vanbroekhoven, A review of the substrates used in microbial fuel cells (MFCs) for sustainable energy production, Bioresour. Technol. 101 (2010) 1533–1543.
[107] J.M. Foley, R.A. Rozendal, C.K. Hertle, P.A. Lant, K. Rabaey, Life cycle assessment of high-rate anaerobic treatment, microbial fuel cells, and microbial electrolysis cells, Environ. Sci. Technol. 44 (2010) 3629–3637.
[108] M.A. Tanni, Md Arifujjam, M.T. Iqbal, Dynamic modeling a of phosphoric acid fuel cell (PAFC) and its power conditioning system, J. Clean. Energy Technol. 1 (2013) 3.

[109] J.F. Shin, W. Xu, M. Zanella, K. Dawson, S.N. Savvin, J.B. Claridge, et al., Self-assembled dynamic perovskite composite cathodes for intermediate temperature solid oxide fuel cells, Nat. Energy 2 (2017) 16214.
[110] S.T. Aruna, L.S. Balaki, S.S. Kuar, B.S. Prakash, Electrospinning in solid oxide fuel cells-a review, Renew. Sustain. Energy Rev. 67 (2017) 673–682.
[111] S. Choi, C. Kucharcxyk, Y. Liang, X. Zhang, I. Takeuchi, H. Ji, et al., Exceptional power density and stability at intermediate temperatures in protonic ceramic fuel cells, Nat. Energy 3 (2018) 202–210.
[112] V.R. Stamenkovic, D. Strmcnik, P.P. Lopes, N.M. Markovic, Energy and fuels from electrochemical interfaces, Nat. Mater. 16 (2017) 57–69.
[113] M. Arenz, T.J. Schmidt, K. Wandelt, P.N. Ross, N.M. Markovic, The oxygen reduction reaction on thin palladium films supported on a Pt(111) electrode, J. Phys. Chem. B 107 (2003) 9813–9819.
[114] D.F. Van der Vliet, et al., Mesostructured thin films as electrocatalysts with tunable composition and surface morphology, Nat. Mater. 11 (2012) 1051–1058.
[115] C. Wang, et al., Monodisperse Pt(3)Co nanoparticles as electrocatalyst: the effects of particle size and pretreatment on electrocatalytic reduction of oxygen, Phys. Chem. Chem. Phys. 12 (2010) 6933–6939.
[116] C. Wang, et al., Multimetallic Au/FePt3 nanoparticles as highly durable electrocatalyst, Nano Lett. 11 (2011) 919–926.
[117] Y. Kang, et al., Multimetallic core/interlayer/shell nanostructures as advanced electrocatalysts, Nano Lett. 14 (2014) 6361–6367.
[118] C. Wang, et al., Correlation between surface chemistry and electrocatalytic properties of monodisperse Pt*x*Ni1-*x* nanoparticles, Adv. Funct. Mater. 21 (2011) 147–152.
[119] C. Chen, et al., Highly crystalline multimetallic nanoframes with three-dimensional electrocatalytic surfaces, Science 343 (2014) 1339–1343.
[120] T. Yoshida, K. Kojima, Toyota MIRAI fuel cell vehicle and progress toward a future hydrogen society, Interface 24 (2015) 45–49.
[121] W.J. Khudhayer, A.U. Shaikh, T. Karabacak, Platinum nanorod arrays with enhanced morphological, crystal, and electrochemical properties for oxygen reduction reaction, Adv. Sci. Lett. 4 (3) (2011) 1.
[122] W.J. Khudhayer, A.U. Shaikh, T. Karabacak, Periodic Pt nanorod arrays with controlled porosity for oxygen reduction reaction, Nanosci. Nanotechnol. Lett. 4 (2012) 1–9.
[123] C. Wang, L. Ma, L. Liao, S. Bai, R. Long, M. Zuo, et al., A unique platinum-graphene hybrid structure for high activity and durability in oxygen reduction reaction, Sci. Rep. 3 (2013) 2580.
[124] C. Wang, N.M. Markovic, V.R. Stamenkovic, Advanced platinum alloy electrocatalysts for the oxygen reduction reaction, ACS Catal. 2 (2012) 891–898.
[125] L. Zou, J. Fan, Y. Zhou, C. Wang, J. Li, Z. Zou, et al., Conversion of PtNi alloy from disordered to ordered for enhanced activity and durability of methanol tolerant oxygen reduction reaction, Nano Res. 8 (8) (1998) 2777–2788. Available from: https://doi.org/10.1007/s12274-015-0784-0.
[126] N. Karthikeyan, B.P. Vinayan, M. Rajesh, K. Balaji, A.K. Subramani, S. Ramaprabhu, Highly durable platinum based cathode electrocatalysts for PEMFC application using oxygen and nitrogen functional groups attached nanocarbon supports, Fuel Cell 15 (2) (2015) 278–287.
[127] K. Sasaki, H. Naohara, Y. Cai, Y.M. Choi, P. Liu, M.B. Vukmirovic, et al., Core-protected platinum monolayer shell high-stability electrocatalysts for fuel-cell cathodes, Angew. Chem. Int. Ed. 49 (2010) 8602–8607.
[128] M. Lefèvre, E. Proietti, F. Jaouen, J.-P. Dodelet, Iron-based catalysts with improved oxygen reduction activity in polymer electrolyte fuel cells, Science 324 (2009) 71–74.
[129] J. Gu, Y.W. Zhang, F.F. Tao, Shape control of bimetallic nanocatalysts through well-designed colloidal chemistry approaches, Chem. Soc. Rev. 41 (2012) 8050–8065.
[130] D. Banham, S. Ye, Current status and future development of catalyst materials and catalyst layers for proton exchange membrane fuel cells: an industrial perspective, ACS Energy Lett. 2 (3) (2017) 629–638.

[131] L. DeSilva Munoz, B. Erable, L. Etcheverry, J. Riess, R. Basseguy, A. Bergel, Combining phosphate species and stainless steel cathode to enhance hydrogen evolution in microbial electrolysis cell (MEC), Electrochem. Com. 12 (2010) 183–186.

[132] S. Cheng, B.E. Logan, High hydrogen production rate of microbial electrolysis cell (MEC) with reduced electrode spacing, Bioresour. Technol. 102 (2011) 3571–3574.

[133] H. Hu, Y. Fan, H. Liu, Hydrogen production using single-chamber membrane-free microbial electrolysis cells, Water Res. 42 (2008) 4172–4178.

[134] B. Genorio, D. Strmcnik, R. Subbaraman, D. Tripkovic, G. Karapetrov, V.R. Stamenkovic, et al., Selective catalysts for the hydrogen oxidation and oxygen reduction reactions by patterning of platinum with calix[4]arene molecules, Nat. Mater. 9 (2010) 998–1003.

[135] G. Chen, Y. Zhao, G. Fu, P.N. Duchesne, L. Gu, Y. Zheng, et al., Interfacial effects in iron-nickel hydroxide–platinum nanoparticles enhance catalytic oxidation, Science 344 (6183) (2014) 495–499.

[136] L. Gan, M. Heggen, S. Rudi, P. Strasser, Core-shell compositional fine structures of dealloyed PtxNi1-x nanoparticles and their impact on oxygen reduction catalysis, Nano Lett. 12 (2012) 5423–5430.

[137] S. Gupta, D. Tryk, I. Bae, W. Aldred, E. Yeager, Heat-treated polyacrylonitrile-based catalysts for oxygen electroreduction, J. Appl. Electrochem. 19 (1) (1989) 19–27.

[138] J.P. Collman, P. Denisevich, Y. Konai, M. Marrocco, C. Koval, F.C. Anson, Electrode catalysis of the four-electron reduction of oxygen to water by dicobalt face-to-face porphyrins, J. Am. Chem. Soc. 102 (19) (1980) 6027–6036.

[139] A.L. Bouwkamp-Wijnoltz, W. Visscher, J.A.R. Van Veen, E. Boellaard, A.M. Van der Kraan, S.C. Tang, On active-site heterogeneity in pyrolyzed carbon-supported iron porphyrin catalysts for the electrochemical reduction of oxygen: an in situ Mössbauer study, J. Phys. Chem. B 106 (50) (2002) 12993–13001.

[140] M. Lefevre, J.P. Dodelet, P. Bertrand, Molecular oxygen reduction in PEM fuel cell conditions: ToF-SIMS analysis of Co-based electrocatalysts, J. Phys. Chem. B 109 (35) (2005) 16718–16724.

[141] U.I. Koslowski, I. Abs-Wurmbach, S. Fiechter, P. Bogdanoff, Nature of the catalytic centers of porphyrin-based electrocatalysts for the ORR: a correlation of kinetic current density with the site density of Fe − N4 centers, J. Phys. Chem. C. 112 (2008) 15356.

[142] A. Garsuch, A. Bonakdarpour, G. Liu, R. Yang, J.R. Dahn, Handbook of fuel cells fundamentals technology and applications 5 (2009) 71–80.

[143] H.A. Gasteiger, N.M. Markovic, Just a dream-or future reality? Science 324 (5923) (2009) 48–49.

[144] D. Li, C. Wang, D. Tripkovic, S. Sun, N.M. Markovic, V.R. Stamenkovic, Surfactant removal for colloidal nanoparticles from solution synthesis: the effect on catalytic performance, ACS Catal. 2 (2012) 1358–1362.

[145] D.H. Chen, C.H. Hsieh, Synthesis of nickel nanoparticles in aqueous cationic surfactant solutions, J. Mater. Chem. 12 (2002) 2412–2415.

[146] S. Srinivasan, R. Mosdale, P. Stevens, C. Yang, Fuel cells: reaching the era of clean and efficient power generation in the twenty-first century, Annu. Rev. Energy Env. 24 (1999) 281.

[147] P. Murray, E. Tsai, S.A. Barnett, A direct-methane fuel cell with a ceria-based anode, Nature 400 (1999) 649–651.

[148] S. Park, J.M. Vohs, R.J. Gorte, Direct oxidation of hydrocarbons in a solid-oxide fuel cell, Nature 404 (2000) 265–267.

[149] Y.P. Zhou, K. Feng, Y. Sun, L. Zhou, A brief review on the study of hydrogen storage in terms of carbon nanotubes, Prog. Chem. 15 (2003) 345–350.

[150] L. Schlappach, A. Zuttel, Hydrogen-storage materials for mobile applications, Nature 414 (2001) 353–358.

[151] A.C. Dillon, et al., Storage of hydrogen in single-walled carbon nanotubes, Nature 386 (1997) 377–379.

[152] A. Chambers, C. Park, R.T.K. Baker, N.M. Rodriguez, Hydrogen storage in graphite nanofibers, J. Phys. Chem. 102 (1998) 4253–4256.

[153] R.T. Yang, Hydrogen storage by alkali-doped carbon nanotubes—revisited, Carbon 38 (2000) 623–626.

[154] V.R. Stemenkovic, B.S. Mun, J.J. Mayrhofer, P.N. Ross, N.M. Markovic, Effect of surface composition on electronic structure, stability, and electrocatalytic properties of Pt-transition metal alloys: Pt-skin versus Pt-skeleton surfaces, J. Am. Chem. Soc. 128 (2006) 8813–8819.
[155] C.J. Tseng, S.T. Lo, S.C. Lo, P.P. Chu, Characterization of Pt-Cu binary catalysts for oxygen reduction for fuel cell applications, Mater. Chem. Phys. 100 (2006) 385–390.
[156] M. Hirscher, M. Becher, Hydrogen storage in carbon nanotubes, J. Nanosci. Nanotechnol. 3 (2003) 3–17.
[157] C.X. Shang, M. Bououdina, Y. Song, Z.X. Guo, Mechanical alloying and electronic simulation of MgH2–M systems (M–Al, Ti, Fe, Ni, Cu, and Nb) for hydrogen storage, Int. J. Hydrog. Energy 29 (2004) 73–80.
[158] J.L. Bobet, E. Grigorova, M. Khrussanova, M. Khristov, P. Peshev, Hydrogen sorption properties of the nanocomposite 90wt.% Mg2Ni–10 wt.% V, J. Alloy. Compd. 356 (2003) 593–597.
[159] R. Atanasoski, 4th Int. Conf. Applications of Conducting Polymers, ICCP-4, Como, Italy Abstract 22, 2004.
[160] M.K. Debe, A.J. Steinbach, K.A. Lewinski, G.M. Haugen, G.D. Vernstrom, R.T. Atanasoski, et al. Activities of Low Pt Loading, Carbon-Less, Ultra-Thin Nanostructured Film-Based Electrodes for PEM Fuel Cells, and Performances in Roll-Good Fabricated MEA's in Single Cells and Stacks, *Fuel Cell Seminar*, Miami, FL, 2003.
[161] M.K. Debe, A.J. Steinbach, G.D. Vernstrom, S.M. Hendricks, M.J. Kurkowski, R.T. Aranasosk, et al., Extraordinary oxygen reduction activity of Pt_3Ni_7, J. Electrochem. Soc. 158 (8) (2011) B910–B918.
[162] M.K. Debe, A.K. Schmoeckel, R. Atanasoski, G.D. Vernstrom, High Voltage Stability of Nano Structured Thin Film Catalysts for PEM Fuel Cells, FC Seminar, Palm Springs, CA, 2005.
[163] A. Atkinson, R.I. Taylor, A.E. Hughes, A quantitative demonstration of the grain boundary diffusion mechanism for the oxidation of metals, Philos. Mag. A 45 (5) (1982) 823–833.
[164] N.M. Marković, P.N. Ross Jr, Surface science studies of model fuel cell electrocatalysts, Surf. Sci. Rep. 45 (4–6) (2002) 117–229.
[165] V.R. Stamenkovic, B. Fowler, B.S. Mun, G. Wang, P.N. Ross, C.A. Lucas, et al., Improved oxygen reduction activity on Pt3Ni (111) via increased surface site availability, Science 315 (5811) (2007) 493–497.
[166] R. Subbaraman, D. Strmcnik Bonakdarpour, V. Stamenkovic, N.M. Markovic, Three phase interfaces at electrified metal – solid electrolyte systems 1. Study of the Pt (hkl) – Nafion interface, J. Phys. Chem. C. 114 (2010) 8414–8422.
[167] R. Subbaraman, D. Strmcnik, A.P. Paulikas, V.R. Stamenkovic, N.M. Markovic, Oxygen Reduction Reaction at Three-Phase Interfaces, Chem. Phys. Chem 11 (2010) 2825–2833.
[168] M.D. Gasda, R. Teki, T.-M. Lu, N. Koratkar, G.A. Eisman, D. Gall, Sputter-deposited Pt PEM fuel cell electrodes: particles vs layers, J. Electrochem. Soc. 156 (5) (2009) B614–B619.
[169] M.D. Gasda, G.A. Eisman, D. Gall, Pore formation by in situ etching of nanorod PEM fuel cell electrodes, J. Electrochem. Soc. 157 (1) (2009) B113.
[170] M.D. Gasda, G.A. Eisman, D. Gall, Nanorod PEM fuel cell cathodes with controlled porosity, J. Electrochem. Soc. 157 (3) (2010) B437.
[171] R. O'Hayre, F.B. Prinz, The air/platinum/Nafion triple-phase boundary: characteristics, scaling, and implications for fuel cells, J. Electrochem. Soc. 151 (5) (2004) A756.
[172] Q. Wang, M. Eikerling, D. Song, Z.S. Liu, Modeling of ultrathin two-phase catalyst layers in PEFCs, J. Electrochem. Soc. 154 (6) (2007) F95.
[173] A. Garsuch, D.A. Stevens, R.J. Sanderson, S. Wang, R.T. Atanasoski, S. Hendricks, et al., Alternative catalyst supports deposited on nanostructured thin film for proton exchange membrane fuel cells, J. Electrochemical Society 157 (2) (2010) B187–B194.
[174] C. Patiz, T. Karabacak, B. Fuhrmaann, B. Rauschenbach, Theoretical and experimental characterization of TiO_2 thin films deposited at oblique angles, J. Appl. Phys. 104 (2008) 94318.
[175] A. Satyaprasad, V. Jain, S.K. Nema, Deposition of superhydrophobic nanostructured Teflon-like coating using expanding plasma arc, Appl. Surf. Sci. 253 (12) (2007) 5462–5466.

[176] D.K. Sarkar, M. Farzaneh, R.W. Paynter, Superhydrophobic properties of ultrathin rf-sputtered Teflon films coated etched aluminum surfaces, Mater. Lett. 62 (8-9) (2008) 1226–1229.

[177] M. Oezaslan, F. Hasche, P. Strasser, Pt-based core–shell catalyst architectures for oxygen fuel cell electrodes, J. Phys. Chem. Lett. 4 (19) (2013) 3273–3291.

[178] G. McWay, US Dept. of Energy Report on Solid State Converseraion Alliance (SECA), US Dep of Energy, 1999.

[179] Y. Wang, S. Song, V. Maragou, P.K. Shen, High surface area tungsten carbide microspheres as effective Pt catalyst support for oxygen reduction reaction, Appl. Catal. B Environ. (2009) 223–228.

[180] A. Jackson, A. Strickler, D. Higgins, T.F. Jaramillo, Engineering Ru@Pt core-shell catalysts for enhanced electrochemical oxygen reduction mass activity and stability, Nanomaterials 8 (2018) 38.

[181] K.C. Wang, H.C. Huang, C.H. Wang, Synthesis of Pd@Pt3Co/C core shell structure as catalyst for oxygen reduction reaction in proton exchange membrane fuel cell, Int. J. Hydrog. Energy 42 (2017) 11771–11778.

[182] R. Nagar, B.P. Vinayan, Metal-semiconductor core-shell nanomaterials for energy applications, Micro Macro Tech. (2017) 99–132. Available from: https://doi.org/10.1016/B978-0-323-44922-9.00005-3.

[183] D. Wang, H.L. Xin, R. Hovden, H. Wang, Y. Yu, D.A. Muller, et al., Structurally ordered intermetallic platinum-cobalt core-shell nanoparticles with enhanced activity and stability as oxygen reduction electrocatalysts, Nat. Mater. 12 (2012) 81–87.

[184] N.N. Kariuki, M.F. Cansizoglu, M. Begum, M. Yurukcu, F.M. Yurtsever, T. Karabacak, et al., SAD-GLAD Pt-Ni@Ni Nanorods as Highly Active Oxygen Reaction Electrocatalysts, ACS Catalysis, 2016.

[185] M.G. Hosseini, R. Mahmoodi, Ni@M(M = Pt, Pd and Ru) core@shell nanoparticle on a Vulcan XC-72R support with superior catalytic toward borohydride oxidation: electrochemical and fuel cell studies, N. J. Chem. (2017) 22.

[186] D. Takioto, T. Ohnishi, J. Nutariya, Z. Shen, Y. Ayato, D. Mochizuki, et al., Ru-core@Pt-shell nanosheet for fuel cell electrocatalysts with high activity and durability, J. Catal. 355 (2017) 185.

[187] D.J. Ham, C. Pak, G.H. Bae, S. Han, K. Kwon, S.A. Jin, et al., Palladium–nickel alloys loaded on tungsten carbide as platinum-free anode electrocatalysts for polymer electrolyte membrane fuel cells, Chem. Commun. 47 (20) (2011) 5792–5794.

[188] W.J. Khudhayer, N. Kariuki, D.J. Myers, A.U. Shaikh, T. Karabacak, GLAD Cr nanorods coated with SAD Pt thin film for oxygen reduction reaction, J. Electrochem. Soc. 159 (6) (2012) 159.

[189] F. Harnisch, U. Schröder, M. Quaas, F. Scholz, Electrocatalytic and corrosion behavior of tungsten carbide in near neutral Ph electrolytes, Appl. Catal. B Environ. 87 (1–2) (2009) 63–69.

[190] T. Karabacak, J.P. Singh, Y.P. Zhao, G.C. Wang, T.M. Lu, Scaling during shadowing growth of isolated nanocolumns, Phys. Rev. B 68 (12) (2003).

[191] W.J. Khudhayer, N.N. Kariuki, X. Wang, Oxygen reduction reaction electrocatalytic activity of glancing angle deposition platinum nanorod arrays, J. Electrochem. Soc. 158 (8) (2011) 1029–1041.

[192] F.M. Yurtsever, M. Begum, M. Yurukcu, T. Karabacak, Stacked Pt-Ni/WC Nanorod Array Electrocatalyst with Gradient Pt-Ni Alloy Composition for ORR in PEM Fuel Cells Abstracts of Papers of the American Chemical Society, 2018, 255.

[193] N.N. Kariuki, W.J. Khudhayer, T. Karabacak, GLAD Pt-Ni alloy nanorods for oxygen reduction reaction, ACS Catal. (2013) 3123–3132.

[194] M. Begum, M. Yurukcu, F. Yurtsever, B. Ergul, N. Kariuki, T. Karabacak, Pt-Ni/WC alloy nanorods arrays as ORR catalyst for PEM fuel cells, ECS Trans. 80 (8) (2017) 919–925.

[195] S. Kim, T. Karabacak, T.M. Lu, Hydrogen Generation Using Ruthenium Nano-Rod Array Electrodes, Smarts and Materials, 2006.

[196] D. Takimoto, T. Ohnishi, J. Nutariya, Z. Shen, Y. Ayato, et al., Ru-core@Pt-shell nanosheet for fuel cell electrocatalyss with high activity and durability, J. Catal. 355 (2017) 185.

[197] R. Jiang, J. Zhang, A review of core-shell nanostructured electrocatalysts for oxygen reduction reaction, J. ENSM 11 (2017) 005.

[198] L. Yang, D. Banham, M. Markiewichz, S. Knights, S. Ye, Development and Application of Core-Shell Cathode Catalysts in PEM Fuel Cel, Electrochem. Soc. MA2018 (01) (2018) 2170. Available from: https://iopscience.iop.org/article/10.1149/MA2018-01/37/2170.
[199] P. Dhanasekaran, S.D. Bhai. Boosting Pt oxygen reduction reaction activity and durability by carbon semi-coated titania nanorods for proton exchange membrane fuel cells, Electrochim. Acta, 263, 596-609.
[200] Y. Rivera-Lugo, M.F. Felix-Navarro, Effect of template, reaction time and platinum concentration in the synthesis of PtCu/CNT catalyst for PEMFC applications, J. Energy 01 (2018) 069.
[201] J. Greeley, I.E.L. Stephens, A.S. Bondarenko, T.P. Johansson, H.A. Hansen, T.F. Jaramillo, et al., Alloys of platinum and early transition metals as oxygen reduction electrocatalysts, Nat. Chem. 1 (2009) 552.
[202] P. Strasser, S. Koh, T. Anniyev, J. Greeley, K. More, C.F. Yu, et al., Lattice-strain control of the activity in dealloyed core—shell fuel cell catalysts, Nat. Chem. 2 (2010) 454.
[203] V.R. Stamenkovic, B.S. Mun, M. Arenz, K.J. Mayrhofer, C.A. Lucas, G. Wang, et al., Trends in electrocatalysis on extended and nanoscale Pt-bimetallic alloy surfaces, Nat. Mater. 6 (3) (2007) 241—247.
[204] Y. Lu, Y. Jiang, X. Gao, X. Wang, W. Chen, Strongly coupled Pd nanotetrahedron/tungsten oxide nanosheet hybrids with enhanced catalytic activity and stability as oxygen reduction electrocatalysts, J. Am. Chem. Soc. 136 (33) (2014) 11687—11697.
[205] Y. Lu, W. Chen, Size effect of silver nanoclusters on their catalytic activity for oxygen electro-reduction, J. Power Sources 197 (2012) 107—110.
[206] Y. Lu, Y. Wang, W. Chen, Silver nanorods for oxygen reduction: strong effects of protecting ligand on the electrocatalytic activity, J. Power Sources 196 (6) (2011) 3033—3038.
[207] X. Xiong, W. Chen, W. Wang, J. Li, S. Chen, Pt-Pd nanodendrites as oxygen reduction catalyst in polymer-electrolyte-membrane fuel cell, Int. J. Hydrog. Energy 42 (40) (2017) 25234—25243.
[208] D.A. Slanac, W.G. Hardin, K.P. Johnston, K.J. Stevenson, Atomic ensemble and electronic effects in Ag-rich AgPd nanoalloy catalysts for oxygen reduction in alkaline media, J. Am. Chem. Soc. 134 (23) (2012) 9812—9819.
[209] S. Liu, C. Zhang, L. Yuan, J. Bao, W. Tu, M. Han, et al., Component-controlled synthesis of small-sized Pd-Ag bimetallic alloy nanocrystals and their application in a non-enzymatic glucose biosensor, Part. Part. Syst. Charact. 30 (6) (2013) 549—556.
[210] Y. Lu, Y. Jiang, X. Gao, X. Wang, W. Chen, Highly active and durable PdAg@ Pd core—shell nanoparticles as fuel-cell electrocatalysts for the oxygen reduction reaction, Part. Part. Syst. Charact. 33 (8) (2016) 560—568.
[211] Y. Shao, G. Yin, Y. Gao, P. Shi, Durability study of Pt/C and Pt/CNTs catalysts under simulated PEM fuel cell conditions, J. Electrochem. Soc. 153 (6) (2006) A1093—A1097.
[212] S.-I.I. Choi, M. Shao, N. Lu, A. Ruditskiy, H.-C. Peng, J. Park, et al., Synthesis and characterization of Pd@Pt-Ni Cre-shell octahedra with high activity toward oxygen reduction, ACS Nano 8 (10) (2014) 10363—10371.
[213] X.-R. Zhang, F.X. Zhang, X. Yang, A.H. Yuan, Theoretical studies of the electronic strucuture and spectrum properties of PtnNim ($n + m = 7, n, m \neq 0$) clusters, J. Clust. Sci. 24 (2013) 945—958.
[214] D.S. Strmcnik, D.V. Tripkovic, D. Van der Vliet, K.C. Chang, V. Komanicky, et al., Unique activity of platinum adislands in the CO electrooxidation reaction, Am. Chem. Soc. 130 (2008) 15332—15339.
[215] S.H. Yang, D.A. Drabold, J.B. Adams, P. Ordejon, K. Glassford, Density functional studies of small platinum clusters, J. Phys. Condens. Matter 9 (1997) L39—L45.
[216] H. Cansizoglu, M. Yurukcu, M.F. Cansizoglu, T. Karabacak, Investigation of physical vapor deposition techniques of conformal shell coating for core/shell structures by Monte Carlo simulations, Thin Solid. Films 583 (2015) 122—128.
[217] M. Yurukcu, H. Cansizoglu, M.F. Cansizoglu, T. Karabacak, Conformality of PVD shell layers on vertical arrays of rods with different aspect ratios investigated by Monte Carlo simulations, MRS Adv. 2 (8) (2017) 465—470.
[218] D.G. Coronel, E.W. Egan, G. Hamilton, A. Jain, R. Venkaraman, B. Weitzman, Monte Carlo simulations of sputter deposition and step coverage of thin films, Thin Solid Films 333 (1998) 77—81.

[219] T. Karabacak, G.C. Wang, T.M. Lu, Physical self-assembly and the nucleation of three-dimensional nanostructures by oblique angle deposition, J. Vac. Sci. Technol. A. 22 (4) (2004) 1778–1784.

[220] M. Yurukcu, An Investigation of Conformality Of the Core-Shell Polymer Electrolyte Membrane (PEM) Fuel Cell Catalysts by Sputter Deposition, Proquest.com, 2018.

[221] D. Song, Q. Wang, Z. Liu, T. Navessin, M. Eikerling, S. Holdcroft, Numerical optimization study of the catalyst layer of PEM fuel cell cathode, J. Power Sources 126 (2004) 104–111.

[222] S. Thomas, A. Bates, S. Park, A.K. Sahu, S.C. Lee, B.R. Son, et al., An experimental and simulation study of novel channel designs for open-cathode high-temperature polymer electrolyte membrane fuel cells, Appl. Energy 165 (2016) 765–776.

[223] S.M. Rao, Y. Xing, Simulation of nanostructured electrodes for polymer electrolyte membrane fuel cells, J. Power Sources 185 (2008) 1094–1100.

[224] Y. Ding, X.T. Bi, D.P. Wilkinson, Numerical investigation of the impact of two-phase flow maldistribution on PEM fuel cell performance, Int. J. Hydrog. Energy 39 (2014) 469–480.

[225] U. Chakraborty, A new stochastic algorithm for proton exchange membrane fuel cell stack design optimization, J. Power Sources 216 (2012) 530–541.

[226] E.L. Miller, D. Papageorgopoulos, N. Stetson, K. Randolph, D. Peterson, K. Cierpik-Gold, et al., U.S. department of energy hydrogen and fuel cells program: progress, challenges and future directions, MRS Adv. 1 (42) (2016) 2839–2855. Available from: https://doi.org/10.1557/adv.2016.495.

[227] C. Wang, M. Chi, D. Li, D. Strmcnik, D. Van der Vliet, G. Wang, et al., Design and synthesis of bimetallic electrocatalyst with multilayered Pt-Skin surfaces, J. Am. Chem. Soc., 133, 2011, pp. 14396–14403.

[228] C. Wang, Q. Zhang, S. Shen, X. Yan, F. Zhu, X. Cheng, et al., The respective effect of under-rib convection and pressure drop of flow fields on the performance of PEM fuel cells, Sci. Rep. 7 (2017) 43447.

[229] A. Ghanbarian, M.J. Kermani, Enhancement of PEM fuel cell performance by flow channel indentation, Energy Convers. Manag. 110 (2016) 356–366.

[230] https://fuelcellsworks.com/news/toyota-tsusho-signs-distribution-agreement-with-canadian-fuel-cell-developer-and-manufacturer.

[231] C. He, S. Desai, G. Brown, S. Bollepalli, PEM fuel cell catalysts: cost, performance and durability, J. Electrochem. Soc. 14 (3) (2005) 41–44.

[232] I. Matanovic, F.H. Garzon, N.J. Henson, Theoretical study of electrochemical processes on Pt-Ni alloys, J. Phys. Chem. 115 (2011) 10640–10650.

[233] N. Wakabayashi, M. Takeichi, H. Uchida, M. Watanabe, Temperature dependence of oxygen reduction activity at Pt-Fe, Pt-Co, and Pt-Ni alloy electrodes, J. Phys. Chem. B 1099 (2005) 5836–5841.

[234] P. Harrop, R. Das, Hybrid and pure electric cars 2009-2019. Research report, http://www.idtechex.com/research/reports/hybrid_and_pure_electric_cars_2009_2019_000227.asp, 2017.

[235] M.K. Debe. Advanced Cathode Catalysts and Supports for PEM Fuel Cells, hydrogen.energy.gov, 2010.

Index

Note: Page numbers followed by "*f*" and "*t*" refer to figures and tables, respectively.

C

D

H

I

K

L

P

Q

R

S

T

U

V

W

X